Advances in Veterinary Science
and Comparative Medicine

Volume 38A

Comparative Vertebrate Exercise Physiology: Unifying Physiological Principles

Advances in Veterinary Science and Comparative Medicine

Edited by

Charles E. Cornelius
Department of Animal Resources
Scripps Research Institute
La Jolla, California

R. R. Marshak
New Bolton Center
University of Pennsylvania
Kennett Square, Pennsylvania

E. C. Melby
SmithKline Beecham
Animal Health Products Research and Development
West Chester, Pennsylvania

Advances in Veterinary Science
and Comparative Medicine

Volume 38A

Comparative Vertebrate Exercise Physiology: Unifying Physiological Principles

Edited by

James H. Jones

Department of Surgical and Radiological Sciences
School of Veterinary Medicine
University of California, Davis
Davis, California

Academic Press

San Diego New York Boston
London Sydney Tokyo Toronto

This book is printed on acid-free paper. ♾

Academic Press, Inc.
A Division of Harcourt Brace & Company
525 B Street, Suite 1900, San Diego, California 92101-4495

United Kingdom Edition published by
Academic Press Limited
24–28 Oval Road, London NW1 7DX

International Standard Serial Number: 0065-3519

International Standard Book Number: 0-12-039238-0

PRINTED IN THE UNITED STATES OF AMERICA

94 95 96 97 98 99 EB 9 8 7 6 5 4 3 2 1

CONTENTS

The Mechanical Design of the Muscular System

LAWRENCE C. ROME

Relating Mechanics and Energetics during Exercise

C. RICHARD TAYLOR

Circulatory Function during Exercise: Integration of Convection and Diffusion

JAMES H. JONES

Respiratory Gas Exchange during Exercise

FRANK L. POWELL, JR.

CONTRIBUTORS

Numbers in parentheses indicate the pages on which the authors' contributions begin.

RUEDI BILLETER, Department of Anatomy, University of Berne, CH-3000 Berne 9, Switzerland (57)

KEVIN E. CONLEY, Magnetic Resonance Laboratory, Department of Radiology, University of Washington Medical Center, Seattle, Washington 98195 (1)

PETER W. HOCHACHKA, Department of Zoology and Sports Medicine Division, University of British Columbia, Vancouver, British Columbia, Canada V6T 2A9 (41)

HANS HOPPELER, Department of Anatomy, University of Berne, CH-3000 Berne 9, Switzerland (57)

JAMES H. JONES, Department of Surgical and Radiological Sciences, School of Veterinary Medicine, University of California, Davis, Davis, California 95616 (217)

HANS OETLIKER, Institute of Physiology, University of Berne, CH-3000 Berne 9, Switzerland (57)

FRANK L. POWELL, JR., Physiology Division, Department of Medicine, University of California, San Diego, La Jolla, California 92093 (253)

LAWRENCE C. ROME, Department of Biology, University of Pennsylvania, Philadelphia, Pennsylvania 19104 (125)

C. RICHARD TAYLOR, Harvard University, Bedford, Massachusetts 01730 (181)

PREFACE

Comparative physiology is an integrative approach to understanding how animals function. Its roots are firmly anchored in "whole animal" physiology, a perspective that is essential for interpreting and comparing data collected from specific organ systems in sometimes very different animals. The demands placed on an animal due to environment, body size, etc., are inevitably met by the integrated responses of multiple systems, none of which is independent of other systems in the body. Only by maintaining an integrative view can many of these relationships be understood and their roles in the adaptation and evolution of the animal be interpreted.

Comparative physiologists have two fundamental goals when investigating how animals function. One goal is to determine what structural and functional elements of animals are relatively invariant among taxa, thereby identifying the fundamental building blocks of animals that have not been particularly malleable during the course of evolution. An alternative, but complementary, goal is to understand the mechanisms that enable certain species to tolerate extreme environments or to achieve exceptional physiological performance and to contrast these extreme cases with other animals. This approach capitalizes on the evolutionary variation in structure and function that occurs in different organ systems and provides a large signal-to-noise ratio in data that are collected. Both of these goals contribute to our understanding of how animals adapt to demands of the environment and, over time, evolve in response to those demands.

During exercise, many organ systems approach or attain their maximal functional capacities, and structural elements may be stressed to their limits. Maximally functioning organs or systems may interact in unique ways, resulting in unusual physiological phenomena. The comparative approach is useful for understanding the basic elements that enable, as well as limit, exercise performance.

This book and a companion volume (Volumes 38A and 38B) apply both of the fundamental research approaches of comparative physiology to the study of exercise. In this volume, principles of structure and function that are fundamental to all vertebrate exercise performance are identified and characterized. These seven chapters deal with the elements of exercise that are common to vertebrates of all taxa: cellular energetics and control, muscle function, translation of metabolic to mechanical power, and oxygen uptake and transport. They define which elements of the vertebrate body constitute nonmalleable building blocks and how these elements are used during exercise. The companion volume explores extremes of adaptation that enable some animals to perform athletic feats that are truly remarkable by human standards.

The contributors to this book are authorities in their fields and provide broad perspectives on questions of interest and significance in the study of exercise performance. I am indebted to each of them for the efforts they have made to identify the key elements in their area that are pertinent for comparing exercise performances of vertebrates.

For many of us, our interest in comparative physiology was either initiated or heightened by reading about phenomena such as how kangaroo rats and camels tolerate desert heat and dryness, how birds breathe, or why small animals metabolize faster than large ones. We have also been deeply touched and inspired by the curiosity, creativity, and ingenuity of one man who asked and answered those questions of how animals function. This book is dedicated to that man, Knut Schmidt-Nielsen.

JAMES H. JONES

Cellular Energetics during Exercise

KEVIN E. CONLEY

Magnetic Resonance Laboratory, Department of Radiology, University of Washington Medical Center, Seattle, Washington 98195

I. Introduction

Vertebrates perform a wide variety of locomotor activities, including hopping, running, swimming, and flying. The speed at which these tasks are accomplished varies from the slow, continuous contractions of maintaining posture to the transient bursts characteristic of ballistic exercise. This diversity of activities means that the speed, force, and duration of contraction generated by striated muscles range widely during exercise among vertebrates (Close, 1972; Fedak *et al.*, 1982; Lindstedt *et al.*, 1985; Taylor *et al.*, 1980). How do muscles accomplish such a variety of tasks?

To answer this question, I examine how the muscle fiber is designed to use ATP in force generation and to supply ATP to meet this contractile demand. My contention is that we can account for the diversity of functions by a remarkably similar set of systems for energy utilization and supply (Kushmerick, 1985, 1987). These common mechanisms achieve a range of muscle tasks as the result of a heterogeneity of the contractile and metabolic properties of the fibers (Kushmerick *et al.*, 1992b; Perry, 1985; Pette, 1985; Pette and Spamer, 1986). Thus, just as a basic set of principles allows gears to vary the speed of an automobile, so too do common principles allow vertebrate muscle to generate the range of performance seen in nature.

1. Contraction Sets Demand, Metabolism Meets Demand

My approach to answering the question of how muscle cells are designed is to examine how energy balance is achieved in each fiber (Kushmerick, 1977, 1983). The concept that the energy used in contraction must be balanced by processes of energy supply comes from some of the earliest myothermal studies of isolated frog muscle (Woledge *et al.*, 1985). These principles were extended to include a biochemically based energy balance by Kushmerick (1983). In these experiments, contractile costs were separated from metabolic costs of recovery to evaluate the basic biochemical and physiological processes of energy supply and demand. I use this same approach of separating the contractile and metabolic processes to evaluate the basic design of muscle fibers and to examine how the diversity of muscle function is accomplished.

The balance between energy demand and supply necessary for muscle contraction over both the short and long terms (Krisanda *et al.*, 1988) is shown as sources and sinks for ATP in the muscle cell in Fig. 1. The demands for ATP during muscle contraction are set by the myosin and the ion pumps, for example, the sarcoplasmic reticulum (SR) (Klug and Tibbits, 1988; Rall, 1985). This demand is met by ATP synthesis in glycolysis and by oxidative phosphorylation in the mitochondria, and is sustained by a continuous supply of carbon fuels and

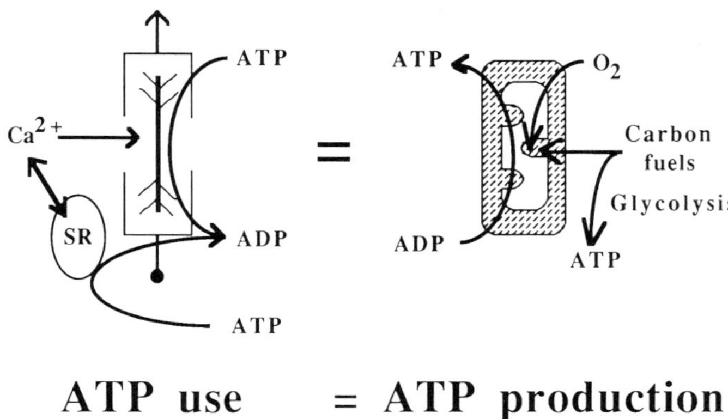

FIG. 1. Diagram of the major sources and sinks for ATP that are involved in muscle energy balance. SR stands for the sarcoplasmic reticulum.

oxygen from intracellular stores and via the cardiovascular system (Taylor *et al.*, 1987; Weber, 1988).

2. *Creatine Kinase Integrates Supply and Demand*

Central to ATP balance in the muscle cell is creatine phosphate (PCr), which maintains constant ATP concentrations in the cell (Ellington, 1989; Meyer *et al.*, 1984). Creatine kinase (CK) is the enzyme that catalyzes the exchange of high-energy phosphate (~P) between ATP and PCr via the near-equilibrium reaction (Lawson and Veech, 1979)

$$PCr + ADP + H^+ \leftrightarrow ATP + Cr. \tag{1}$$

At the onset of exercise, PCr is broken down to regenerate ATP via the CK reaction and eventually plateaus at a stable concentration when oxidative ATP synthesis balances the contractile demand. Once exercise has ended, this energy buffer is refilled to its original concentration by PCr resynthesis via oxidative phosphorylation (Kushmerick and Paul, 1976). Separation of breakdown from the synthesis processes in a human arm is shown in Fig. 2. Contractile ATP use is determined by eliminating oxygen delivery during contraction with imposed ischemia. The resultant linear breakdown of PCr and stability of ATP in the absence of oxidative synthesis is apparent during contraction. The

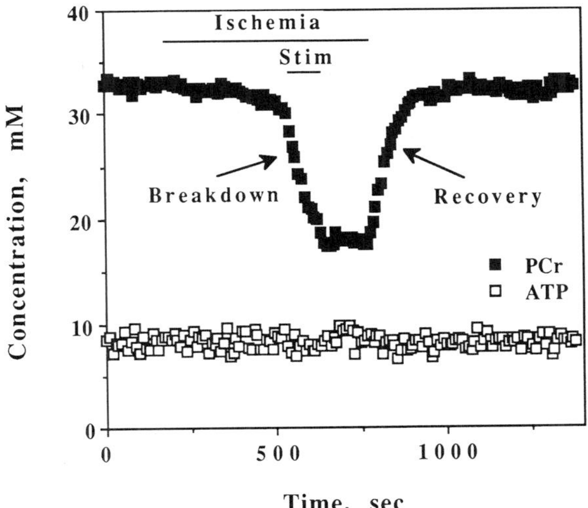

FIG. 2. PCr changes at constant ATP concentration during ischemic stimulation and aerobic recovery in human arm muscle. Stim stands for the period of electrical stimulation of the muscle.

aerobic recovery of PCr to its original concentration occurs when the tissue is reoxygenated. In the following pages, I will show how one can utilize the properties of the near-equilibrium creatine kinase reaction to determine the contractile cost of exercise and the metabolic capacity for ATP supply.

II. Setting Demand: Energy Utilization

What consumes energy in contracting muscle? Two tasks use the majority of the ATP during contraction: generating and maintaining force, and turning the muscle on and off (Rall, 1985). The maximal force generated per cross-sectional area of muscle is similar among vertebrates, but the frequency of activation and the rate at which this force is generated varies greatly among animals (Heglund and Taylor, 1988; Taylor *et al.*, 1980). The following section deals with the costs associated with force generation by the cycling of actomyosin cross-bridges, and with the activation and relaxation of the muscle by cycling calcium.

1. Exercise Can Increase ATP Utilization 1000-Fold

Resting muscle uses ATP to maintain ion gradients and enable the excitability of the sarcolemma, as well as to carry out basic processes, for example, protein synthesis and degradation. With the onset of contraction, additional recruited cellular ATPases provide a sink for ATP. The interaction between actin and myosin uses ATP in the cycling of cross-bridges, by which the muscle generates force. Activation and relaxation of the muscle also require ATP for the sarcoplasmic reticulum to sequester the Ca^{2+} that was released to trigger contraction. In the classic preparation for muscle studies, frog muscle at 0°C, the cross-bridge and calcium cycling during tetanic stimulation increases the rate of ATP utilization by 1000-fold (Kushmerick, 1983). However, because skeletal muscle has higher basal ATP needs at homeothermic body temperatures, this increase in energy use during contraction in mammals ranges from 40-fold in human muscle (Blei *et al.*, 1993a) to 400-fold in hummingbird muscle (Hochachka and Matheson, 1992). The major portion of the cost is due to cross-bridge cycling (~70–85%) and turning the contractions on and off (calcium cycling, ~15–30%) (Crow and Kushmerick, 1983; Rall, 1985), whereas the pumping of ions across the sarcolemmal membrane consumes a considerably smaller fraction of total ATP.

2. *ATPase Rate Is Proportional to Maximal Speed*

Bárány (1967) examined a wide range of animal muscles from invertebrates to mammals and found that the maximal rate of ATP use varied in proportion to maximal shortening velocity over a 200-fold range of speeds. For vertebrate fibers in the same animal, the range of maximal speeds and ATPase rates is closer to 3- to 5-fold (Chase and Kushmerick, 1988; Crow and Kushmerick, 1983; Elzinga *et al.*, 1987). These results show that ATP utilization is directly related to contraction velocity and, therefore, to the rate of cross-bridge cycling.

The ATPase rate and contraction speed are often used to classify vertebrate fibers. Histochemical stains in combination with functional measurements reveal classes of fibers that broadly correspond to the underlying myosin isoform and metabolic properties of each fiber. At one extreme are the highest ATPase rates and fastest myosins, which are termed fast-twitch, or Type II, fibers. At the other extreme are lower ATPase rates and slower myosin isoforms in slow-twitch, or Type I, fibers. These classifications are often subdivided to include metabolic properties, for example, fast-twitch glycolytic (FG), fast-twitch oxidative (FOG), or slow-twitch oxidative (SO). I use these categories to describe the basic design properties of fibers with the caveat that single-fiber analysis reveals a continuum of myosin isoform and metabolic properties among these discrete classifications (Pette, 1985; Pette and Spamer, 1986). Given the range of maximal speeds and ATP costs found among fibers, do fiber types differ in the cost of a single contraction?

3. *Measuring Cost/Twitch*

Characterizing contractile costs requires segregating out the metabolic processes associated with resynthesizing energy stores (Kushmerick, 1983). A separation of the contractile from the metabolic components of the energy balance was first accomplished in frog muscle (Woledge *et al.*, 1985). Cooling this muscle to 0°C slows the ATP synthesis processes to a negligible rate during contraction and permits measurement of energy use in the form of the heat released during contraction. We have separated these components in the human arm *in vivo* by halting oxygen delivery to the muscle during contraction (Fig. 2; Blei *et al.*, 1993a). In these experiments, inflation of a cuff around the upper arm occludes blood flow to the forearm and eliminates oxygen delivery. Once the muscle oxygen stores are depleted by basal metabolism, oxidative resynthesis of ATP during contraction is neglig-

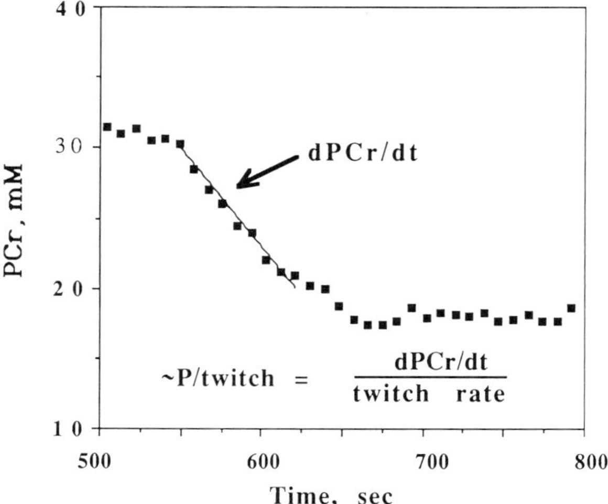

FIG. 3. Creatine phosphate (PCr) breaks down linearly during ischemic stimulation in the human arm and provides a measure of the contractile cost per twitch (~P/twitch).

ible, and PCr breakdown rate then measures the ATP cost of contraction (Fig. 3). As in the frog experiment, the forearm muscles are electrically stimulated with supramaximal voltages to activate all fibers. The glycolytic contribution is small in these experiments (6% in frog and <15% in human forearm) and PCr breakdown during contraction is linear. Typically, this breakdown continues with no change in ATP concentration until PCr is almost exhausted, as expected if PCr is in near equilibrium with ATP via the CK reaction. We have found that cost/twitch is constant over stimulation frequencies of 0.5 to 2.0 Hz, before the twitches fuse into a tetanus (3.0 Hz). This protocol has allowed us to measure directly the ATP cost per twitch *in vivo*.

4. *Contractile Cost Depends on Mechanics*

The mechanical conditions under which a muscle is activated have been found to strongly affect the chemical energy costs of muscle contraction. The importance of mechanical state was best demonstrated in the classic studies of Fenn on frog muscle, in which energy cost increased during contraction if a load was applied (Rall, 1982). This result, termed the "Fenn effect," demonstrated that muscle contraction is not a viscoelastic process that releases energy like a recoiling spring, that is, independent of the load; instead, the mechanical conditions of contraction affect the muscle's energetic cost. Not only is the Fenn effect important in terms of understanding the basic mechanisms of muscle contraction, but also because a number of mechanical states have been used to study muscle, including isometric (constant length) and isotonic (constant force) conditions (Rall, 1982; Woledge *et al.*, 1985). Isometric contractions provide a measure of muscle economy as the energy cost measured as the integral of the tension developed over the time of the twich (time–tension integral or TTI). Isotonic conditions permit determination of muscle efficiency because the muscle changes in length and thereby performs mechanical work. These two mechanical conditions impose very different energetic costs of contraction. A faster cross-bridge cycling rate results in a higher cost/TTI in an isometric contraction (i.e., lower economy), but efficiency does not differ greatly among fibers because the cycling rate of cross-bridges is reflected in the work performed (Heglund and Cavagna, 1985, 1987). Which mechanical condition is relevant to how muscles perform *in vivo*?

The key to answering this question is to determine whether muscles generate force with little change in length or if they shorten during contraction. The implication for energetics is that force generation with only slight shortening or while lengthening makes possible the elastic and nonelastic storage of energy that can be recovered later in the limb cycle (Biewener *et al.*, 1981; Cavagna *et al.*, 1985). Muscle length changes have been examined contemporaneously with electrical activity (electromyography) during contraction in locomoting animals (Goslow *et al.*, 1981). Over a range of gaits, limb muscles of the dog were found to shorten little during a significant fraction of the period of electrical activity, and several muscles were found to be stretched prior to or during contraction. Stretching muscle and tendons during locomotion is consistent with the elastic storage of kinetic energy that can be recovered later in the stride. The limited extent of muscle shortening during contraction and the storage of elastic energy both indicate that muscles operate during locomotion *in vivo* primarily by generating isometric force (Rall, 1985). The significance of the mechanical state to contractile energetics is discussed in the following.

5. *Cost/Twitch Varies with Fiber Type (Myosin Isoforms)*

Single-fiber analysis has shown that maximal ATPase rate and shortening velocity vary among myosin isoforms (Chase and Kushmerick, 1988; Elzinga *et al.*, 1987). The dependence of muscle contractile demand on the isoform has also been shown by examination of muscles composed primarily of a single fiber type by Crow and Kushmerick (1982). These researchers took advantage of the principle of energy balance to measure the contractile energetics of slow-twitch mouse *soleus* and fast-twitch mouse *EDL* (*extensor digitorum longus*) using the recovery oxygen consumption after a brief tetanus. They measured a greater oxygen consumption per TTI for the fast-twitch *EDL* compared to the slow-twitch *soleus*, consistent with there being a greater ATP utilization with the higher cross-bridge cycling rate. A similar result is found in muscles transformed in fiber type. The *EDL* in mice fed the creatine analogue, β-guanidinopropionate (β-GPA), shift from predominantly fast-twitch to slow-twitch myosin (Moerland *et al.*, 1989). The relationship between oxygen consumption and TTI in the transformed *EDL* shifted to that found in the slow-twitch *soleus* (Moerland and Kushmerick, 1993). These results confirm that the myosin isoform strongly influences the cost of contraction. Thus, the greater ATPase activity and maximal speed of fast-twitch fibers are reflected in a larger ATP utilization per TTI (cost/TTI) under isometric conditions compared to slow-twitch fibers.

This dependence of contractile cost on fiber type can be used *in vivo* to evaluate the fiber type composition of muscles (Blei *et al.*, 1993a). Surgical biopsies of the wrist muscle, *flexor carpi radialis*, from healthy human subjects show that fast-twitch composition varies among individuals from 35 to 85% (Mizuno *et al.*, 1992). Our ischemic stimulation protocol activates all fibers in a muscle so that the cost/twitch represents the summed ATPase rate of the underlying fibers. This procedure reveals that the cost/twitch reproducibly varies among subjects for the wrist flexor muscles by a 1.6-fold range (0.29–0.45% PCr/twitch; Blei *et al.*, 1993b), similar to the range of variation in fiber types in this muscle. These results suggest that the differences in fiber composition of the muscle are reflected in the cost/twitch.

Faster myosin isoforms not only cycle cross-bridges more rapidly, but must also cycle calcium more quickly so that contraction and relaxation can keep pace (Klug and Tibbits, 1988; Rall, 1985). A rapid rate of cross-bridge cycling therefore requires an equally rapid rate of Ca^{2+} release and resequestration. The cost of this activation and relaxation has been determined by stretching the sarcomeres beyond any overlap of actin and myosin so that only calcium cycles with each stimulation (Crow and Kushmerick, 1983; Rall, 1985). For the few muscles studied, both cross-bridge and calcium cycling rates increase proportionally with maximum shortening velocity. Comparison of fast- and slow-twitch mouse muscles differing in maximal speed by a factor of three reveals a similar relative contribution of cross-bridge and Ca^{2+} cycling to total energetic costs (i.e., 15–30% for Ca^{2+} cycling). The fraction of contractile cost associated with cross-bridge and with calcium cycling is, therefore, similar for a variety of fiber types.

Activation costs have not been systematically examined across muscles or animals, and a number of examples exist where calcium cycling costs may be a larger proportion of contractile costs than in mammalian muscle. An extreme example of higher cycling costs is shown by the low efficiency of locust flight muscle, which operates at 20 Hz (Josephson and Stevenson, 1991). Despite extensive elastic storage, this muscle has an efficiency of about 6%, as compared to 15–30% in vertebrate muscles (Heglund and Cavagna, 1985, 1987). Josephson and Stevenson (1991) suggest that this low efficiency is due in part to the high activation costs associated with the 20-Hz wing beat rate. With the assumption that cross-bridge cycling has an efficiency of 20%, these workers estimate that activation costs are two to four times higher in the locust than those measured in vertebrate muscle. Support for the contention that the fraction of contractile costs associated with activation increases at high twitch rates comes from the rattlesnake tail-shaker muscle, which can maintain an even faster twitch frequency than the locust (i.e., 90 Hz; Chadwick and Rahn, 1954). The large proportion of sarcoplasmic reticulum (27% by volume versus <10% in mammals; Schaeffer and Lindstedt, 1992) in this muscle indicates a high activation cost relative to cross-bridge cycling. Thus, the fixed proportion of cross-bridge to calcium cycling costs found in mammalian fibers may not hold for muscles that sustain high twitch rates.

6. *Cost Depends on Body Size*

The mechanics of locomotion change systematically with body size (Heglund *et al.*, 1982). By examining animals of different body masses, we can determine the mechanics under which muscle generates force and the mechanical basis of the costs of contraction. These allometric studies have revealed that the cost of locomotion increases as body size decreases in terrestrial animals, while mass-specific cost/step is constant (Heglund and Taylor, 1988). What explains the size dependence of locomotory costs in the face of the size independence of cost/step?

A major mechanical factor that scales with body size is the stride frequency required for each speed. As body size decreases, more strides are needed to cover a given distance. The cost of locomotion increases as more strides are taken because muscles must generate force and be turned on and off more frequently. The inverse relationship between cost per distance and body size holds for animals as large as several hundred kilograms, such as running horses, all the way down to tens of milligrams, such as insects (Full, 1989).

Also associated with the higher stride frequency of smaller animals is a greater intrinsic (or maximum) velocity of the muscle fibers (Heglund and Taylor 1988; Lindstedt *et al.,* 1985; Rome *et al.,* 1990). Smaller animals not only take more steps per distance but also cycle cross-bridges at a greater rate than larger animals. How can the constancy of the mass-specific cost per step be reconciled with this greater cross-bridge cycling rate per step?

The answer to this paradox comes from measurements made by Kram and Taylor (1990) and a scaling analysis of Rome (1992) concerning the time the foot is in contact with the ground. Because foot contact time decreases with body size, the time-tension integral during which the muscle generates force also decreases. The lower stride frequency of smaller animals is therefore accompanied by a reduced foot contact time and time-tension integral of force generation as compared with larger animals. These results indicate that there is a trade-off between the cross-bridge cycling rate and the period of force generation so that mass-specific cost per step is constant across body sizes.

The cellular basis of energy demand during contraction originates from two major activities: activation (turning the muscle on and off) and force generation. The resultant cost of exercise increases as body size decreases because smaller animals must activate and generate force more frequently per unit distance than larger animals.

III. Meeting Demand

If demand can increase 100- to 1000-fold with exercise, how is energy mobilized in so short a time? An excellent synthesis of the design of biochemical energy supply pathways throughout the animal kingdom is provided by Hochachka (1985) and addresses this question in broad terms. In this section, I will review how immediate and long-term energy requirements for contraction are met in vertebrate striated muscle cells. The first question examined is how energy supply is related to the time course of energy demand. Then the possible mechanisms for matching long-term energy supply to sustained contractile demand are presented.

A. Energy Supply

Muscle contraction can be sustained as long as energy balance is achieved, or until fatigue sets in owing to buildup of inhibitory byproducts of contraction and/or excitation failure (Westerblad *et al.,*

1991; Wilkie, 1986). Energy use and the onset of fatigue are both related to the pathway of energy supply. For short-term needs, PCr and glycolysis can support a large ATP flux rate, but they generate by-products that cause fatigue (i.e., H^+ and inorganic phosphate, Pi). Longer-term ATP needs are sustained at a lower rate by oxidative phosphorylation, avoiding the buildup of inhibitory metabolites by using the end products of PCr utilization and glycolysis to generate ATP.

Short-term, rapidly mobilized energy is provided by the energy store in the cell, PCr (Kushmerick and Meyer, 1985), and by the glycolytic synthesis of ATP (Connett, 1987). The near-equilibrium nature of the creatine kinase reaction means that PCr is an immediate source of ~P for maintenance of ATP concentration during contraction (Meyer *et al.*, 1984). Glycolysis mobilizes energy on the time scale of seconds during exercise by catabolizing glucose or glycogen (Connett, 1987). These sources typically meet ATP needs for periods on the order of seconds to minutes, but result in the accumulation of metabolic by-products in the absence of oxidative phosphorylation. The use of PCr generates Pi and, in the absence of oxidation, glycolysis generates lactate, causing pH to drop. Both by-products have effects on muscle that facilitate fatigue (Wilkie, 1986), such as attenuating Ca^{2+} cycling (Westerblad *et al.*, 1991) and inhibiting the interaction between actin and myosin (Chase and Kushmerick, 1988). Muscle contraction at rates above the oxidative capacity is, therefore, limited by immediate energy supply (PCr and ATP), by the buildup of metabolic by-products, and the ability to cycle Ca^{2+}.

To sustain muscle contraction for periods exceeding a few minutes, other sources of ATP are needed, and the by-products of energy supply must be used to avoid inhibition of muscle activation and contraction. Oxidation of the glycolytic end products by the mitochondria provides a longer-term supply of ATP and reduces the buildup of metabolites. This oxidative ATP synthesis requires sustained supply of oxygen and carbon fuels, such as carbohydrate and fat, to continue ATP generation to meet demand (Weber, 1988). How the fuel and oxygen supply needs of the mitochondria are met is discussed in the final sections of this article.

1. Metabolic Properties Are Heterogeneous

Just as contractile demands vary among fibers depending on the myosin and SR ATPases, the capacity of energy supply pathways also differ among muscle fibers (Chi *et al.*, 1986). Two extremes characterize the relationship between energy demand and supply: high force

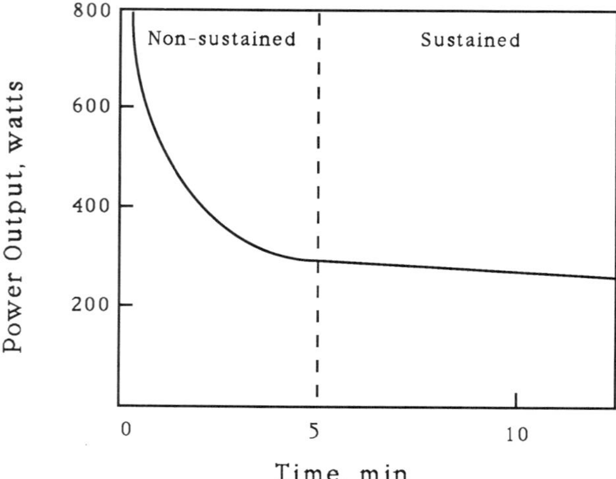

FIG. 4. Maximal power output of human legs as a function of time during bicycling. This plot illustrates that high power output decreases rapidly, whereas lower power output can be sustained. Adapted from Wilkie (1985).

generation with limited duration, and low force generation that can be sustained. The consequence of this demand/supply trade-off is illustrated in Fig. 4 by the duration of power output as a function of time for human bicycling (Wilkie, 1985). The higher power outputs generated by fast-twitch fibers are short term and cannot be sustained for more than a few minutes ("nonsustained" in Fig. 4). These fast-twitch fibers have high energy needs and rely on short-term energy supply pathways such as PCr and glycolysis, which generate inhibitory byproducts (e.g., H^+ and Pi). For example, fast glycolytic fibers (Type IIB) with the highest power outputs also have the largest energy needs and fatigue first (Armstrong and Laughlin, 1985). The fast oxidative fibers (Type IIA) with lower power outputs and energy needs are able to contract for longer periods. An animal can achieve a constant power output that can be sustained using by slow-twitch oxidative, or Type I fibers. These fibers generate less power and have a lower energy demand that can be met by long-term energy supply. Thus, several factors are involved in determining the duration of muscle contractile activity:

1. the nature of the demand process—high versus low rate of ATP use;
2. the nature of the supply process—short- versus long-term ATP synthesis;
3. whether inhibitory products are generated in the supply process—glycolysis versus oxidative phosphorylation.

TABLE I

METABOLIC PROPERTIES OF MOUSE MUSCLES ILLUSTRATING THE *TWITCH NOW, PAY LATER* (*EDL*) AND *PAY AS YOU GO* (*SOLEUS*) STRATEGIES[a]

Property	Fast twitch (*EDL*)	Slow twitch (*soleus*)	*EDL*/*soleus*
ATP demand			
Cost/TTI ($\Delta \sim$ Pg N^{-1} m^{-1} sec^{-1})	23 (11)[b]	9	2.5 (1.2)[b]
ATP supply			
PCr content (μmol g wet^{-1})	24	13	1.8
Glycolytic capacity (μmol g^{-1} s^{-1})	4.3	1.3	3.3
Oxidative capacity (μmol g^{-1} s^{-1})	0.3	0.9	0.3

[a] Data are taken from Crow and Kushmerick (1982) and Kushmerick *et al.* (1992a).

[b] Results in parentheses indicate the decreased cost/TTI at high time-tension integrals in the *EDL*.

This trade-off between power output (energy demand) and the pathway for energy supply is exemplified in Table I for representative muscles of the mouse containing primarily fast-twitch (*EDL*, 99% FT) or slow-twitch (*soleus*, 76% ST) fibers. These muscles represent the two strategies for meeting energy demand that Kushmerick has termed the *Twitch now, pay later* (*EDL*) and the *Pay as you go* (*soleus*) approaches (Kushmerick, 1985, 1987).

2. *Twitch Now, Pay Later*

Fast-twitch myosin isoforms are used in ballistic or sprint exercise and have a large ATP demand, but a low capacity for sustained ATP supply (Armstrong and Laughlin, 1985). These fibers typically use rapidly mobilized, but short duration energy sources, resulting in nonsustainable power output (Fig. 4). For example, the fast-twitch *EDL* has a large PCr buffer and high glycolytic enzyme activity for ATP supply, but low oxidative capacity for ATP resynthesis. Thus, contractile demand in this muscle is sustained by breakdown of PCr and by glycolysis during the short term, resulting in a debt that is repaid after exercise by oxidation of lactate and resynthesis of the energy store (PCr). Because PCr is rapidly depleted and glycolytic ATP synthesis is accompanied by the buildup of H^+, contraction duration is limited by both energy supply and the fatigue that results from the accumulation of inhibitory metabolites.

3. *Pay as You Go*

Fibers that must pay as they go have a limited short-term buffer (lower PCr content) and reduced glycolytic activities, but a larger long-term supply capability inherent in their higher oxidative capacities. The small PCr buffer and large oxidative capacity mean that ATP breakdown during contraction is quickly matched by oxidative ATP synthesis. The slow-twitch mouse *soleus* is an example of a muscle with reduced maximal power output capacity (glycolytic capacity), but high aerobic capacity for oxidation of glycolytic end products and recycling of phosphorus metabolites (Pi and ADP) for sustained ATP synthesis.

Heart muscle is an excellent example of the tight linkage between contractile demand and ATP supply that is inherent in the pay as you go strategy. Cardiac muscle typically has a PCr content lower than that of slow-twitch fibers and has little excess glycolytic capacity relative to that needed for oxidative phosphorylation (Kushmerick, 1987). Thus, over the full operating range of the heart, the ATP demand of each beat must be met by oxidative ATP synthesis.

These two design strategies for muscle fibers are also seen in nonmammalian muscles. Single-fiber analysis of muscle from the amphibian *Xenopus laevis* shows a range of fiber properties similar to that of the fast- and slow-twitch muscles of the mouse (Elzinga *et al.*, 1987). Differences between fast- and slow-twitch fibers in cost/twitch (twofold), ATPase rate (fourfold), and oxidative capacity (fivefold) correspond to the trade-off between speed and oxidative capacity seen in mammalian fibers (Table I). These design strategies represent the extremes of fiber properties, within which muscle fibers form a continuum of myosin isoform and metabolic properties (Pette, 1985; Pette and Spamer, 1986). The independence of contractile and metabolic properties can be seen in the results of muscle adaptation studies. In endurance-trained human muscle, mitochondrial volume can increase in each of the three fiber types (Howald *et al.*, 1985). Similarly, cold acclimation in fish results in increased mitochondrial volume densities in both the white and red muscles (Egginton and Sidell, 1989). These results suggest that mitochondrial content and myosin properties are independently regulated traits of a fiber.

Some muscles do not trade contractile speed for oxidative capacity, but instead have a rapid contraction rate and high oxidative capacity. The rattlesnake tail-shaker muscle (90 Hz; Chadwick and Rahn, 1954), hummingbird *pectoralis* (30 Hz), and locust flight muscle (20 Hz; Josephson and Stevenson, 1991) all sustain these rapid contractions for extended periods. These muscles are characterized by fast-twitch fibers with high oxidative capacity for long-term energy supply (Schaeffer and Lindstedt, 1992; Suarez *et al.*, 1991). Initial results for the rattlesnake muscle show that in addition to its large long-term supply capacity, it also possesses a large energy buffer capacity (PCr = 32 m*M*) typically associated with fast-twitch fibers. Thus, a number of specialized strategies in addition to the two defined here are used by muscle fibers to balance energy supply and contractile demand.

B. Sustaining Supply to Meet Demand

A balance between energy supply and energy demand is critical for sustained muscle performance. There are two primary components underlying the energy balance in aerobic muscle metabolism (Fig. 5). First, ATP synthesis by the mitochondria must match the breakdown of ATP by myosin and SR ATPases during muscle contraction. Second, oxygen and substrate supplies must meet the oxidative demands of the mitochondria to sustain ATP synthesis.

1. Contractile Demand and Oxidative Supply

The first factors that must be balanced for aerobic exercise to be sustained are contractile ATP demand and ATP supply by oxidative phosphorylation. Many factors and mechanisms have been implicated in the control of respiration, as several reviews have expertly discussed (Balaban, 1990). My discussion is limited to the mechanisms by which increased contractile demand could signal an elevation in ATP synthesis. To determine what links energy supply to demand, I will focus on those factors governing oxidative phosphorylation that are also involved in the contractile demand for ATP.

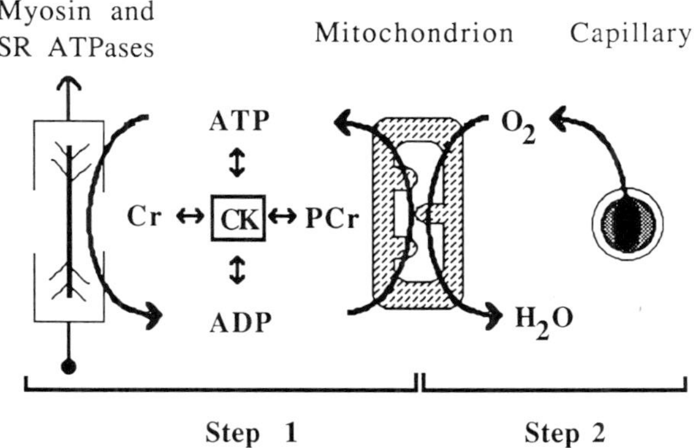

FIG. 5. Scheme of the two steps in the energy balance underlying sustained muscle contraction. Step 1 is the ATP balance in which mitochondrial ATP supply meets contractile ATP demand. Creatine kinase (CK) maintains ATP concentrations via PCr. Step 2 involves the delivery and consumption of oxygen and carbon fuels to sustain ATP supply.

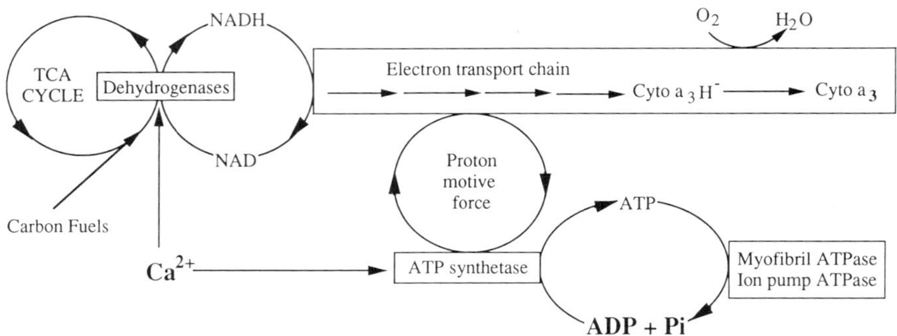

FIG. 6. Diagram of the major steps in oxidative phosphorylation in the mitochondrion. Factors related to contraction are also involved in oxidative phosphorylation: Ca^{2+}, ADP, and Pi. TCA stands for the Tricarboxylic Acid Cycle.

The synthesis of ATP involves not only the metabolites of muscle contraction, ADP and Pi, but also carbon substrates and oxygen, as shown in Fig. 6 (Nicholls and Ferguson, 1992). Carbon fuels provide the reducing equivalents (e.g., NADH) for the electron transport chain (ETC). According to the chemiosmotic theory, the electrons provided by NADH pass down the ETC and provide the energy to transpose protons across the inner mitochondrial membrane. Oxygen acts as the terminal electron acceptor that maintains this electron flow. The electrochemical gradient or membrane potential generated by the transposition of protons provides the proton motive force for the ATP synthetase to phosphorylate ADP, forming ATP. Thus, a number of factors must be supplied for oxidative phosphorylation, but of these, only ADP and Pi are directly related to the process of muscle contraction.

Calcium is a second factor involved in oxidative phosphorylation that is also related to contraction, as it is the trigger for the interaction between actin and myosin (Denton and McCormack, 1990; Hansford, 1985; McCormack *et al.*, 1990). Release of Ca^{2+} alters the activity of mitochondrial dehydrogenases, increasing NADH production. Calcium may also activate the ATP synthetase to increase ATP generation during contraction. Thus, two links are apparent between muscle contraction and ATP supply (Fig. 6). The first link is ADP and Pi, which are the substrates for phosphorylation and ADP is a reactant in the creatine kinase reaction. These metabolites act as feedback signals that are generated by contraction and that trigger an increase in supply. A second link is calcium, which may act as a feed-forward signal by simultaneously turning on the demand for and supply for ATP. Although many signals are possible, these metabolites provide a clear link between the contractile demand for ATP and the oxidative supply of ATP by the mitochondria.

2. Feedback Control of Energy Supply

The long-term matching of ATP supply to demand is related to the drop in PCr during exercise in skeletal muscle (Fig. 2) (Funk *et al.*, 1990; Meyer, 1988). This PCr change represents a shift in the muscle's energy state as a result of increased contractile demand. In control theory, this shift represents an error signal that indicates a mismatch of energy supply and demand (Chance *et al.*, 1985, 1986b). The feedback circuit operating in muscle uses the error signal to eliminate the mismatch between energy supply and demand by activating ATP synthesis to balance demand. A new steady-state is reached and net PCr breakdown cease during exercise once this signal elevates ATP synthesis enough to meet contractile energy needs. How does the change in PCr generate a signal that activates mitochondrial ATP supply?

3. Feedback Control Model

A link between the increased contractile demand signaled by the decrease in PCr and mitochondrial respiration occurs via the creatine kinase reaction. The drop in PCr changes the CK equilibrium so that ADP and Pi increase (Kushmerick, 1983). Decades ago ADP was found to control oxidative phosphorylation in isolated liver mitochondria experiments by Chance and Williams (1956). Respiration state was characterized in these experiments using saturating concentrations of substrates and oxygen and limiting levels of ADP. Similar conditions hold in normally functioning skeletal muscle because:

1. the oxygen partial pressure required to saturate the terminal oxidases is low ($<$0.5 Torr);
2. intracellular stores of glycogen and lipid are present to meet substrate needs;
3. ADP is low and potentially limiting to respiration.

Conditions in the muscle cell differ from the *in vitro* mitochondria studies in the Pi level. Since these early studies, Pi has been found to be a regulator of mitochondrial respiration and ^{31}P MRS measurements have shown Pi to be low and potentially limiting in resting muscle.

Are the ADP and Pi levels that result from the PCr changes during exercise sufficient to stimulate mitochondrial respiration in a feedback control mechanism (Chance *et al.*, 1986b; Kushmerick *et al.*, 1992a)? Exercise causes changes in Pi that can exceed 20 mM but ADP levels remain in the micromolar range, which are too low to be detected in most experiments. High-resolution ^{31}P spectroscopy shows that the ADP values calculated via the CK reaction correspond with measured ADP in smooth muscle confirming our measurements of metabolite levels and that the CK reaction is near equilibrium (Fisher and Dillon, 1988). However, in order to evaluate whether these changes in ADP and Pi operate in feedback control, we need to know the relationship between these metabolites and phosphorylation rate *in vivo*.

4. Feedback Control Models

The mechanisms linking the change in ADP and Pi to oxidative phosphorylation remain under intense investigation. Both kinetic and thermodynamic control models for muscle incorporate ADP and Pi as feedback signals. The thermodynamic scheme is based on the chemiosmotic control of oxidative phosphorylation via the mitochondrial membrane potential. This model incorporates ADP and Pi into the cytosolic phosphorylation potential [ln(ATP/ADP*Pi)] which, along with the membrane potential, sets the driving force for oxidative phosphorylation. The kinetic scheme uses ADP and Pi in a general reactant-limited process (Palsson and Lightfoot, 1984) which is represented by a hyperbolic relationship between phosphorylation rate and substrate level. This scheme is termed the Michaelis–Menten relationship in biochemistry and an expanded form of this relationship is used to represent the role for both ADP and Pi in mitochondrial respiration (Chance *et al.*, 1986a,b).

$$v = V_{max}(\mathrm{ADP*Pi})/(K_m{}'*K_m{}'')/(1 + \mathrm{ADP}/K_m{}' + \mathrm{Pi}/K_m{}'' + (\mathrm{ADP*Pi})/(K_m{}'*K_m{}'')), \qquad (2)$$

where v is the phosphorylation rate, V_{max} is the maximum phosphorylation rate, and $K_m{}'$ and $K_m{}''$ are the binding constants for ADP (0.025 μM) and Pi (1 mM), respectively.

Since both the thermodynamic and kinetic schemes use ADP and Pi as feedback signals, distinguishing which model operates in respiratory control has not been possible (Connett and Honig, 1989; Connett, 1988) and recent analyses suggest that both schemes may simply be different means of expressing the same underlying processes (Westerhof and van Dam, 1987). The key feature of both of these models is that feedback control of oxidative phosphorylation via ADP and Pi is consistent with the mechanisms for regulating mitochondrial respiration. Further, these models illustrate that as ADP and Pi increase with exercise, phosphorylation rate increases to V_{max} thereby providing a means of determining the oxidative phosphorylation capacity of a muscle.

5. *Measuring* V_{max}

We have recently taken advantage of the role of ADP and Pi in respiration to measure the maximum phosphorylation rate *in vivo* in human arm. The essence of this procedure is to raise ADP and Pi to saturating levels by exercise to elicit the maximum phosphorylation rate (i.e., V_{max}). The key steps in this procedure appear in the upper panels of Fig. 7.

The experiment works as follows. During the ischemic stimulation protocol, NAD is fully reduced to NADH and [ADP] and Pi rise to $>175\ \mu M$ and $>20\ \mu M$, respectively. Both substrates are at least 7×

Physiological

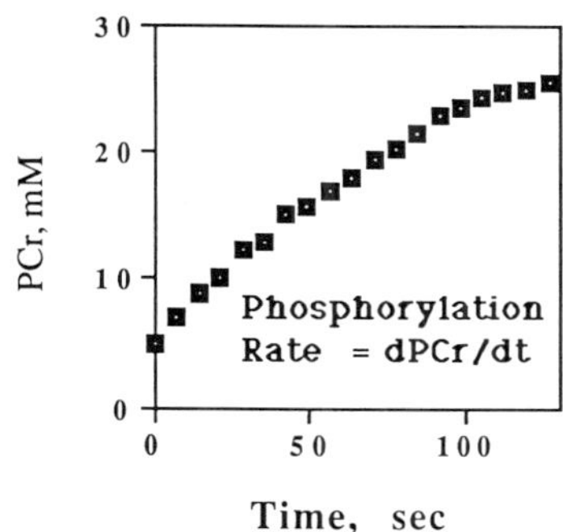

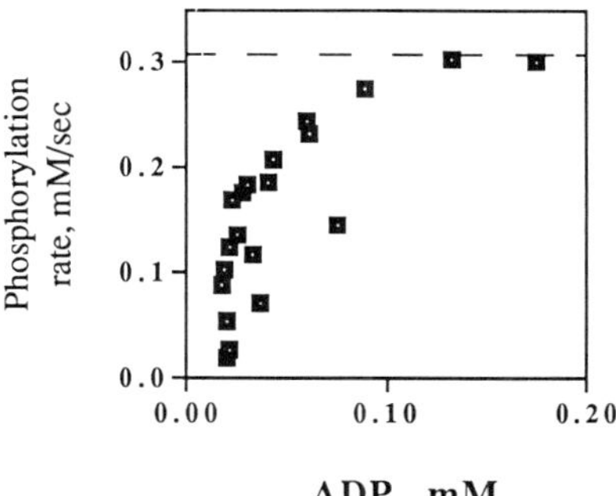

Morphological

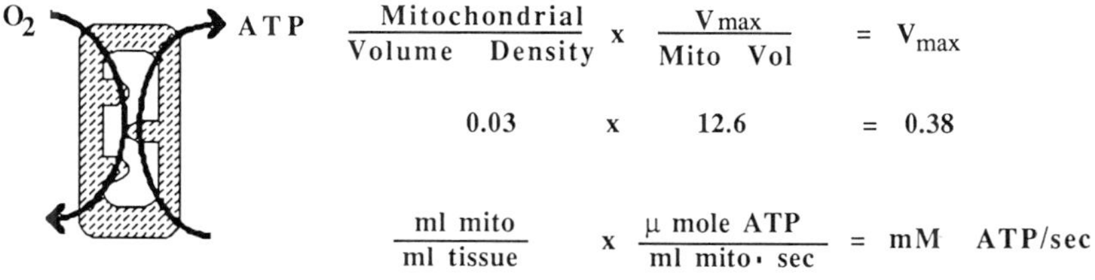

FIG. 7. Scheme showing two methods for independently estimating the oxidative capacity of muscle (V_{max}). The physiological method uses the aerobic PCr recovery following exercise, as determined by ^{31}P nuclear magnetic resonance spectroscopy to measure V_{max}. The morphological method uses the mitochondrial properties of the forearm and the oxidative capacity of isolated mitochondria.

their K_m according to Eq. 2. These high substrate levels leave oxygen as the sole limiter of respiration (Fig. 6). With release of the ischemic cuff, reperfusion rapidly supplies sufficient oxygen (>0.5 Torr P_{O_2}) to remove the limitation to oxidative phosphorylation. Once sufficient oxygen is supplied, ADP, Pi, and NADH activate oxidative phosphorylation and PCr recovery begins (upper left panel in Fig. 7). This recovery allows us to determine the phosphorylation rate as a function of ADP as is shown in the upper right panel in Fig. 7. The asymptote of this curve is the maximum phosphorylation rate at saturating substrate concentrations—V_{max} by definition (Chance *et al.*, 1985, 1986b). The dotted line in Fig. 7 (left panel) defines the V_{max} under these conditions as 0.31 m*M*/sec, which is 40-fold higher than resting (i.e., 0.008 m*M*/sec; Blei *et al.*, 1993a). How does this *in vivo* physiological measure of V_{max} compare with an independent estimate from mitochondria?

An estimate of V_{max} in the *flexor carpi radialis* can be derived from the mitochondrial volume of this forearm muscle as shown in the lower panel of Fig. 7. This morphological measure is based on the *in vitro* oxygen consumption rate determined for maximally respiring isolated mitochondria at 30°C (Schwerzmann *et al.*, 1989), which requires additional information to be applied to the arm: muscle temperature, P/O_2 and mitochondrial volume density. The temperature in muscle following ischemia has been found to be 33.5°C (Harris *et al.*, 1975), which elevates the *in vitro* mitochondrial rate by a factor of 1.3 (assuming that respiration rate doubles with each 10 degrees, i.e., $Q^{10} = 2$). Conversion of oxygen consumption to ATP synthesis is uncertain because the P/O_2 ranges from 4.6 in isolated hearts (Sako *et al.*, 1988) and 5 in isolated mitochondria (Nicholls and Ferguson, 1992) to 6 in isolated muscles (Crow and Kushmerick, 1982). Using a P/O_2 of 5 yields a maximal phosphorylation rate of 0.38 m*M*/s for the reported mitochondrial content of the *flexor carpi radialis* (i.e., 0.03 ml mito/ml tissue, Jerusalem *et al.*, 1975) corrected for temperature and intracellular water content (0.7 ml water/ml tissue, Meyer *et al.*, 1985; Sjøgaard and Saltin, 1982). Despite the assumptions incorporated into this calculation, the estimate of 0.38 m*M*/s for mitochondrial V_{max} in forearm muscle is reasonably close to our measured value of 0.31 m*M*/s.

Several independent methods also indicate muscle can operate at the V_{max} set by the mitochondria. A similar maximum oxygen consumption ($\dot{V}_{O_2\ max}$) was found during contraction and by uncoupling the mitochondria of single fibers of amphibian muscle (Elzinga and Laarse, 1988). For animals exercising at their aerobic capacity, the average maximal oxygen consumption of mitochondria of the musculature are strikingly similar among mammals to the $\dot{V}_{O_2\ max}$ elicited from isolated mitochondria (Hoppeler, 1990; Schwerzmann *et al.*, 1989). These data indicate that the V_{max} defined here represents the *in vivo* capacity for oxidative phosphophorylation. The functional relationship and V_{max} defined in Fig. 7 allow us to ask whether a simple feedback control of respiration can meet the contractile costs during steady-state exercise in skeletal muscle.

6. *Feedback Control during Exercise*

Evaluation of feedback control usually involves making [ADP] measurements on a portion of muscle while determining the phosphorylation rate from the oxygen consumption of the whole muscle (e.g., Kushmerick *et al.*, 1992a). Our exercise protocol allows us to eliminate errors due to differences in reference frame by using the same volume of muscle to measure ADP and Pi, phosphorylation rate, and the cost of contraction (Blei *et al.*, 1993a,b). During sustained, steady-state exercise at 2 Hz stimulation, the contractile cost for the subject in Fig. 7 was 0.14 m*M* ~P twitch^{-1}, or 0.28 m*M* ~P s^{-1} (as determined in Fig. 3) and the steady-state ADP was 0.070 m*M*. The phosphorylation rate versus ADP curve in the upper right panel in Fig. 7 yields a phosphorylation rate of ~0.25 m*M* ~P s^{-1}, which is close to that needed to balance the contractile cost during steady-state exercise (i.e., 0.28 m*M* ~P s^{-1}). Thus, the relation between phosphorylation rate and ADP determined under transient conditions predicts an ATP synthesis rate during steady-state exercise that is similar to the measured contractile cost. These results indicate that a simple feedback control model can therefore account for the energy balance achieved during steady-state exercise.

A limitation to feedback control is that muscle and locomotory performance cannot be sustained at the maximum level defined by V_{max}. For example, $\dot{V}_{O_2\ max}$ is maintained for only a few minutes in most animals. Accumulation of lactate in the blood (Seeherman *et al.*, 1981) occurs at exercise levels well below $\dot{V}_{O_2\ max}$ (i.e., <90%), which indicates net glycolytic energy production prior to reaching the aerobic capacity. Similar results are seen in single muscles or small muscle groups where sustained muscle contractions reach their limit at ADP levels (0.08 m*M*) below that necessary to elicit V_{max} (Kushmerick *et al.*, 1992a). Our observations indicate that the inability to attain steady-state at higher work levels is due to the role that glycolysis plays in ATP supply at work rates above 85% of V_{max}. When glycolysis exceeds that needed to sustain aerobic metabolism, lactate and protons accumulate and the pH drops. This proton accumulation reduces the ADP level for a given PCr level according to the CK reaction (Eq. 1), which reduces the signal for mitochondrial respiration and decreases the sustainable ATP supply. The result is that the increased glycolysis limits steady-state performance to below the limit of mitochondrial respiration. These results do not negate a feedback control of respiration, but illustrate that we do not completely understand the control of metabolic pathways such as glycolysis in cellular metabolism.

With these limitations to feedback control of oxidative phosphorylation in mind, it nonetheless appears that a link between contractile ATP demand and mitochondrial ATP supply resides in two features of the change in PCr with exercise. The short-term energy needs of contraction are met by the drop in PCr at the onset of exercise. This PCr drop also elevates ADP and Pi, which increases the longer-term synthesis of ATP to sustain contraction.

7. *Uniform Signal for Respiration*

The feedback model illustrates how mitochondria respond to a signal for increased ATP synthesis. However, muscles vary in metabolite levels (PCr and ATP; Fig. 8A), as well as in the ATPase activities that generate feedback signals. In addition, muscles also differ in their volumes of mitochondria and therefore the ATP synthesis rate controlled by these feedback signals (Hoppeler *et al.*, 1981). How can feedback control uniformly regulate mitochondrial respiration when metabolite levels, ATPase rates, and the capacity for generating ATP range so widely in skeletal muscle?

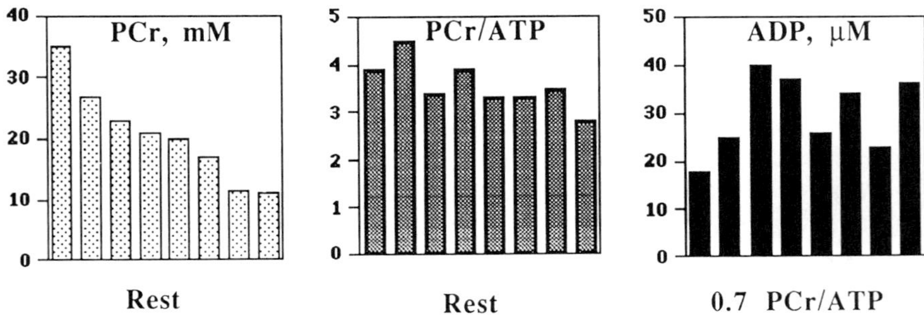

FIG. 8. High-energy phosphate concentrations in isolated mouse and rat muscles at 25°C. The individual muscles are ranked in each plot according to their resting PCr concentration in the left panel. These plots illustrate the large range in PCr but similar ratios of PCr/ATP at rest. Decreasing PCr/ATP by 30% results in ADP concentrations near the concentration that elicits half-maximal respiration in mitochondria (*i.e.*, K_m or 0.5 V_{max}). Data from Kushmerick et al. (1992b).

An answer to this question comes from apparent equilibrium between the creatine and adenylate pools (Connett, 1988). This equilibrium between the two pools results in a similar ratio of total creatine to ATP in a variety of tissues despite their 150-fold differences in absolute concentrations (Conley *et al.*, 1989a; Connett, 1988), and is also reflected in a similar PCr/ATP in resting muscle (Conley and Kushmerick, 1993). Figure 8 shows that the 4-fold range of PCr concentration in rat and mouse muscles reduces to less than a 1.5-fold range when expressed as PCr/ATP (Kushmerick *et al.*, 1992b). We propose that the parallel scaling of the adenylate and creatine pools adjusts for the differences in PCr and ATP concentrations between muscle fibers and provides a similar ADP signal for controlling mitochondrial ATP synthesis. How does this similar PCr/ATP among muscles at rest result in a uniform signal for respiration during exercise?

Two factors enable the PCr/ATP ratio to function as a uniform signal for respiration during exercise:

1. the PCr/ATP ratio determines the ADP concentration in the cell;
2. ADP along with Pi determines the relative activity of mitochondrial oxidative phosphorylation (Fig. 7).

To illustrate this system, Fig. 8 shows that the ADP concentration is between 20 and 40 μM when PCr/ATP is decreased by 30%, according to the CK equilibrium. These ADP concentrations correspond to the K_m for muscle oxygen consumption (25–50 μM; Kushmerick *et al.*, 1992a). Thus, the same drop in PCr/ATP, independent of the absolute decrease in PCr results in an ADP necessary to elevate muscle respiration to 50% of maximum (i.e., $K_m = 1/2\ V_{max}$). This example illustrates that it is the ratio of PCr to ATP, not the absolute levels alone, that is important to respiratory control. The large differences in PCr among muscles do not pose a problem for the feedback control of respiration since PCr/ATP generates the regulatory signal for controlling oxidative phosphorylation.

This normalized respiratory control scheme allows us to evaluate a criticism of feedback control that there is an insufficient dynamic range in these models to account for the ATP synthesis rates observed in muscle (Hochachka and Matheson, 1992). An example against feedback control is that the 400-fold range between rest and $\dot{V}_{O_2\ max}$ in the hummingbird cannot be accounted for by current control models. Few studies have measured the phosphorylation rate from rest to V_{max} in the same muscle *in vivo*; however, this range of rates is available for the wrist flexor muscles. The feedback model describes in these muscles a 40-fold range from 0.008 mM ~P sec^{-1} at rest (Blei *et al.*,) to 0.31 mM ~P sec^{-1} at V_{max} for a tissue with a 3% mitochondrial volume. To evaluate this model for the hummingbird, we scale the feedback control model according to the limit to oxidative phosphorylation (V_{max}). Increasing V_{max} by 10-fold in proportion to the 30% mitochondrial volume in hummingbird muscle yields a V_{max} of 3.1 mM ~P sec^{-1} and a rest to V_{max} range of 387-fold! For the same range of ADP from 0.012 mM at rest to >0.1 mM at V_{max} (Blei *et al.*, 1993a; Fig. 7), ATP synthesis rates can be regulated via feedback control over the 40-fold range in human arm defined by a 3% mitochondrial volume or an almost 400-fold range in hummingbird defined by a 30% mitochondrial volume. Thus even without correcting for the differences in temperature between the human arm (33.5°C) and the hummingbird (>40°C), there is sufficient range in the feedback control model to account for the rest to ATP synthesis capacity range observed in even the most aerobic muscles.

Recent studies have suggested that heart may be an exception to feedback control (Heineman and Balaban, 1990). Heart has a low PCr level and a large oxidative capacity, which results in high respiratory gain so that very small changes in PCr result in large changes in ATP synthesis (Kushmerick and Conley, 1991). Little change in PCr and ADP have been observed over the range of respiration rates achievable in open-chested, anesthetized heart preparations. The highest oxygen consumption rate in dog hearts is typically 15 μmole g^{-1} min^{-1} (Heineman and Balaban, 1990), while direct measurement of cardiac respiration in exercising dogs is 3 times higher (45 μmol g^{-1} min^{-1}; Huang and Feigl, 1988). As can be seen in Fig. 7, a large change in phosphorylation rate accompanies a small change in ADP (and PCr) over the first third of the curve. Thus it may be difficult to detect the change in PCr in the open chested, anesthetized preparation over the reported range of oxygen consumption and phosphorylation rates. Nonetheless, these studies have raised the question concerning what role other regulators play in controlling mitochondrial respiration. Particularly exciting possibilities are those signals that operate via feedforward control.

8. *Feed-Forward Control Occurs without an Error Signal*

Regulation of energy balance by a feed-forward mechanism would occur if the metabolic signal that increases the rate of ATP usage were to also increase the rate of oxygen consumption (Hansford, 1985; McCormack *et al.*, 1990). Calcium is a good candidate for this feed-forward signal, as it simultaneously activates muscle contraction (ATP demand) as well as the mitochondrial enzymes that supply NADH to the electron transport chain and synthesize ATP (ATP supply). The most likely enzyme complexes that could be sites of feed-forward control are mitochondrial dehydrogenases, which produce NADH from substrate oxidation, and mitochondrial ATP synthetase (F_1-ATPase), which couples the proton motive force (Δp) to ATP synthesis (Fig. 6). How could this enzyme activation by calcium stimulate respiration without a change in ADP?

One suggestion is that elevation of NADH increases the mitochondrial proton motive force and the driving force behind ATP synthesis (Koretsky and Balaban, 1987). The Δp is the sum of the larger electrochemical membrane potential ($\Delta\Psi$) and smaller ΔpH gradient across the inner mitochondrial membrane. The diffusion of protons down their electrochemical gradient ($\Delta\Psi$) into the mitochondrial matrix, via the F_1-ATPase, drives the phosphorylation of ADP to form ATP (Nicholls and Ferguson, 1992). Increases in $\Delta\Psi$ should cause increased ATP synthesis at a constant ADP concentration, but these increases probably play a minor role in respiratory control, as NADH changes are small relative to the change in oxygen consumption.

A more likely site for respiratory control is the F_1-ATP synthetase itself (Das and Harris, 1990a,b). In heart submitochondrial particles, calcium concentrations above 1 m*M* cause increased ATP synthetase activity by removing the action of an endogenous, low-molecular-weight, calcium-sensitive protein inhibitor (Yamada and Huzel, 1989). This concentration of Ca^{2+} is within the physiological range of heart mitochondrial matrix concentrations (0.5–2 m*M*) (Wier, 1990). In the *in vivo* heart, ADP is often found at concentrations that are not limiting to respiration and do not change with oxygen consumption rate (Heineman and Balaban, 1990). With excess ADP, Ca^{2+} could stimulate respiration by activating the F_1-ATP synthetase, thereby increasing ADP phosphorylation to ATP.

Respiration involves many factors that act to coordinate substrate, oxygen, and electron fluxes, as well as regulate the overall respiratory rate (Fig. 6). Kascar and Burns have presented a scheme whereby the "control strength" of these factors may vary, depending on the conditions examined (Groen *et al.*, 1982; Kascar and Burns, 1973). Feedback and feed-forward control models illustrate differences in the "strength" of ADP versus Ca^{2+} in respiratory control in skeletal versus heart muscle. For example, in skeletal muscle the control may reside in ADP because it is the limiting factor for oxidative phosphorylation, whereas in heart, ADP appears not to be limiting *in vivo*, and Ca^{2+} may control respiration (Balaban, 1990). Examination of other conditions may reveal that the two simplifications of control presented here are not the exclusive means by which ATP supply is regulated. Instead, feed-forward and feedback mechanisms may simply represent two possibilities for linking contractile demand to oxidative ATP supply.

9. Substrate and Oxygen Supply

Carbon fuels and oxygen comprise the final step in the cellular energy balance between contractile demand and ATP supply (Figs. 5 and 6). Both glycolysis and oxidative phosphorylation produce ATP in the catabolism of substrates. Glycolysis generates ATP through substrate-level phosphorylation but has a low ATP yield (2–3 ATP/glucose). Oxidative phosphorylation has a much higher yield (36 ATP/glucose) by fully catabolizing the substrates to CO_2 and water using oxygen. The supply of carbon fuels and oxygen for metabolism comes from intracellular stores and from extracellular sites after being transported to the cell. Common to the transport of all of these substrates are the uptake of the substrate into the blood, its circulation to the muscle, delivery into the cell, and diffusion to the site of use. Despite these common features, the mechanisms of transport and potential limitations to utilization are different for carbon substrates and oxygen.

10. Carbon Fuels: Several Substrates Are Used for Energy Supply

A number of carbon fuels are available for generating ATP, depending on the needs of the cell (Weber, 1988). Carbohydrates (CHO) and fat are the primary substrates used by the muscle cell and are transported to the cell as well as stored intracellularly. Two types of intracellular stores are present:

1. glycogen, which is broken down to yield glucose;
2. lipids, which supply free-fatty acids (FFA).

Cells also readily take up and oxidize lactate from the blood, or utilize lactate during aerobic recovery after an exercise bout. Why so many different substrates and storage locations?

11. Substrate Selection Related to Exercise Intensity

Substrates are used by different fiber types according to a strategy that parallels the range of energy demands in these fibers (Weber, 1988). Each substrate has unique properties suited for different roles. Glucose has the advantage that it can enter directly into glycolysis in the cytosol and does not need to diffuse to mitochondria. It can generate ATP at a two-fold higher rate than fat and can be rapidly taken up from the blood or mobilized from glycogen (Saltin and Gollnick, 1988).

In contrast, FFA transport in blood requires complexing with albumin, but this substrate has an advantage over glycogen in its higher energy storage density (129 ATP/mol versus 36 ATP/mol for glucose; Gollnick and Saltin, 1988).

Carbohydrates are useful for short-term, high-intensity exercise because glycogen can be rapidly mobilized and has a high maximum energy yield rate (Saltin and Gollnick, 1988). The ATP for contraction in exercise at or above the oxidative capacity is derived almost exclusively from PCr and utilizes intracellular glycogen stores via glycolysis. At the highest exercise intensities, the energy yield per molecule is low for glycogen because it is only partially catabolized to lactate. This lactate is transported via the blood to heart and skeletal muscles, where it is readily oxidized (Mazzeo *et al.*, 1986).

Longer-term energy supply is achieved with FFA because of their higher energy density (ATP/mol) but lower maximum rate of ATP generation (Gollnick and Saltin, 1988). They are well suited for the more economical, sustained performance characteristic of slow-twitch fibers. For example, in vertebrate hearts, organs that require sustained substrate supply, the capacity for FFA utilization increases with ATPase activity whereas the glycolytic capacity changes little (Driedzic *et al.*, 1987). In humans at rest, 50% of the substrate oxidized is FFA and this value increases to about 80% at moderate exercise intensities. Extracellular stores meet this need for FFA at moderate exercise intensities, but at work rates >70% $\dot{V}_{O_2\ max}$, intracellular CHO generates an increasing proportion of the energy supply, until it is the predominant substrate at $\dot{V}_{O_2\ max}$ and above.

The size of the intracellular stores of these substrates reflects the pattern of energy use in fibers. Glycolytic fibers tend to have larger glycogen deposits, whereas oxidative fibers have larger lipid stores. The supply of substrate does not appear to limit performance in the short term, but long-term performance in humans does appear to be limited by glycogen availability, despite the superiority of FFA as a fuel for sustained contractions. Thus, increased utilization of FFA by fibers might prolong endurance because it decreases the rate at which glycogen is used and extends the period before this store is emptied. Although lipid supplies may still be available, once glycogen is depleted, athletic performance cannot be maintained (Saltin and Gollnick, 1988).

12. Transport Problems Favor Intracellular Substrate Use

Why are the intracellular stores of substrates so important if large extracellular depots are available? The many steps involved in substrate transport may provide insight into an answer. Under exercise conditions, the mobilization and uptake of substrates into the blood does not appear to be a limiting step for substrate transport. The physical properties of glucose and lactate make them easily dissolved in blood, whereas FFA are carried as complexes with albumin in the blood. Because these substrates remain at high concentrations in the venous blood, convective transport by the circulation would also not appear to be a factor limiting substrate uptake. Weber (1988) argues that the limit to maximal flux rate resides in transport into the cell. This uptake is carrier-mediated for each substrate, and these carriers appear to be saturated at flux rates below the maximal cellular consumption rates.

An additional step in the movement of substrates to their sites of utilization is intracellular diffusion. Glucose is phosphorylated upon uptake and passes into the glycolytic pathway in the cytosol, but both lactate and FFA must diffuse to the mitochondria to be oxidized. Oxidative tissues with large fiber diameters tend to have mitochondria clustered at the sarcolemmal membrane, as well as dispersed throughout the cell (Hoppeler *et al.*, 1981). This subsarcolemmal localization reduces the diffusion distance for blood-borne substrates and is found to increase disproportionately with elevations in oxidative capacity following endurance training (Hoppeler *et al.*, 1985) or cold exposure (Buser *et al.*, 1982). Further evidence supporting a transport limitation to maximal fluxes of substrates are results from training experiments. The increased cellular energy demand following endurance training is met predominantly by an elevated intracellular utilization store (Buser *et al.*, 1982) and utilization of lipid (Jansson and Kaijser, 1987). Independent of what limits extracellular substrate flux, these training results indicate a preference for utilization of intracellular energy stores over extracellular substrates that must be transported to the cell (Weber, 1988).

13. Oxygen

The final component in the pathway of energy from contractile demand to oxidative phosphorylation is oxygen. Oxygen stores are present in the cell in the form of oxymyoglobin and dissolved oxygen. Unlike the intracellular substrate depots, this oxygen store is small relative to the oxidative needs of most fibers, so that transport of oxygen to the mitochondria is critical to sustain oxidative phosphorylation. This transport involves a cascade of steps from the lungs via the circulation to the capillaries (Weibel, 1984). Once in the cell, oxygen diffuses through the tissue to mitochondria as molecular O_2 or by facilitated diffusion, bound with myoglobin. A series of integrated studies conducted by Taylor, Weibel, and collaborators provide a comprehensive structural and functional analysis of the design of this oxygen transport cascade in mammals (Taylor *et al.*, 1987; Taylor and Weibel, 1981). My treatment will focus on the terminal structures of the respiratory system that are critical to oxygen delivery and uptake: the capillaries and the mitochondria.

14. Heterogeneity of Properties

The variation in metabolic and structural properties of muscle fibers discussed in the foregoing affects the delivery and consumption of oxygen as well. The range of capillary and mitochondrial densities and cell sizes found among vertebrate muscles is reflected in the transport capacities, diffusion distances, and oxidative capacities of these fibers. Capillaries and mitochondria fulfill a variety of roles in muscle in addition to oxygen delivery and consumption, such as the supply and catabolism of substrates for metabolism and the removal of metabolic wastes. How then does oxygen supply meet demand given this diversity of properties and the many roles served by these structures?

To separate the importance of these roles, we tested the hypothesis that capillaries and mitochondria form a structural and functional unit in the respiratory system. We asked three questions to determine whether oxygen is the primary factor determining the properties of capillaries and mitochondria. First, are capillaries and mitochondria related in structural terms relevant to oxygen delivery and consumption? Second, does this structural relationship result in a similar oxidative capacity of mitochondria independent of the muscle or animal? Third, is there any evidence that cellular oxygen availability limits oxidative phosphorylation? The alternative to this hypothesis is that oxygen is only one of a number of important factors governing capillary and mitochondrial properties. If this alternative hypothesis were true, then no fixed relationship should be observed between these structures when evaluated across a range of muscle oxidative capacities.

15. Capillary and Mitochondrial Unit

Quantitative indices of capillaries and mitochondria are clearly related to the oxidative capacities of individual muscles and of the whole-body musculature in mammals (Else and Hulbert, 1985; Hoppeler, 1990). Both the capillary lengths and mitochondrial volumes of muscles scale in direct proportion to $\dot{V}_{O_2\,max}$ in mammals ranging in size from 10 g to 500 kg (Mathieu *et al.*, 1981). In addition, the capillary density and mitochondrial volume density are related in a similar

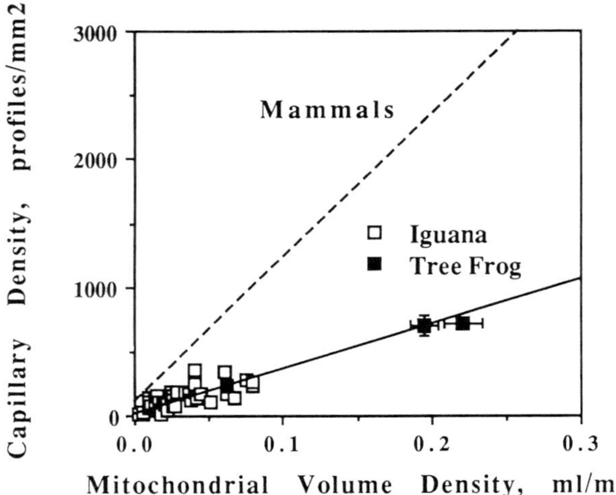

FIG. 9. Densities of capillary and mitochondrial volumes in striated muscles. The dashed line is the regression for mammalian muscles (Conley *et al.*, 1987) and the solid line is the regression for the Cuban iguana, *Cyclura nubila* (open squares; Conley *et al.*, 1989b). Also included are the means (±SE) reported for the grey tree frog, *Hyla versicolor* (solid squares; Marsh and Taigen, 1987).

manner in muscles ranging from glycolytic skeletal muscle to entirely oxidative cardiac muscle, independent of the body size, species, or aerobic capacity of the mammal examined (Conley *et al.*, 1987). We evaluated whether this relationship is unique to mammals by examining a reptile that maintains the same body temperature, but a lower aerobic capacity, as a similarly sized mammal. Figure 9 shows that capillaries and mitochondria are in lower densities in muscles of the Cuban iguana (*Cyclura nubila*) than in mammals, as expected from its 10-fold lower $\dot{V}_{O_2\,max}$ (Conley *et al.*, 1989b). However, the slope of the regression indicates that the iguana has fewer capillaries per mitochondrial volume than are found in mammalian muscle. Muscles of the grey tree frog (*Hyla versicolor*) have mitochondrial volume densities as high as mammals but are related to capillary density in a similar manner as in the iguana, which indicates that this relationship is not unique to reptiles (Marsh and Taigen, 1987).

Structural density per se is not representative of the properties relevant to oxygen supply or uptake because the properties of blood, capillaries, and mitochondria all differ between mammals and other vertebrates (Conley *et al.*, 1989b; Pough, 1980). What is relevant in making these comparisons is the structural capacity to deliver oxygen as compared to the biochemical capacity to use this oxygen. For an animal exercising at $\dot{V}_{O_2\ max}$, the hemoglobin (Hb) mass in the capillary sets the amount of oxygen available to the cell. The limit for oxygen consumption resides on the inner mitochondrial membrane area (S_{im}) where oxidative phosphorylation occurs. Thus, the relevant terms are Hb within the capillary available to supply oxygen to a given inner mitochondrial membrane area (S_{im}). When expressed in these terms, this reptile has 5 mg Hb m^{-2} inner membrane versus 8–13 mg Hb m^{-2} for mammals. Expression of the structural properties of the muscles in terms relevant to oxygen delivery and uptake yields a relationship in agreement between the iguana and mammals within a factor of two as compared to the ten-fold difference in the $\dot{V}_{O_2\ max}$.

The capillary and mitochondrial volume densities of muscles of a wide variety of vertebrates have been reported (Fitch *et al.*, 1984), but the structural and blood properties necessary for calculating oxygen balance have generally not been measured. Such a comprehensive study of the muscle and blood characteristics is necessary to confirm whether the properties of capillaries and mitochondria are designed primarily so that their oxygen delivery capacity matches the mitochondrial oxidative capacity. The results reported here do indicate that capillaries and mitochondria are similarly related in a variety of vertebrate muscles when expressed in physical terms relevant to oxygen delivery and uptake.

16. Functional Capacity of Structural Unit

The second criterion that would suggest that capillaries and mitochondria form a unit is if mitochondria respire at the same maximal rate independent of muscle properties. The average oxidative capacity per unit inner membrane surface area ($\dot{V}_{O_2\ max}/S_{im}$) determined *in situ* for the Cuban iguana falls within the range observed in mammals when the differences in body temperature during exercise are taken into account (Conley and Kushmerick, 1993). Preliminary results in isolated mitochondria indicate that these values are similar to those determined for the cat (3 μmol O_2 min^{-1} m^{-2}; Schwerzmann *et al.*, 1989) and the Teju lizard, *Tupinambis sp.* (~2 μmol O_2 min^{-1} m^{-2}). Thus, in addition to the similar structural relationship between capillaries and mitochondria, the oxidative capacity per unit inner membrane area agrees among these species.

17. Evidence against an Oxygen Limitation

The third criterion supporting the hypothesis of the existence of a capillary and mitochondrial unit is that if supply capacity is matched to demand capacity, then cellular oxygen partial pressure (P_{O_2}) cannot be limiting to oxidative phosphorylation. There is some evidence that P_{O_2} may be limiting to respiration in isolated myocytes and the isolated heart (Heineman *et al.*, 1992; Rumsey *et al.*, 1990); however, in both cases this limitation may be related to the artificial nature of the preparation. An oxygen limitation is incompatible with several other lines of evidence. The first is that the terminal oxidase is saturated with oxygen at P_{O_2} less than 0.5 Torr (Gayeski *et al.*, 1987). Measurements of myoglobin saturation in muscle frozen in liquid nitrogen while stimulated near its oxidative capacity indicate that P_{O_2} does not drop below ~1 Torr (Gayeski and Honig, 1991; Honig *et al.*, 1992). Although none of these studies is definitive, together they indicate that an oxygen limitation to oxidative phosphorylation is unlikely.

Thus, three lines of evidence support the hypothesis that capillaries and mitochondria form a structural and functional unit designed principally for the delivery and uptake of oxygen. A number of examples provide quantitative exception to this relationship but support the conclusion of the central role of oxygen delivery and uptake in determining the properties of these structures. For example, muscle from fish found in cold waters have a larger content of mitochondria relative to capillaries compared to more temperate water species (Fitch *et al.*, 1984). Acclimation to lower temperature (25 to 5°C) in fish species from temperate waters also results in an increased mitochondrial volume relative to capillaries (Egginton and Sidell, 1989). An explanation for the higher mitochondrial volume is that the addition of more enzyme adjusts for the reduced V_{max} at these temperatures. Hence, the increased mitochondrial content may be compensation for the low reaction rates, but similar power output requirements, for swimming at low temperatures. These exceptions support the concept that the structural properties of capillaries and mitochondria are adaptable, but additional species must be studied to determine whether this match between the capacities for oxygen delivery and consumption found in mammals and the Cuban iguana is a ubiquitous feature of vertebrate muscles.

IV. Summary

How is the muscle fiber designed to accomplish the diversity of tasks performed by striated muscle? Basically, a common contractile mechanism and a similar organization of metabolism in striated muscles are used to generate a wide spectrum of speeds and durations of contraction. The speed of contraction ranges from manyfold within an animal to over a hundred-fold between animals, owing to variation in the intrinsic velocities of the myosin isoforms. Recruiting fibers that contain the myosin isoform that contracts at the appropriate velocity varies the speed of locomotion at minimal cost. The magnitude and duration of the energy supply required to meet this contractile demand depends on the size of the cellular energy buffer and the capacities of the metabolic pathways. The faster the contractile speed, the larger the PCr pool and the greater the glycolytic capacity to meet a high rate of ATP use. Slower-contracting fibers have a smaller buffer for the short term, but an increased oxidative capacity for continuous energy supply to maintain energy balance over the long term. In general, fibers trade contractile speed for duration of performance, but a number of exceptions exist where rapid contractions are maintained for extended periods.

In the face of this heterogeneity of properties, common features are found that assure an energy balance. The PCr/ATP buffer system offers a simple mechanism of feedback control of energy supply despite the wide range of high-energy phosphate concentrations and oxidative capacities found in skeletal muscle. An oxygen balance system also appears to be present in the terminal structures of the respiratory system, the capillaries, and mitochondria. Despite the diversity of these structures, a rather constant ratio of oxygen delivery capacity to mitochondrial oxidative capacity is found in vertebrate striated muscles.

Finally, a critical feature of muscle energy balance that remains unresolved is (are) the mechanism(s) controlling mitochondrial respiration in heart. Feedback control appears to account for linking ATP supply to demand in skeletal muscle, but the mechanisms governing respiratory control in heart are still under vigorous investigation. Thus, the links between contractile demand and oxidative phosphorylation are still unresolved in this tissue, which may indicate that a key element is missing in our understanding of the cellular energetics of exercise.

Acknowledgments

This review owes a great debt to my frequent and lively discussions over the years with Martin J. Kushmerick. Valuable comments on the text were offered by M.J.K., Rick Hedges, and Sharon Jubrias. I was supported while writing this review by American Heart Association Grant-in-Aid 901369, U.S. National Institutes of Health R01 AGAR 10853, and the University of Washington Department of Radiology.

REFERENCES

Armstrong, R., and Laughlin, M. (1985). Metabolic indicators of fibre recruitment in mammalian muscles during locomotion. *J. Exp. Biol.* **115,** 201–213.

Balaban, R. (1990). Regulation of oxidative phosphorylation in the mammalian cell. *Am. J. Physiol.* **258,** C377–C389.

Bárány, M. (1967). ATPase activity of myosin correlated with speed of muscle shortening. *J. Gen. Physiol.* **50,** 197–216.

Biewener, A., Alexander, R. M., and Heglund, N. (1981). Elastic energy storage in the hopping of kangaroo rats (*Dipodomys spectabilis*). *J. Zool.* **195,** 369–383.

Blei, M., Conley, K., and Kushmerick, M. (1993a). Separate measures of ATP utilization and recovery in human skeletal muscle. *J. Physiol. (London)* **465,** 203–222.

Blei, M., Conley, K., Esselman, P., Odderson, I., Cress, M., and Kushmerick, M. (1993b). Individual variation in contractile cost and recovery in a human skeletal muscle. *Proc. Natl. Acad. Sci. U.S.A.* **90,** 7396–7400.

Buser, K. S., Kopp, B., Gehr, P., Weibel, E. R., and Hoppeler, H. (1982). Effect of cold environment on skeletal muscle mitochondria in growing rats. *Cell Tissue Res.* **225,** 427–436.

Cavagna, G., Mazzanti, M., Heglund, N., and Citterio, G. (1985). Storage and release of mechanical energy by active muscle: A non-elastic mechanism? *J. Exp. Biol.* **115,** 79–87.

Chadwick, L. E., and Rahn, H. (1954). Temperature dependence of rattling frequency in the rattlesnake, *Crotalus Viridis*. *Science* **119,** 442–443.

Chance, B., and Williams, C. (1956). The respiratory chain and oxidative phosphorylation. *Advances in Enzymology* **17,** 65–134.

Chance, B., Leigh, J., Clark, B., Maris, J., Kent, J., Nioka, S., and Smith, D. (1985). Control of oxidative metabolism and oxygen delivery in human skeletal muscle: A steady-state analysis of the work/energy cost transfer function. *Proc. Natl. Acad. Sci. U.S.A.* **82,** 8384–8388.

Chance, B., Leigh, J., Kent, J., and McCully, K. (1986a). Metabolic control principles and ^{31}P-NMR. *Fed. Proc., Fed. Am. Soc. Exp. Biol.* **45,** 2915–2920.

Chance, B., Leigh, J., Kent, J., McCully, K., Nioka, S., Clark, B., Maris, J., and Graham, T. (1986b). Multiple controls of oxidative metabolism in living tissues as studied by phosphorus magnetic resonance. *Proc. Natl. Acad. Sci. U.S.A.* **83,** 9458–9462.

Chase, P. B., and Kushmerick, M. J. (1988). Effects of pH on contraction of rabbit fast and slow skeletal muscle fibers. *Biophys. J.* **53,** 935–946.

Chi, M. M., Hintz, C. S., Henriksson, J., Salmons, S., Hellendahl, R. P., Park, J. L., Nemeth, P. M., and Lowry, O. H. (1986). Chronic stimulation of mammalian muscle: Enzyme changes in individual fibers. *Am. J. Physiol.* **251,** C633–C642.

Close, R. (1972). Dynamic properties of mammalian skeletal muscles. *Physiol. Rev.* **52,** 129–197.

Conley, K., and Kushmerick, M. (1993). Unifying principles in aerobic muscle energetics. *In* "The Vertebrate Gas Transport Cascade: Adaptations to Environment and Mode of Life" (J. E. P. W. Bicudo, ed.), pp. 304–313. CRC Press, Boca Raton, FL.

Conley, K., Kayar, S., Rösler, K., Weibel, E. R., Hoppeler, H., and Taylor, C. R. (1987). Adaptive variation in the mammalian respiratory system in relation to energetic demand. IV. Capillaries and their relationship to oxidative capacity. *Respir. Physiol.* **69,** 47–64.

Conley, K., Baudinette, R., and Kushmerick, M. (1989a). Muscle high-energy phosphate compounds: ATP/creatine is constant in marsupials. *Am. Zool.* **29,** 156A.

Conley, K. E., Christian, K. A., Hoppeler, H., and Weibel, E. R. (1989b). Capillary and mitochondrial unit in muscles of a large lizard. *Am. J. Physiol.* **256,** R982–R988.

Connett, R. J. (1987). Glycolytic regulation during an aerobic rest-to-work transition in dog gracilis muscle. *J. Appl. Physiol.* **63,** 2366–2374.

Connett, R. J. (1988). Analysis of metabolic control: New insights using scaled creatine kinase model. *Am. J. Physiol.* **254,** R949–R959.

Connett, R. J., and Honig, C. R. (1989). Regulation of $\dot{V}_{O_2}$ in red muscle: Do current biochemical hypotheses fit *in vivo* data? *Am. J. Physiol.* **256,** R898–R906.

Crow, M. T., and Kushmerick, M. J. (1982). Chemical energetics of slow- and fast-twitch muscles of the mouse. *J. Gen. Physiol.* **79,** 147–166.

Crow, M. T., and Kushmerick, M. J. (1983). Correlated reduction of velocity of shortening and the rate of energy utilization in mouse fast-twitch muscle during a continuous tetanus. *J. Gen. Physiol.* **82,** 703–720.

Das, A., and Harris, D. (1990a). Control of mitochondrial ATP synthase in heart cells: Inactive to active transitions caused by beating or positive inotropic agents. *Cardiovasc. Res.* **24,** 411–417.

Das, A., and Harris, D. (1990b). Regulation of the mitochondrial ATP synthase in intact rat cardiomyocytes. *Biochem. J.* **266,** 355–361.

Denton, R., and McCormack, J. (1990). Ca^{2+} as a second messenger within mitochondria of heart and other tissues. *Annu. Rev. Physiol.* **52,** 451–466.

Driedzic, W., Sidell, B., Stowe, D., and Branscombe, R. (1987). Matching of vertebrate cardiac energy demand to energy metabolism. *Am. J. Physiol.* **252,** R930–R937.

Egginton, S., and Sidell, B. (1989). Thermal acclimation induces adapative changes in subcellular structure of fish skeletal muscle. *Am. J. Physiol.* **256,** R1–R9.

Ellington, W. R. (1989). Phosphocreatine represents a thermodynamic and functional improvement over other muscle phosphagens. *J. Exp. Biol.* **143,** 177–194.

Else, P. L., and Hulbert, A. J. (1985). An allometric comparison of the mitochondria of mammaliam and reptilian tissues: The implications for the evolution of endothermy. *J. Comp. Physiol.* **156,** 3–11.

Elzinga, G., and Laars, V. D. (1988). Oxygen consumption of single muscle fibres of *Rana temporaria* and *Xenopus laevis* at 20°C. *J. Physiol. (London)* **399,** 405–418.

Elzinga, G., Lännergren, J., and Stienen, G. (1987). Stable maintenance heat rate and contractile properties of different single muscle fibers from *Xenopus laevis* at 20°C. *J. Physiol. (London)* **393,** 399–412.

Fedak, M., Heglund, N., and Taylor, C. (1982). Energetics and mechanics of terrestrial locomotion. II. Kinetic energy changes of the limbs and body as a function of speed and body size in birds and mammals. *J. Exp. Biol.* **97,** 23–40.

Fisher, M., and Dillon, P. (1988). Direct determination of ADP in hypoxic porcine carotid artery using ^{31}P NMR. *NMR in Biomed.* **1,** 121–126.

Fitch, N., Johnston, I., and Wood, R. (1984). Skeletal muscle capillary supply in a fish that lacks respiratory pigments. *Respir. Physiol.* **57,** 201–211.

Full, R. (1989). Mechanics and energetics of terrestrial locomotion: Bipeds to polypeds. *In* "Energy Transformations in Cells and Organisms" (W. Weiser and E. Gnaiger, eds.), pp. 175–182. Thieme, Stuttgart.

Funk, C., Clark, A. J., and Connett, R. J. (1990). A simple model of aerobic metabolism: Applications to work transitions in muscle. *Am. J. Physiol.* **258,** C995–C1005.

Gayeski, T., and Honig, C. (1991). Intracellular P_{O_2} in individual cardiac myocytes in dogs, cats, rabbits, ferrets, and rats. *Am. J. Physiol.* **260,** H522–H531.

Gayeski, T. E., Connett, R. J., and Honig, C. R. (1987). Minimum intracellular P_{O_2} for maximum cytochrome turnover in red muscle *in situ*. *Am. J. Physiol.* **252,** H906–H815.

Gollnick, P., and Saltin, B. (1988). Fuel for muscular exercise: Role of fat. *In* "Exercise, Nutrition, and Energy Metabolism" (E. S. Horton and R. L. Terjung, eds.), pp. 72–88. Macmillan, New York.

Goslow, G., Seeherman, H., Taylor, C., McCutchin, M., and Heglund, N. (1981). Electrical activity and relative length changes of dog limb muscles as a function of speed and gait. *J. Exp. Biol.* **94,** 15–42.

Groen, A., Wanders, R., Westerhoff, H., van der Meer, R., and Tager, J. (1982). Quantification of the contribution of various steps to the control of mitochondrial respiration. *J. Biol. Chem.* **257,** 2754–2757.

Hansford, R. (1985). Relation between mitochondrial calcium transport and control of energy metabolism. *Rev. Physiol. Biochem. Pharmacol.* **102,** 1–72.

Harris, R., Hultman, E., Kaijser, L., and Nordesjo, L.-O. (1975). The effect of circulatory occlusion on isometric exercise capacity and energy metabolism of the quadriceps muscle in man. *Scand. J. Clin. Lab. Invest.* **35,** 87–95.

Heglund, N., and Cavagna, G. (1985). Efficiency of vertebrate locomotory muscles. *J. Exp. Biol.* **115,** 283–292.

Heglund, N., and Cavagna, G. (1987). Mechanical work, oxygen consumption and efficiency in isolated frog and rat striated muscle. *Am. J. Physiol.* **253,** C22–C29.

Heglund, N., and Taylor, C. (1988). Speed, stride frequency, and energy cost per stride: How do they change with body size and gait. *J. Exp. Biol.* **138,** 301–318.

Heglund, N., Fedak, M., Taylor, C., and Cavagna, G. (1982). Energetics and mechanics of terrestrial locomotion. IV. Total mechanical energy changes as a function of speed and body mass in birds and mammals. *J. Exp. Biol.* **97,** 57–66.

Heineman, F. W., and Balaban, R. S. (1990). Control of mitochoncrial respiration in the heart *in vivo*. *Annu. Rev. Physiol.* **52,** 523–542.

Heineman, F., Kupriyanov, V., Marshall, R., Fralix, T., and Balaban, R. S. (1992). Myocardial oxygenation in the isolated working rabbit heart as a function of work. *Am. J. Physiol.* **262,** H255–H267.

Hochachka, P. (1985). Fuels and pathways as designed systems for support of muscle work. *J. Exp. Biol.* **115,** 149–164.

Hochachka, P., and Matheson, G. (1992). Regulation ATP turnover rates over broad dynamic work ranges in skeletal muscle. *J. Appl. Physiol.* **73,** 1697–1703.

Honig, C., Connett, R., and Gayeski, T. (1992). O_2 transport and its interaction with metabolism: A systems view of aerobic capacity. *Med. Sci. Sports Exercise* **24,** 47–53.

Hoppeler, H. (1990). The different relationship of $\dot{V}_{O_2\ max}$ to muscle mitochondria in humans and quadrupedal animals. *Respir. Physiol.* **80,** 137–146.

Hoppeler, H., Mathieu, O., Krauer, R., Claassen, H., Armstrong, R., and Weibel, E. R. (1981). Design of the mammalian respiratory system. VI. Distribution of mitochondria and capillaries in various muscles. *Respir. Physiol.* **44,** 87–111.

Hoppeler, H., Howald, H., Conley, K., Lindstedt, S. L., Claassen, H., Vock, P., and Weibel, E. R. (1985). Endurance training in humans: Aerobic capacity and structure of skeletal muscle. *J. Appl. Physiol.* **59,** 320–327.

Howald, H., Hoppeler, H., Claassen, H., Mathieu, O., and Straub, R. (1985). Influences of endurance training on the ultrastructural composition of the different muscle fiber types in humans. *Pflüegers Arch.* **403,** 369–376.

Huang, A., and Feigl, E. (1988). Adrenergic coronary vasoconstriction helps maintain uniform transmural blood flow distribution during exercise. *Circ Res.* **62,** 286–298.

Jansson, E., and Kaijser, L. (1987). Substrate utilization and enzymes in skeletal muscle of extremely endurance-trained men. *J. Appl. Physiol.* **62,** 999–1005.

Jerusalem, F., Engel, A., and Peterson, H. (1975). Human muscle fiber fine structure: Morphometric data on controls. *Neurology* **25,** 127–134.

Josephson, R., and Stevenson, R. (1991). The efficiency of a flight muscle from the locust *Schistocerca americana. J. Physiol. (London)* **442,** 413–429.

Kascar, H., and Burns, J. (1973). The control of flux. *Semin. Ser.—Soc. Exp. Biol.* **27,** 64–104.

Klug, G., and Tibbits, G. (1988). The effect of activity on calcium-mediated events in striated muscle. *Exercise Sport Sci. Rev.* **16,** 1–59.

Koretsky, A. P., and Balaban, R. S. (1987). Changes in pyridine nucleotide levels alter oxygen consumption and extra-mitochondrial phosphates in isolated mitochondria: A ^{31}P-NMR and NAD(P)H fluorescence study. *Biochim. Biophys. Acta* **893,** 398–408.

Kram, R., and Taylor, C. (1990). Energetics of running: a new perspective. *Nature (London)* **346,** 265–267.

Krisanda, J., Moerland, T., and Kushmerick, M. (1988). ATP supply and demand during exercise. *In* "Exercise, Nutrition and Energy Metabolism" (E. S. Horton and R. L. Terjung, eds.), pp. 27–44. Macmillan, New York.

Kushmerick, M. (1977). Energy balance in muscle contraction: A biochemical approach. *Curr. Top. Bioenerg.* **6,** 1–36.

Kushmerick, M. (1983). Energetics of muscle contraction. *In* "Handbook of Physiology" (L. D. Peachey, R. H. Adrian, and S. Geiger, eds.), sect. 10, pp. 189–236. Williams & Wilkins, Baltimore, MD.

Kushmerick, M., and Conley, K. (1991). An analysis of regulation of respiration in the heart. *In* "Imaging, Measurements and Analysis of the Heart" (S. Sideman and R. Beyar, eds.), pp. 289–298. Hemisphere, New York.

Kushmerick, M., Meyer, R., and Brown, T. (1992a). Regulation of oxygen consumption in fast- and slow-twitch muscle. *Am. J. Physiol.* **263,** C598–C606.

Kushmerick, M., Moerland, T., and Wiseman, R. (1992b). Contents of PCr, Pi, ATP in mammalian fiber types. *Proc. Natl. Acad. Sci. U.S.A.* **89,** 7521–7525.

Kushmerick, M. J. (1985). Patterns in mammalian muscle energetics. *J. Exp. Biol.* **115,** 165–177.

Kushmerick, M. J. (1987). Energetics studies of muscles of different types. *Basic Res. Cardiol.* **2,** 17–30.

Kushmerick, M. J., and Meyer, R. A. (1985). Chemical changes in rat leg muscle by phosphorus nuclear magnetic resonance. *Am. J. Physiol.* **248,** C542–C549.

Kushmerick, M. J., and Paul, R. J. (1976). Aerobic recovery metabolism following a single isometric tetanus in frog sartorius muscle at 0 degrees C. *J. Physiol. (London)* **254,** 693–709.

Lawson, J., and Veech, R. (1979). Effects of pH and free Mg^{2+} on the K_{eq} of the creatine kinases reaction and other phosphate hydrolyses and phosphate transfer reactions. *J. Biol. Chem.* **254,** 6528–6537.

Lindstedt, S., Hoppeler, H., Bard, K., and Thronson, H. (1985). Estimate of muscle shortening rate during locomotion. *Am. J. Physiol.* **249,** R699–R703.

Marsh, R., and Taigen, T. (1987). Properties enhancing aerobic capacity of calling muscles in gray tree frogs *Hyla versicolor. Am. J. Physiol.* **252,** R786–R793.

Mathieu, O., Krauer, R., Hoppeler, H., Gehr, P., Lindstedt, S. L., Alexander, R. M., Taylor, C. R., and Weibel, E. R. (1981). Design of the mammalian respiratory system. VII. Scaling mitochondrial volume in skeletal muscle to body mass. *Respir. Physiol.* **44,** 113–128.

Mazzeo, R., Brooks, G., Schoeller, D., and Budinger, T. (1986). Disposal of blood [1-^{13}C]lactate in humans during rest and exercise. *J. Appl. Physiol.* **60,** 232–241.

McCormack, J., Halestrap, A., and Denton, R. (1990). Role of calcium ions in regulation of mammalian intramitochondrial metabolism. *Physiol. Rev.* **70,** 391–425.

Meyer, R. (1988). A linear model of muscle respiration explains mono-exponential phosphocreatine changes. *Am. J. Physiol.* **257,** C1149–C1157.

Meyer, R. A., Sweeney, H. L., and Kushmerick, M. J. (1984). A simple analysis of the "phosphocreatine shuttle." *Am. J. Physiol.* **246,** C365–C377.

Meyer, R. A., Brown, T. R., and Kushmerick, M. J. (1985). Phosphorus nuclear magnetic resonance of fast- and slow-twitch muscle. *Am. J. Physiol.* **248,** C279–C287.

Mizuno, M., Secher, N., and Quistorff, B. (1992). ^{31}P-MRS, electromyographic activity, and fiber type composition of human forearm flexor muscles. *Soc. Mag. Res. Med.* **11,** 779.

Moerland, T. S., and Kushmerick, M. J. (1993). Contractile economy and aerobic recovery metabolism in skeletal muscle adapted to creatine depletion. *Am. J. Physiol.* (in press).

Moerland, T. S., Wolf, N., and Kushmerick, M. J. (1989). Administration of a creatine analogue induces isomyosin transitions in muscle. *Am. J. Physiol.* **257,** C810–C816.

Nicholls, D., and Ferguson, S. (1992). "Bioenergetics 2." Academic Press, London.

Palsson, B., and Lightfoot, E. (1984). Mathematical modeling of dynamics and control in metabolic networks. I. On Michaelis–Menten kinetics. *J. Theor. Biol.* **111,** 273–302.

Perry, S. (1985). Properties of muscle proteins: A comparative approach. *J. Exp. Biol.* **115,** 31–42.

Pette, D. (1985). Metabolic heterogeneity of muscle fibres. *J. Exp. Biol.* **115,** 179–189.

Pette, D., and Spamer, C. (1986). Metabolic properties of muscle fibers. *Fed. Proc., Fed. Am. Soc. Exp. Biol.* **45,** 2910–2914.

Pough, F. (1980). Blood oxygen transport and delivery in reptiles. *Am. Zool.* **20,** 173–185.

Rall, J. (1982). Sense and nonsense about the Fenn effect. *Am. J. Physiol.* **242,** H1–H6.

Rall, J. (1985). Energetic aspects of skeletal muscle contraction: Implications of fiber type. *Exercise Sport Sci. Rev.* **13,** 33–74.

Rome, L. (1992). Scaling of muscle fibres and locomotion. *J. Exp. Biol.* **168,** 243–252.

Rome, L., Sosnicki, A., and Goble, D. (1990). Maximum velocity of shortening of three fibre types from horse soleus muscle: Implications for scaling with body size. *J. Physiol. (London)* **431,** 173–185.

Rumsey, W., Schlosser, C., Nuutinen, E., Robiolio, M., and Wilson, D. (1990). Cellular energetics and the oxygen dependence of respiration in cardiac myocytes isolated from adult rat. *J. Biol. Chem.* **265,** 15392–15399.

Sako, E. Y., Kingsley, H. P., From, A. H., Foker, J. E., and Ugurbil, K. (1988). ATP synthesis kinetics and mitochondrial function in the postischemic myocardium as studied by ^{31}P NMR. *J. Biol. Chem.* **263**(22), 10600–10607.

Saltin, B., and Gollnick, P. (1988). Fuel for muscular exercise: Role of carbohydrate. *In* "Exercise, Nutrition, and Energy Metabolism" (E. S. Horton and R. L. Terjung, eds.), pp. 45–71. Macmillan, New York.

Schaeffer, P., and Lindstedt, S. (1992). Structure–function coupling in the fastest-contracting vertebrate muscle: The rattlesnake tail-shaker muscle. *Physiologist* **35,** 224.

Schwerzmann, K., Hoppeler, H., Kayar, S. R., and Weibel, E. R. (1989). Oxidative capacity of muscle and mitochondria: Correlation of physiological, biochemical, and morphometric characteristics. *Proc. Natl. Acad. Sci. U.S.A.* **86,** 1583–1587.

Seeherman, H., Taylor, C., Maloiy, G., and Armstrong, R. (1981). Design of the mammalian respiratory system. II. Measuring maximum aerobic capacity. *Respir. Physiol.* **44,** 11–24.

Sjøgaard, G., and Saltin, B. (1982). Extra- and intracellular water spaces in muscles of man at rest and dynamic exercise. *Am. J. Physiol.* **243,** R271–R280.

Suarez, R., Lighton, J., Brown, G., and Mathieu-Costello, O. (1991). Mitochondrial respiration in hummingbird flight muscle. *Proc. Natl. Acad. Sci. U.S.A.* **88,** 4870–4873.

Taylor, C. R., and Weibel, E. R. (1981). Design of the mammalian respiratory system. *Respir. Physiol.* **44,** 1–164.

Taylor, C., Heglund, N., McMahon, T., and Looney, T. (1980). Energetic cost of generating muscular force during running. *J. Exp. Biol.* **86,** 9–18.

Taylor, C. R., Karas, R. H., Weibel, E. R., and Hoppeler, H. (1987). Adaptive variation in the mammalian respiratory sytem in relation to energetic demand. *Respir. Physiol.* **69,** 1–127.

Weber, J.-M. (1988). Design of exogenous fuel supply systems: Strategies for endurance locomotion. *Can. J. Zool.* **66,** 1116–1121.

Weibel, E. (1984). "The Pathway for Oxygen." Harvard Univ. Press, Cambridge, MA.

Westerblad, H., Lee, J., Lännergren, J., and Allen, D. (1991). Cellular mechanisms of fatigue in skeletal muscle. *Am. J. Physiol.* **261,** C195–C209.

Westerhof, H., and van Dam, K. (1987). "Thermodynamics and control of biological free-energy transduction." Amsterdam, Elsevier.

Wier, W. (1990). Cytoplasmic [Ca^{2+}] in mammalian ventricle: Dynamic control by cellular processes. *Annu. Rev. Physiol.* **52,** 467–485.

Wilkie, D. (1985). Muscle function: A personal view. *J. Exp. Biol.* **115,** 1–13.

Wilkie, D. (1986). Muscular fatigue: Effects of hydrogen ions and inorganic phosphate. *Fed. Proc., Fed. Am. Soc. Exp. Biol.* **45,** 2921–2923.

Woledge, R., Curtin, N., and Homsher, E. (1985). "Energetic Aspects of Muscle Contraction," Monogr. Physiol. Soc. No. 41. Academic Press, London.

Yamada, E., and Huzel, N. (1989). Calcium-binding ATPase inhibitor protein of bovine heart mitochondria. Role of ATP synthesis and effect of Ca^{2+}. *Biochemistry* **28,** 9714–9718.

Solving the Common Problem: Matching ATP Synthesis to ATP Demand during Exercise

PETER W. HOCHACHKA

Department of Zoology and Sports Medicine Division, University of British Columbia, Vancouver, British Columbia, Canada V6T 2A9

I. Introduction

It has been well known for years that under aerobic steady-state exercise conditions, ATP utilization rates must be closely balanced with ATP synthesis rates. Otherwise, the cell's supply of ATP would be rapidly depleted. The higher the mass-specific metabolic rate of the tissue, the more crucial is this requirement for close coupling between energy demand and energy supply. In humans, the highest muscle aerobic metabolic rates (about 100 μmol ATP g^{-1} min^{-1}) could be sustained by endogenous ATP for only about 4 s, and by endogenous creatine phosphate (PCr) for, at most, about another half minute (Anderson and Saltin, 1985). During hovering flight in the hummingbird, the breast muscles must be able to turn over ATP at the rate of about 500 μmol g^{-1} min^{-1} (Suarez *et al.*, 1990); endogenous ATP could meet this demand for about 0.5 s, while endogenous PCr could extend balanced function for an additional 1–2 s. Energy demands for insect flight muscles can exceed the hummingbird metabolic rates by at least three- or fourfold; endogenous supplies of ATP plus phosphagen could sustain these flight muscle energy demands for only a second or less (Rowan and Newsholme, 1979). Although it is not as severe a problem as in small, very active animals, even in humans it is clear that ATP utilization rates must be effectively balanced with ATP synthesis rates. Our main task in this paper is to critically review how this balance may be achieved.

II. Energy Coupling Quantification

Although the concept of energy demand–energy supply coupling is an old one in biology, the issue usually has been considered only in rather qualitative terms. Most previous metabolic biochemistry insights into this problem (Arthur *et al.*, 1992a,b; Atkinson, 1990; Balban, 1990; Chance *et al.*, 1986; Connett, 1988; Connett and Honig, 1989; Cooke and Pate, 1990; Dobson *et al.*, 1987, 1988; Dudley *et al.*, 1987; From *et al.*, 1990; Funk *et al.*, 1990; Graham *et al.*, 1990; Hochachka *et al.*, 1991a,b; Meyer *et al.*, 1984; Nioka *et al.*, 1992; Rumsey *et al.*, 1990) came from metabolite concentration changes, especially changes in [ATP], [ADP], [AMP], [Pi], [PCr], [creatine] (Cr), [H^+], and [lactate] to obtain estimates of the degree of imbalance in ATPase and ATP synthase fluxes. If energy coupling is very tight, ATPase fluxes are very similar to ATP synthase fluxes, and [ATP] will not change. If energy coupling is loose, ATPase fluxes exceed ATP synthase, and ATP concentrations clearly must fall. Thus, one measure of coupling efficiency (Hochachka *et al.*, 1991b; Matheson *et al.*, 1991) is the change in [ATP]/total ATP turnover, and the percent imbalance in energy coupling can be defined operationally as (both numerator and denominator being expressed in μmol g^{-1} over any given work time period)

$$\% \text{ imbalance} = \frac{\Delta[\text{ATP}] \times 100}{\text{total ATP turnover}}.$$

The lower the percent imbalance in ATPase fluxes compared to ATP synthase fluxes, the more efficient the coupling of the two pathways.

This approach suffers from the problem that even during intense exercise, in mammalian skeletal muscles [ATP] almost never declines below about 50% of rest values; thus, in many cases, estimates of [ATP] change are based on differences between two fairly large and similar values. That is why inosine monophosphate (IMP) may, at times, be a better measure of the degree to which ATPase fluxes are out of balance with ATP synthase fluxes. This is because in vertebrate fast muscles, IMP is formed from AMP in a reaction catalyzed by AMP deaminase, and its accumulation in human muscles (Graham *et al.*, 1990), as in all vertebrates (Mommsen and Hochachka, 1988), is stoichiometric with ATP depletion. Because in nonworking muscle IMP occurs at very low concentrations, its percent change during intense exercise is proportionately much greater than the percent change in [ATP]. The stoichiometry between ATP and IMP implies, however, that a change in the concentration of either species (as a fraction of overall ATP turnover rate) yields essentially the same information on the percent imbalance in ATPase and ATP synthesis rates over any particular work period.

Because creatine phosphokinase (CPK) is generally assumed to function under near-equilibrium conditions (Connett, 1988; Kushmerick, 1985; Meyer *et al.*, 1984; Nioka *et al.*, 1992) and to "buffer" changes in [ATP], decreases in [PCr] could supply another means of estimating percent imbalance in energy coupling; that is, percent energy imbalance could be defined as $((\Delta[\text{ATP}] + \Delta[\text{PCr}]) \times 100)/\text{ATP}$ turnover. Though acceptable, this interpretation may be less appropriate than ours because, at the enzymatic level, energy coupling concerns ATP, not related pathway intermediates. In terms of stabilizing energy coupling, the glycolytic and mitochondrial metabolic pathways also clearly serve in "buffering" changes in [ATP], although by different mechanisms than in the case of CPK-based ATP "buffering." Furthermore, in fast-twitch muscles, Δ[IMP] supplies the most unequivocal measure of energy coupling, but its concentration changes are not stoichiometric with [PCr] changes, while they clearly mirror [ATP] changes (Graham *et al.*, 1990; Schulte *et al.*, 1992). Thus, we marginally prefer to quantify efficiency of energy coupling by focusing directly on ATP or IMP, assuming PCr to be an ATP source analogous to other ATP-yielding pathways. Our overall argument, however, would not change significantly if we used [PCr] changes as well.

III. Energy Coupling in Human Muscles

The highest mass-specific O_2 uptake rates in human muscles occur during work of small muscle masses, because of the high perfusion rates possible. In such cases, aerobic metabolic rates equivalent to about 105 μmol ATP turned over g^{-1} skeletal muscle min^{-1} can be sustained for about 10 min (Andersen and Saltin, 1985); when glycolytic plus PCr contributions are added (Graham *et al.*, 1990), the total ATP turnover during the exercise protocol is 1217 μmol ATP g^{-1} muscle 10 min^{-1}. During the exercise, ATPase fluxes modestly exceed ATP synthase fluxes, so that when fatigue is reached, [ATP] is reduced by about 2.1 μmol ATP g^{-1} muscle. For the total ATP turnover, the percent imbalance is about 2.1 × 100/1217, or 0.17%, indicating coupling efficiency up to 10 times higher than observed in muscles that are fast-twitch glycolytic type (Arthur *et al.*, 1992b; Moyes *et al.*, 1992; Schulte *et al.*, 1992). This coupling efficiency is close to that for rat muscle during treadmill exercise, in which the percent imbalance in fast-twitch oxidative (FOG) muscle (driven almost solely by O_2-based metabolism) is estimated to be 0.06–0.12% (Hochachka *et al.*, 1991b). For O_2-based work, the efficiency of energy coupling can be adjusted with training, acute hypoxia, chronic hypoxia acclimation, or long-term altitude adaptation (Hochachka *et al.*, 1991a); nevertheless, in all of these cases the precision of flux regulation in the two directions is impressively high. It is even higher at lower work rates, where this kind of quantification of energy coupling is difficult, because [ATP] changes are too small to be accurately measured. However, for subjects near 70% of their aerobic exercise maxima, IMP concentrations increase by about 0.2 μmol g^{-1} over a 120-min exercise period, during which the muscle ATP turnover rate can be shown to be about 4860 μmol g^{-1} (Spencer *et al.*, 1991). Under these submaximal exercise conditions, the percent imbalance in energy coupling is less than 0.005%—over 10 times tighter coupling than at maximum aerobic ATP turnover rates!

Enzymes interact with their substrates and modulators in molecular, not molar, terms. Biologists and physiologists, therefore, may find it useful to phrase the problem in terms "perceived" by the enzymes involved. For such illustrative purposes, Matheson *et al.* (1991) calculate that during a 15-min calf muscle exercise protocol in Andean natives, before fatigue is reached about 3.3 × 10^{20} ATP molecules g^{-1} muscle are converted to ADP and Pi, then back again, at cycling rates that are over 100 times faster than at rest. In the process, only 20–25% of the ATP pool is lost to IMP (Matheson *et al.*, 1991); as noted, the situation is probably more than 10 times better at low work rates!

How is such precision of rate control achieved?

IV. Regulation of ATP Demand

One way to address this question is through control circuitry models with enormous loop gains and super-sensitive feedback mechanisms;

however, to metabolic biochemistry, such control models are useful only if they are consistent with known catalytic requirements at any given step in metabolism. That is why it may be useful in addressing this question to begin instead with first principles. Classically, enzyme kinetics and regulation are analyzed in terms of reaction velocities (v) as functions of substrate concentrations, [s]. For Michaelis–Menten kinetics, enzyme activation usually involves increasing, whereas enzyme inhibition involves decreasing, apparent enzyme–substrate affinities, with or without effects on V_{max}; kinetic order cannot exceed 1. For allosteric or regulatory enzymes displaying sigmoidal substrate saturation curves, activation and inhibition similarity is largely based on enzyme–substrate affinity changes, but kinetic order can greatly exceed unity. Usually it is assumed that [s] is the independent parameter and v is the dependent parameter, which may be valid for many tissues, but for working muscle the reverse may be the case.

In an intact organism, a working muscle's ATP demand is thought to be mainly determined by (a) recruitment of motor units and (b) rate coding (Hoffer *et al.*, 1987). Once a motor unit is recruited, energy demand is seemingly determined largely, if not solely, by the firing frequency (activation intensity), which at the cellular level means Ca^{2+} activation tempered by product inhibition. This is well illustrated in a recent study of isolated dog muscle preparation. The advantage of such a preparation (Arthur *et al.*, 1992a; Hogan *et al.*, 1992) is that the circulation can be isolated to allow arterial–venous gradients and flow to be quantified as a function of work rate (tension-time integral, TTI). Maximal and submaximal ATP turnover rates (during isometric contractions) can be elicited by sciatic nerve stimulation, and both work and metabolic performance can be quantified at severe and moderate hypoxia, and compared to values at normoxia.

Under normoxic conditions, nerve stimulation at 0.5 Hz leads to near half-maximal activation of TTI, considered proportional to actomyosin ATPase activity. Moderate hypoxia at 0.5 Hz leads to a down-regulation of ATPase—*even if the muscle retains the capacity for normoxic ATPase rates*. This appears to be a regulated response, because higher-intensity stimulation (1 Hz) in moderate hypoxia elicits an ATPase activity essentially equivalent to that in normoxia at 0.5 Hz. During severe hypoxia similar, but more exaggerated, patterns are observed as in moderate hypoxia. So it appears that under any given conditions, a set or balance point (between activation and inhibition) determines the effective ATPase activity and, therefore, the work rate achievable.

It is probable that stimulus-dependent Ca^{2+} release serves as a crucial activating signal for myosin ATPase, whereas end products such as diprotonated phosphate serve in attenuating ATPase activity (Arthur *et al.*, 1992a). A key point is that activation cannot depend on a rise in the concentration of the substrate, ATP. The affinity of Mg^{2+} myosin ATPase for ATP is so high (affinity constant in the nM range) that the enzyme is, in effect, fully saturated with respect to ATP under most physiological conditions (Hackney and Clark, 1985). That also describes the situation for actin, a kind of cosubstrate for myosin ATPase, that is moved rather than chemically converted during enzyme function (actin determining the direction of sliding of filaments and myosin determining the rate; Sellers and Kachar, 1990).

In functional terms then, myosin in resting muscle behaves like a latent catalyst—*latent but continuously primed for catalytic action, controlled by an exogenously generated activation signal.* As the process of ATP turnover is initiated by ATPase activation via an exogenous signaling path extending essentially all the way back into the central nervous system, most biologists would accept that this clearly represents a proactive control mechanism. For any working situation faced by the animal, this activation cascade represents the means for setting the ATP demand of muscular exercise, albeit tempered by product (especially diprotonated phosphate) inhibition (Arthur *et al.*, 1992a; Cooke and Pate, 1990; Hogan *et al.*, 1992; Wilkie, 1986).

We argue that *for any metabolic enzyme involved in ATP regeneration, the required flux is set by the ATP demand and substrate concentrations are adjusted accordingly; that is, reaction velocity (v) is the independent parameter, whereas substrate concentration [s] is the dependent variable.* In working muscle, then, setting the ATP demand involves balancing activation and inhibitory effects on an enzyme system that is, even at rest, catalytically primed and ready. V_{max} for myosin ATPase ($=k_{cat} \cdot e_0$) is regulated mainly by varying e_0, whereas the effective rate of enzyme turnover is the more minor, tempering part of control. How does the ATP supply system respond?

V. Regulation of ATP Supply

It is, of course, clear that activating muscular work leads to metabolic activation. Empirically, we can illustrate the way in which ATP synthesis rates are balanced with ATPase rates under normoxic and hypoxic conditions by using the isolated dog muscle example considered earlier. In this case, 0.5-Hz stimulation under normoxia leads to about half-maximal activation of oxygen consumption ($\dot{V}_{O_2}$), whereas under moderate hypoxia metabolic rate is down-regulated to about one-third of maximum. This appears to be a regulated response, because further (1 Hz) stimulation at moderate hypoxia achieves the same $\dot{V}_{O_2}$ as in normoxia at 0.5 Hz (Arthur *et al.*, 1992a; Hogan *et al.*, 1992). Thus as expected, the ATP supply part of the system is very closely coupled to the ATP utilization part. Although most earlier studies argue that changes in metabolite concentrations are the proximate causes of such metabolic rate transitions, it has been difficult to show this over wide ranges of muscle work for two reasons:

1. the form of v versus s plots is often incompatible with the Michaelis–Menten kinetic models assumed;
2. the kinetic order of the v versus s response often far exceeds unity.

This can be illustrated with ADP, although the arguments hold for essentially all other parameters thus far measured or calculated for working skeletal muscles.

A. ADP: A Potential Regulator of ATP Supply

For over three decades (see Nioka *et al.*, 1992) ADP has been considered to be a potential activator of mitochondrial respiration *in vivo*. To satisfy this role, at least three provisions must be met:

1. resting concentrations of free ADP ([ADPf]) must be subsaturating;
2. as work rates rise, [ADPf] must rise (in 1:1 ratio according to Michaelis–Menten models);
3. as work rates are modified by other exogenous means, such as by hypoxia, the concentrations of free ADP must change in a predictable way.

Data on several systems indicate that ADPf indeed meets some of these criteria and thus, on first glance, may be taken as a reasonable candidate for "driving" mitochondrial metabolism (Arthur *et al.*, 1992a; Hogan *et al.*, 1992). However, where v versus [ADPf] can be fitted to Michaelis–Menten relationships, only a small (two- to four-fold) dynamic range typically is used; over this range, v versus s plots could be fitted to Michaelis–Menten or to sigmoidal kinetic models.

The problem is illustrated with hummingbird flight muscle functioning at its extreme. During hovering, the flight muscles of hummingbirds sustain metabolic rates equal to nearly 500 μmol ATP g^{-1} min^{-1}, rates that are perhaps 400 times higher than at rest. No one has ever reported, nor ever expects to find, metabolite concentration changes of similar magnitude. In fact, as a general rule, whenever dynamic range is broad (Hochachka *et al.*, 1991b), the relationship between v and s usually cannot be accommodated by Michaelis–Menten kinds of kinetic models because:

1. the shapes of the curves are inappropriate;
2. the relation between change in [ADPf] in the subsaturating range and change in metabolic rate is not even close to the required 1:1 ratio.

In isolated dog muscle, for example, a 3-fold change in [ADPf] could not drive an observed (Arthur *et al.*, 1992a) 18-fold increase in $\dot{V}_{O_2}$ by Michaelis–Menten kinetics. Such data could be accommodated by allosteric models and sigmoidal ADP saturation of mitochondrial function, but this would require a negative modulator to account for large left-shifts in v versus s plots under hypoxic work conditions (Arthur *et al.*, 1992a). Unfortunately, at this time no such modulator is known. What is more, if we assume allosteric kinetics, we need to know which enzyme is being titrated by increasing ADP concentrations. Again, we cannot be certain from *in vivo* data; however, ATP synthase, when studied in isolation, shows cooperative kinetics for both the forward and back reactions (Matsuno-Yagi and Hatefi, 1985). Hence, it is possible that complex ADP saturation curves arise from complex regulatory and catalytic properties of ATP synthase per se, and not from effects at other sites, such as the adenylate translocase. In any event, in our context at this time the important conclusion is that simple Michaelis–Menten models of metabolic regulation simply cannot account for observed ATP turnover rates during large rest-to-work transitions.

This disturbing conclusion does not apply only to our dog muscle data (Arthur *et al.*, 1992a; Hogan *et al.*, 1992). The same conclusion for data on phosphate metabolites (and derived parameters) arises from numerous other systems, including various rat muscles (Dudley *et al.*, 1987; Leijendekker *et al.*, 1983), rabbit *gastrocnemius* (Nioka *et al.*, 1992), dog *gracilis* (Connett, 1988; Connett and Honig, 1989), greyhound *biceps femoris* (Dobson *et al.*, 1988), thoroughbred horse leg muscles (Rose *et al.*, 1988; Snow *et al.*, 1985), and human knee extensor (Andersen and Saltin, 1985; Graham *et al.*, 1990).

Our recent noninvasive ^{31}P-NMRS studies of human *gastrocnemius* in high-altitude-adapted Andean natives compared to three differently trained subject groups (Hochachka *et al.*, 1991a; Matheson *et al.*, 1991; McKenzie *et al.*, 1991) drive home this point. These measurements show, in contrast to expectations from classical control models, that the highest muscle ATP turnover rates are sustained (in endurance-trained and Andean subjects) with the lowest changes in metabolite (especially ADPf) concentrations; that is, v versus s plots for these subjects are right-shifted compared to sedentary or power-trained individuals, indicating controlling processes not easily accommodated by Michaelis–Menten models. Moreover, the muscles of these subjects sustain more than a 15-fold increase in ATP turnover rate (working metabolic rate compared to resting metabolic rate, RMR), but only about a 3-fold increase in free [ADP] or [phosphate].

For these kinds of reasons, we have proposed (Hochachka and Matheson, 1992) that the time has come to face up to a perhaps unpalatable situation: namely, that phosphate intermediates (such as the adenylates, PCr, Pi, and derived parameters from them) usually considered as key metabolic regulators cannot account for the dynamic operating range of skeletal muscle metabolism. Even when change of one of these species is in the right direction, it is not of adequate magnitude. Thus, something is missing in our current understanding of muscle metabolic regulation. One possibility that we need to consider is other pathway metabolites.

B. Other Pathway Metabolites

Unfortunately, when considering the possibility of other metabolites as regulators, difficulties similar to those encountered with phosphate intermediates also arise. To consider a specific example, at near-maximum running exercise in the rat, the concentrations of glycolytic intermediates in the *gastrocnemius*, the *soleus*, and the *plantaris* typically change by no more than 2- to 3-fold (some going up, some going down) whereas ATP turnover rates increase by at least 100-fold (Hochachka *et al.*, 1991b).

The problem is perhaps best clarified by considering net flux through a specific glycolytic step, that catalyzed by pyruvate kinase (PK):

$$\text{phosphoenolpyruvate (PEP)} + \text{ADP} \rightarrow \text{ATP} + \text{pyruvate}.$$

In usual transitions from rest to work, change in muscle [PEP] is often almost immeasurable, whereas [ADP] increases two- to threefold; at the same time, [ATP] may decrease modestly, whereas [pyruvate] increases some two- to threefold (Hochachka *et al.*, 1991b). No combination of these substrate and product concentration changes can possibly account on its own for up to two orders of magnitude increase in net flux through the PK reaction.

Such an observation is not at all unusual in muscle exercise studies, nor is it restricted to mammalian muscle. In fish white muscle, at ATP turnover rates again some two orders of magnitude higher than at rest, [glycolytic intermediates] typically change by less than two- to threefold (Dobson *et al.*, 1987). In a linear path like glycolysis, change in [substrate] for a given enzyme in the pathway means the same relative change in [product] for the enzyme preceding it, making it all the harder to reconcile observed minor changes in [pathway intermediates] with the large flux changes that obviously must occur at every step in the pathway.

A dramatic example of ATP demand versus change in [pathway intermediates] arises from studies of insect flight muscle in which glycolytic and Krebs cycle intermediates, as well as adenylates, are assayed (Rowan and Newsholme, 1979; Sacktor and Hurlbut, 1966; Sacktor and Worsmer-Shavit, 1966). These indicate that the relationships between flux and pathway [substrates] do not fit Michaelis–Menten models. They likewise show that up to several 100-fold increases in ATP turnover rate are sustainable with much more modest (sometimes immeasurable) changes in substrate concentrations.

C. Effective Enzyme Concentration as a Control Parameter

For allosteric enzymes, we could get around such large discrepancy by assuming equally large modulator effects on enzyme–substrate affinities or on maximum velocities. We can also accommodate the metabolite data for enzymes such as glycogen phosphorylase and pyruvate dehydrogenase, which are known to be regulated by covalent modification. Many metabolic enzymes, however, display Michaelis–Menten kinetics with no known modulator effects. We can conceive of only one way to explain the discrepancy in such cases: up-regulating effective enzyme concentration. *The only way that flux changes of 100-fold or more can occur for simple enzyme catalysts with modest (less than 2-fold) change in [pathway intermediates] is to increase the effective concentrations of pathway enzymes.* For these metabolic enzymes, as for myosin ATPase, where $V_{max} = k_{cat} \cdot e_0$, the main (or coarse level) regulation mechanism must alter the ratio of active to inactive enzyme molecules (e_0), whereas fine-tuning is achieved by adjustments of catalytic activity (through changes in [s] or other pathway intermediates).

The missing regulatory element that we propose is required in the resting state, and is the existence of a latent pool of enzymes—a pool of enzymes that are inaccessible to (or nonreactive with) their substrates. In our view (Hochachka and Matheson, 1992), *muscle activation would involve an effective exposure or release of enzymes from this latent pool so as to allow large changes in flux while maintaining near-normal (near-resting) concentrations of pathway intermediates.*

Our model of metabolic pathway regulation is formally analogous to that for the *in vivo* regulation of myosin ATPase. In resting muscle, myosin is, in effect, a latent ATPase, the catalytic potential of which is released only when properly activated from the outside. We propose that a similar process may be occurring in metabolic enzyme activation in muscular work, but may have been overlooked by all earlier workers because of:

1. dogma (that effective enzyme concentration is the same under rest and work conditions and is not a short-term regulated parameter); and
2. the destructive techniques normally used for isolating enzymes and analyzing their activities.

In this framework, v versus s plots for muscle metabolism take on a new and exciting meaning. In the case of human calf muscle, for example (Matheson *et al.*, 1991), our interpretation assumes that the 15-fold increase in metabolic and work rates requires a 5-fold increase in effective enzyme concentration (e_0 adjustment) plus a 3-fold increase in free [ADP], a kinetic fine-tuning of flux. In the dog muscle (Arthur *et al.*, 1992a; Hogan *et al.*, 1992), under normoxia an observed 18-fold change in $\dot{V}_{O_2}$ is explained by a 6-fold change in active enzyme concentration (e_0 regulation) plus a 3-fold change in free ADP concentration. Finally, to extend the pyruvate kinase example given earlier, the increase in net PK flux by two orders of magnitude is explained by about a 50-fold change in e_0 with about 2- to 3-fold substrate- and product-mediated kinetic adjustments. In this view, *the mechanisms of classical control theories add on to an equally fundamental e_0 compensation for the increased energy demand.*

It is gratifying that a similar control process (unmasking of latent catalytic potential) is described by Low and his co-workers (1990; Harrison *et al.*, 1991) for the regulation of glycolysis in red blood cells. Their concept begins with the observation that several glycolytic enzymes (especially glyceraldehyde-3-phosphate dehydrogenase, GAPDH) and other proteins specifically and reversibly associate with the N-terminus tail of the cytoplasmic domain of band 3 protein, the chloride–bicarbonate (anion) exchanger that occurs at high concentrations in membranes of erythrocytes as well as in many other cells. GAPDH is totally inhibited when bound, whereas its full catalytic activity reappears when it is released into the cytosol.

Admittedly, the evidence for analogous mechanisms on rest–work transitions in skeletal muscle metabolic activation is indirect. Yet it is worth emphasizing that a regulatory system controlling a pool of catalytically active enzymes displays three advantageous features:

1. It acts like an on–off switch;
2. It can achieve very large flux changes at nearly constant substrate concentrations, because of the enormous turnover number for most metabolic enzymes; for example, assuming product is removed as it is formed, an inactive–active state transition of one enzyme molecule with a turnover number of 10^3 s^{-1} would yield a proportional increase in flux through that specific step in the pathway even if no change in [*s*] were allowed. Exposure of such enormous catalytic capacity makes it easy with this model to explain how it is possible to increase metabolic flux by over 100-fold with negligible change in [pathway intermediates];
3. With this kind of coarse, but potent, on–off control mechanism in place, the role of [substrate] and [modulator] changes is mainly a means for fine-tuning enzyme catalytic rates over subsaturating concentration ranges, where *v* is most sensitive to change in [*s*].

VI. Current Status and the Next Issues

One conclusion from this analysis seems inescapable: most current attempts to probe mechanisms controlling ATP turnover rates, directed as they are at mainstream pathway intermediates (such as the adenylates, phosphagen, Pi, pH, and so forth), are somewhat misdirected and incomplete. The data considered suggest that ATPases are a proactive component, whereas energy metabolism is a reactive component in the control of ATP turnover rate. For the proactive component, the main control signal (Ca^{2+}) is an exogenous activator, its effects tempered by product inhibition. For the reactive component, flux behaves as an independent parameter, whereas changes in [pathway intermediates] are minimized and, in themselves, cannot account for pathway flux adjustments.

What is needed next, therefore, is insight into the nature or identity of the "activators," "deinhibitors," and or "inhibitors" of the reactive control component of ATP turnover, all or any of which could be exogenous to the mainstream pathways they regulate. This need was already illustrated in 1979 in studies of control requirements for insect muscle mitochondrial metabolism (Rowan and Newsholme, 1979). They analyzed locust flight muscle that metabolizes at rates of 600–800 μmol ATP g^{-1} min^{-1}; assuming usual RMR values for muscle, this rate approaches three orders of magnitude up-regulation of flux with almost immeasurable concentration changes in Krebs cycle intermediates. Sometime earlier, Sacktor and co-workers (Sacktor and Hurlbut,

1966; Sacktor and Worsmer-Shavit, 1966) had already seen a similar situation for the glycolytic pathway in blowfly flight muscles, and we encountered the same problem during high work rates in rat leg (Hochachka *et al.*, 1991b) and hummingbird flight muscles (Suarez *et al.*, 1990). These studies all recognized, implicitly or explicitly, our main conclusion: *key control elements of energy-consuming and of energy-yielding pathways must lie external to the pathways per se, because the concentrations of the intermediates do not change by a sufficiently large factor to directly cause the flux change.* The problem is that the nature of such controlling elements is unknown; our analysis favors the unconventional concept of metabolic activation dominated by exogenously stimulated release of latent enzyme catalytic capacities. It is the only way we can explain large flux changes with minor changes in concentrations of pathway intermediates.

One particularly satisfying aspect of this concept of regulation of ATP turnover rate is a kind of energy coupling symmetry, with both the ATPase and ATP synthesis pathways being dependent on the exposure of latent catalytic capacity. In the fundamental equation $V_{max} = k_{cat} \cdot e_0$, for pathways of both ATP utilization and ATP synthesis, the dominant (or coarse level) control is focused on e_0, whereas fine-tuning involves kinetic adjustments through substrate, product, and modulator effects. The latter are equivalent to the classical loop gains and feedback mechanisms of physiological control theory, and to the allosteric or covalent modification controls of metabolic biochemistry.

VII. Summary

Biologists have long known that rates of adenosine triphosphate utilization and production need to be extremely closely balanced in all cells at all work rates. To put it in molecular rather than molar terms, in human muscle engaged in a 15-min work protocol, some 3.3×10^{20} ATP molecules g^{-1} are used and resynthesized at close to 100 times the resting cycling rates before fatigue occurs. However, during this interval only a 20–25% decrease in the ATP pool is sustained. Review of recent studies as to how such remarkable regulatory precision is achieved suggests that in resting muscle, myosin behaves as a latent catalyst for which the full catalytic capacity is (a) realized with the arrival of its activator signal (Ca^{2+}) and (b) attenuated with reaction products. These proactive controls, initiated at the onset of ATP utilization, set the required flux through ATP-producing pathways. For any given enzyme step in ATP-producing pathways, reaction velocity (v)

becomes the independent parameter, whereas substrate concentration ([s], the dependent parameter) must be adjusted accordingly.

Because the dynamic range for muscles (change from resting to maximum ATP turnover rates) can exceed 100-fold, in many studies of working muscle the percentage change in ATP turnover rate usually greatly exceeds the percentage change in substrate concentrations. These kinds of observations are not easily explained by current metabolic regulation models, but are consistent with pathway enzymes behaving as latent catalysts in resting muscle. In this view, the unmasking of such latent catalytic potential is the main explanation for how large changes in v can be achieved with modest (sometimes immeasurable) changes in [s]. In terms of the basic enzyme velocity equation,

$$V_{max} = k_{cat} \cdot e_0$$

for pathways of ATP utilization and of ATP synthesis, the dominant (coarse level) regulation is achieved through e_0 adjustments, whereas fine-tuning of flux is achieved by effective k_{cat} adjustment.

Acknowledgments

This work was supported by NSERC (Canada). The first kernels for this paper arose from an assignment for a symposium in Lake Louise in February, 1991. This initial challenge to classical control dogma was fine-tuned for the International Congress of Sports Medicine and Human Performance in Vancouver held in April, 1991, and a symposium on vertebrate oxygen transport held in São Sebastiao, Brazil, in September, 1991. I wish to thank the organizers of these various conferences (especially G. Matheson and J. Sutton) for forcing me to think long and hard about how muscle energy coupling is achieved.

References

Andersen, P., and Saltin, B. (1985). Maximal perfusion of skeletal muscle in man. *J. Physiol. (London)* **366,** 233–249.

Arthur, P. G., Hogan, M. C., Bemout, D. E., Wagner, P. D., and Hochachka, P. W. (1992a). Modelling the effects of hypoxia on ATP turnover in exercising muscle. *J. Appl. Physiol.* **73,** 737–742.

Arthur, P. G., West, T. G., Brill, R. W., Schulte, P. M., and Hochachka, P. W. (1992b). Recovery metabolism of skipjack tuna (*Katsuwonus pelamis*) white muscle: Rapid and parallel changes of lactate and phosphocreatine after exercise. *Can. J. Zool.* **70,** 1230–1239.

Atkinson, D. E. (1990). An experimentalist's view of control analysis. *In* "Control of Metabolic Processes" (A. Cornish-Bowden and M. L. Cardenas, eds.), pp. 413–427. Plenum, New York.

Balban, R. S. (1990). Regulation of oxidative phosphorylation in the mammalian cell. *Am. J. Physiol.* **258,** C377–C389.

Chance, B., Leigh, J. S., Kent, J., and McCully, K. (1986). Metabolic control principles and ^{31}P NMR. *Fed. Proc., Fed. Am. Soc. Exp. Biol.* **45,** 2915–2920.
Connett, R. J. (1988). Analysis of metabolic control: New insights using scaled creatine kinase model. *Am. J. Physiol.* **254,** R949–R959.
Connett, R. J., and Honig, C. R. (1989). Regulation of $\dot{V}_{O_2\ max}$. Do current biochemical hypotheses fit *in vivo* data? *Am. J. Physiol.* **256,** R898–R906.
Cooke, R., and Pate, E. (1990). *In* "Biochemistry of Exercise" (A. W. Taylor, ed.), pp. 59–72. Human Kinetics Books, Champaign, IL.
Dobson, G. P., Parkhouse, W. S., and Hochachka, P. W. (1987). Regulation of anaerobic ATP-generating pathways in trout fast-twitch skeletal muscle. *Am. J. Physiol.* **253,** R186–R194.
Dobson, G. P., Parkhouse, W. S., Weber, J. M., Stuttard, E., Harman, J., Snow, J. H., and Hochachka, P. W. (1988). Metabolic changes in skeletal muscle and blood in greyhounds during 800 m track sprint. *Am. J. Physiol.* **255,** R513–R519.
Dudley, G. A, Tullson, P. C., and Terjung, R. L. (1987). Influence of mitochondrial content on the sensitivity of respiratory control. *J. Biol. Chem.* **262,** 9109–9114.
From, A. H. L., Zimmer, S. D., Michurski, S. P., Mohanakrishnan, P., Ulstad, V. K., Thomas, W. J., and Ugurbil, K. (1990). Regulation of oxidative phosphorylation in the intact cell. *Biochemistry* **29,** 3733–3743.
Funk, C. I., Clark, A., Jr., and Connett, R. J. (1990). A simple model of metabolism: Applications to work transitions in muscle. *Am. J. Physiol.* **258,** C995–C1005.
Graham, T. E., Bangsbo, J., Gollnick, P. D., Juel, C., and Saltin, B. (1990). Ammonium metabolism during intense dynamic exercise and recovery in humans. *Am. J. Physiol.* **259,** E170–E176.
Hackney, D. D., and Clark, P. K. (1985). Steady-state kinetics at high enzyme concentration. The myosin Mg-ATPase. *J. Biol. Chem.* **260,** 5505–5510.
Harrison, M. L., Rathinavelu, P., Arese, P., Geahlen, R. L., and Low, P. S. (1991). Role of band 3 tyrosine phosphorylation in the regulation of erythrocyte glycolysis. *J. Biol. Chem.* **266,** 4106–4111.
Hochachka, P. W., and Matheson, G. O. (1992). Regulating ATP turnover rates over broad dynamic work range in skeletal muscles. *J. Appl. Physiol.* **73,** 1697–1703.
Hochachka, P. W., Stanley, C., Matheson, G. O., McKenzie, D. C., Allen, P. S., and Parkhouse, W. S. (1991a). Metabolic and work efficiencies during exercise in Andean natives. *J. Appl. Physiol.* **70,** 1720–1730.
Hochachka, P. W., Bianconcini, M., Parkhouse, W. S., and Dobson, G. P. (1991b). Role of actomyosin ATPase in metabolic regulation during intense exercise. *Proc. Natl. Acad. Sci. U.S.A.* **88,** 5764–5768.
Hoffer, J. A., Sugano, N., Loeb, G. E., Marks, W. B., O'Donovan, M. J., and Pratt, C. A. (1987). Cat hindlimb motoneurons during locomotion. II. Normal activity patterns. *J. Neurophysiol.* **57,** 530–553.
Hogan, M. C., Arthur, P. G., Bebout, D. E., Hochachka, P. W., and Wagner, P. D. (1992). The role of O_2 in regulating tissue respiration in dog muscle working *in situ*. *J. Appl. Physiol.* **73,** 728–736.
Kushmerick, M. J. (1985). Patterns of mammalian muscle energetics. *J. Exp. Biol.* **115,** 165–177.
Leijendekker, W. J., van Hardenveld, C., and Kassenaar, A. A. H. (1983). The influence of the thyroid state on energy turnover during tetanic stimulation in the fast twitch (mixed type) muscle of rats. *Metab., Clin. Exp.* **32,** 615–621.
Low, P. S., Geahlen, R. L., Mehler, E., and Harrison, M. L. (1990). Extracellular control of erythrocyte metabolism mediated by a cytoplasmic tyrosine kinase. *Biomed. Biochim. Acta* **49,** S135–S140.

Matheson, G. O., Allen, P. S., Ellinger, D. C., Hanstock, C. C., Gheorghiu, D., McKenzie, D. C., Stanley, C., Parkhouse, W. S., and Hochachka, P. W. (1991). Skeletal muscle metabolism and work capacity: A ^{31}P-NMR study of Andean natives and lowlanders. *J. Appl. Physiol.* **70,** 1963–1976.

Matsuno-Yagi, A., and Hatefi, Y. (1985). Studies on the mechanism of oxidative phosphorylation. Catalytic site cooperativity in ATP synthesis. *J. Biol. Chem.* **260,** 14424–14427.

McKenzie, C. D., Goodman, L., Matheson, G. O., Davidson, B., Nath, C. C., Parkhouse, W. S., Allen, P. S., and Hochachka, P. W. (1991). Cardiovascular adaptations in Andean natives after 6 weeks of exposure to sea level. *J. Appl. Physiol.* **70,** 2650–2655.

Meyer, R. A., Sweeny, H. L., and Kushmerick, M. J. (1984). A simple analysis of the "phosphocreatine shuttle." *Am. J. Physiol.* **246,** C365–C377.

Mommsen, T. P., and Hochachka, P. W. (1988). The purine nucleotide cycle as two temporally separated metabolic units: A study of trout muscle. *Metab., Clin. Exp.* **37,** 552–556.

Moyes, C. C., Schulte, P. M., and Hochachka, P. W. (1992). Recovery metabolism of trout white muscle: The role of the mitochondria. *Am. J. Physiol.* **262,** R295–R304.

Nioka, S., Argov, Z., Dobson, G. P., Forster, R. E., Subramanian, H. V., Veech, R. L., and Chance, B. (1992). Substrate regulation of mitochondrial oxidative phosphorylation in hypercapnic rabbit muscle. *J. Appl. Physiol.* **72,** 521–528.

Rose, R. J., Hodgson, D. R., Kelso, T. B., McCutcheon, L. J., Reid, T. A., Bayly, W. M., and Gollnick, P. D. (1988). Maximum O_2 uptake, O_2 debt and deficit and muscle metabolites in thoroughbred horses. *J. Appl. Physiol.* **64,** 781–788.

Rowan, A. N., and Newsholme, E. A. (1979). Changes in the contents of adenine nucleotides and intermediates of glycolysis and the citric acid cycle in flight muscle of the locust upon flight and their relationship to the control of the cycle. *Biochem. J.* **178,** 209–216.

Rumsey, W. L., Schlosser, C., Nuutinen, E. M., Robiollo, M., and Wilson, D. F. (1990). Cellular energetics and the oxygen dependence of respiration in cardiac myocytes isolated from adult rat. *J. Biol. Chem.* **265,** 15392–15402.

Sacktor, B., and Hurlbut, E., (1966). Regulation of metabolism in working muscle *in vivo.* II. Concentration of adenine nucleotide, arginine phosphate, and inorganic phosphate in insect flight muscle during flight. *J. Biol. Chem.* **241,** 632–634.

Sacktor, B., and Worsmer-Shavit, E. (1966). Regulation of metabolism in working muscles *in vivo*, concentration of glycolytic intermediates. *J. Biol. Chem.* **241,** 626–633.

Schulte, P. M., Moyes, C. D., and Hochachka, P. W. (1992). Integrating metabolic pathways in postexercise recovery of white muscle. *J. Exp. Biol.* **166,** 181–195.

Sellers, J. R., and Kachar, B. (1990). Polarity and velocity of sliding filaments: Control of direction by actin and of speed by myosin. *Science* **249,** 406–408.

Snow, D. H., Harris, R. C., and Gash, S. P. (1985). Metabolic responses of equine muscle to intermittent maximal exercise. *J. Appl. Physiol.* **58,** 1689–1697.

Spencer, M. K., Yan, Z., and Katz, A. (1991). Carbohydrate supplementation attenuates IMP accumulation in human muscle during prolonged exercise. *Am. J. Physiol.* **261,** C71–C76.

Suarez, R. K., Lighton, J. R. B., Moyes, C. D., Brown, G. S., Gass, S. P., and Hochachka, P. W. (1990). Fuel Selection in hummingbirds: Ecological implications of metabolic biochemistry. *Proc. Natl. Acad. Sci. U.S.A.* **87,** 9207–9210.

Wilkie, D. R. (1986). Muscular fatigue: Effects of hydrogen ions and inorganic phosphate. *Fed. Proc., Fed. Am. Soc. Exp. Biol.* **45,** 2921–2923.

ADVANCES IN VETERINARY SCIENCE AND COMPARATIVE MEDICINE, VOL. 38A

Structural Basis of Muscle Performance

RUEDI BILLETER,* HANS OETLIKER,† AND HANS HOPPELER*

Department of Anatomy and †Institute of Physiology, University of Berne, CH-3000 Berne 9, Switzerland

I. Skeletal Muscles and Muscle Fibers

This paper presents an overview of the structural elements that allow for muscle contractility. A description of the contractile apparatus (Section I), its control elements (Section II), and its energy supply (Section III) will be given. The main focus of this paper is on cross-striated vertebrate muscle, the muscle type for which the most information is available. For other muscle tissues, the reader is referred to the appropriate review literature (e.g., Bullard, 1983; Beenakkers *et al.*, 1984; Pringle, 1976; Kamm and Stull, 1985; Marston, 1989).

In vertebrates, the principal function of muscles is locomotion, however, they also serve as protein stores and can generate heat (Block, 1991). A muscle represents an ordered assembly of contractile fibers arranged as syncytia, with up to several thousand nuclei that are located preferentially at the fiber's periphery. Each skeletal muscle fiber is linked through one or more synaptic endings to a motor neuron that has its perikayron located in the spinal cord. Fibers vary greatly in length and diameter, from a few millimeters long and about 10 μm in diameter (e.g., mouse extraocular muscles) to several centimeters long and up to 80 μm in diameter (e.g., human *m. vastus lateralis*). Every muscle fiber is surrounded by a basal lamina, and generally the fibers are grouped together in bundles surrounded by perimysial connective tissue (Fig. 1). The arrangement of the fibers can vary greatly between different muscles, from the simple "strap" arrangement of *m. sartorius* through unipennate (e.g., *m. extensor digitorum longus*) to multipennate (e.g., *m. deltoideus*).

At their ends, muscle fibers are joined to fibrous tissue via the myotendinous junction, where fingerlike invaginations of the sarcolemma and basal lamina greatly increase the surface area (Eisenberg and Milton, 1984). For muscles longer than approximately 10 cm in humans, the fibers form end to end anastomoses (Zenker *et al.,* 1990) and are innervated by several end plates.

A. The Sarcomere

The contractile material of a muscle fiber, the myofibrils, amounts to some 80% of its volume in most fibers. Myofibrils are more or less cylindrical in shape, measure 1–2 μm in diameter, and often extend along the entire length of a fiber. They consist of serially linked sarcomeres (Figs. 1 and 2), the contractile units, that are built of longitudinal thick and thin filaments precisely aligned between perpendicular structures called Z-discs. The Z-discs, and probably also the M-bands of neighboring myofibrils, are linked together by cytoskeletal proteins, as described in the following. This linkage keeps them in register and is responsible for the cross-striation pattern of the tissue seen in light microscopy, the hallmark feature of skeletal and cardiac muscle tissue. In relaxed, unstretched fibers, the spacing of the Z-discs is about 2.2 μm in most mammalian muscle fibers.

The sacromeres contract when thick and thin filaments slide relative to each other (Huxley and Hanson, 1954), pulling their Z-discs closer together, as shown in the electron micrographs of Fig. 2. At maximum contraction, the Z-discs are spaced 1.6 μm apart, at maximum exten-

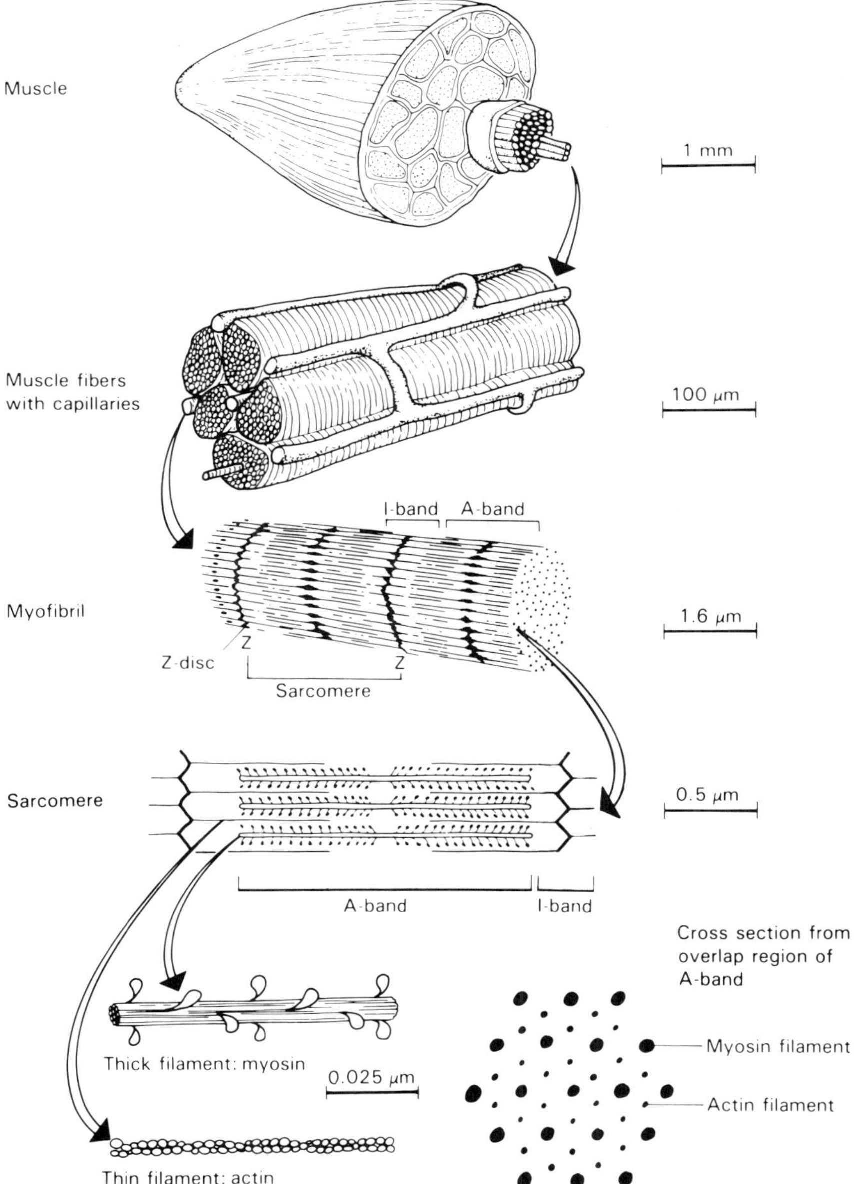

FIG. 1. Schematic representation of the different structural elements in muscle tissue. One micrometer (μm) corresponds to 1/1000th of a millimeter (mm) or 0.0000397 in. Reprinted from Billeter and Hoppeler (1992); adapted from Warwick and Williams (1973), with permission of Churchill-Livingstone.

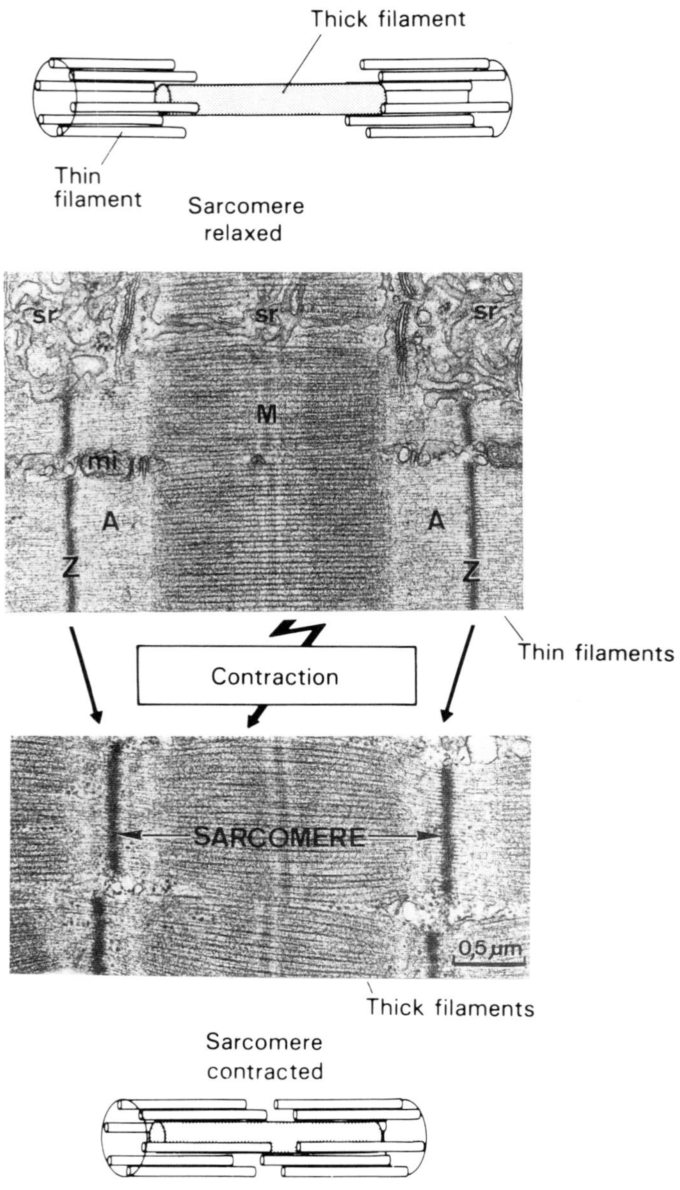

FIG. 2. The "sliding filament" model of muscle contraction. In the relaxed sarcomere, the Z-lines are spaced about 2.2 μm apart. Thick and thin filaments overlap only partially. In the contracted state, the Z-lines have moved closer, and thick and thin filaments overlap over most of their lengths. A = A-band; Z = Z-line; M = M-line; mi = mitochondria; sr = sarcoplasmic reticulum. The drawings give an enlarged, simplified view of the filaments. Reprinted from Billeter and Hoppeler (1992); adapted from Alberts *et al.* (1989), with permission of Garland Publishing, Inc.

sion about 3 μm. In skeletal muscle, the signal initiating contraction is an increase in the intracellular Ca^{2+} concentration (described in Section II). The simultaneous sliding in series of a muscle fiber's sarcomeres (up to half a million) generates considerable length changes and leads to force development.

1. The Thick Filament: Myosin

The main protein of the thick filament is myosin. Sarcomeric myosin molecules are hexamers and are composed of two heavy chains and four light chains (Fig. 3). The long tails of the heavy chains are intertwined in a left-handed "coiled coil" helix (McLachlan and Karn, 1982). Isolated myosin molecules can form filaments spontaneously because the tails have a tendency to aggregate with each other. This process is thought to be controlled by various accessory proteins *in vivo* (Epstein and Fischman, 1991), resulting in the formation of the backbone of the thick filament, with a bipolar structure and composed of about 300 myosin molecules per half filament. There is a "bare" zone in the middle of the filament, where the tails are aligned in an antiparallel fashion, such that no heads are found in this region (Fig. 4).

The pear-shaped heads (Cooke, 1989) of the myosin molecule are the force-generating sites in the muscle. They contain the ATPase site, at which the energy for contraction is derived from the hydrolysis of ATP and ADP and phosphate. An *in vitro* assay has been developed in which whole myosin molecules, or just their head segments, are immobilized on a nitrocellulose-coated glass plate. Upon the addition of sufficient ATP, these molecules can be shown to move fluorescent-labeled actin

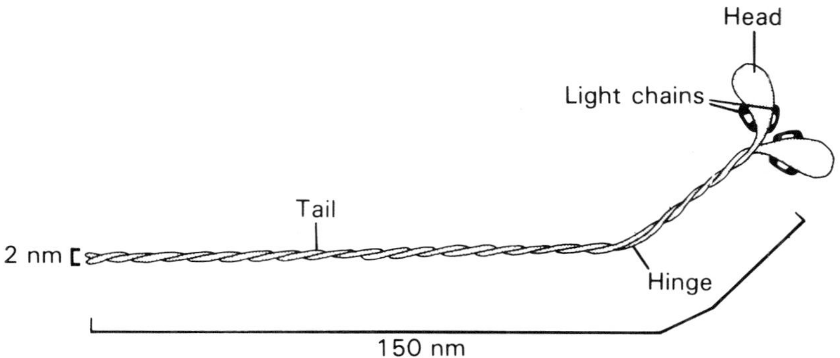

Fig. 3. A muscle myosin molecule consists of six protein chains: two intertwined heavy chains and four light chains. The ATPase site is located in the heads. 1 nm equals 1/1,000,000th of a mm or 0.00000003937 in. Reprinted from Billeter and Hoppeler (1992); adapted from Alberts *et al.* (1989), with permission.

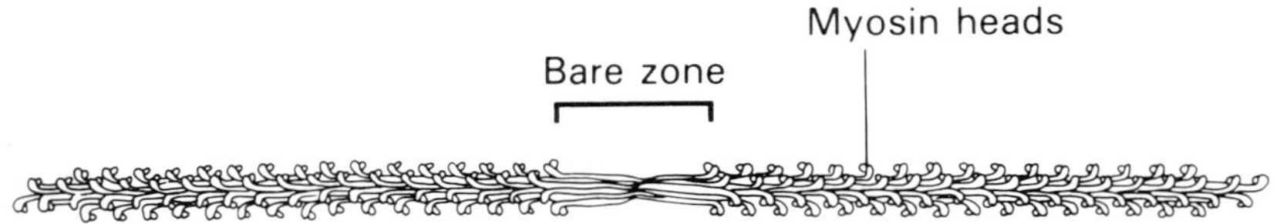

FIG. 4. Schematic representation of a thick filament, composed of myosin molecules aligned with their tails in an orderly fashion, resulting in their heads sticking out of the filament in a regular spacing. The bare zone in the middle, where tails are arranged in an antiparallel fashion, is without heads but binds M-line proteins in the context of the sarcomere. Reprinted from Billeter and Hoppeler (1992); adapted from Alberts *et al.* (1989), with permission.

filaments in a directed manner (reviewed in Warrick and Spudich, 1987).

The ATPase of the myosin head (Mg^{2+}-ATPase) has an activity of about 0.02 s^{-1}. Interaction of the head with filamentous actin in the "activated" state (see the following) increases this ATPase activity approximately a thousand-fold. Therefore, myosin has an actin-activated ATPase activity (reviewed in Taylor, 1979).

The function of the light chains of myosin is not well understood. Two subgroups of myosin light chains are distinguished on the basis of their binding sites on the heads and the reagents that are used to detach them from the heavy chain. The alkali light chains (those that can be removed from the heads by alkali) occur in at least five different skeletal muscle isoforms. All but one of them (LC 3f) interact with actin upon binding of the head (reviewed in Amos, 1985; Vibert and Cohen, 1988). The DTNB light chains (removed from the head with 5,5′-dithiobis-2-nitrobenzoic acid in the presence of EDTA) can be phosphorylated by myosin light chain kinase. In smooth muscle, contraction is controlled by DTNB light chain phosphorylation (Ito and Hartshorne, 1990); in skeletal muscle, regulation of contraction is linked to the thin filaments (see the following and Section II), but phosphorylation of these light chains could be involved in the phenomenon of post-tetanic potentiation (Moore and Stull, 1984; Metzger *et al.*, 1989).

The myosins and actins found in myofibrils are specialized forms of ubiquitous proteins that occur in every eukaryotic cell (Kiehart, 1990). In nonmuscle cells, a special myosin is involved in activities such as intracellular transport, cytokinesis, and surface receptor capping. In addition, there is another myosin involved in amoeboid movement and other activities (Hammer, 1991; Scholey, 1991). The components of both systems have the same basic structure (Vale and Goldstein, 1990).

The motor proteins (myosin, dynein, kinesin) all have conserved, force-generating globular domains that contain an ATPase activity. This globular "head" domain is joined to different tail segments that enable attachment to specific structures, such as the tails of neighboring myosins in the case of the myofibrillar myosins described earlier. The motor proteins produce force and movement along the cytoskeletal filaments of actin (actin filaments, unidirectional) or tubulin (microtubules, bidirectional).

2. *The Thin Filament: Actin, Troponin, Tropomyosin*

The backbone of the thin filaments (Fig. 5) is filamentous actin (F-actin). F-actin consists of two intertwined strands of adjoining actin molecules (globular actin, G-actin). G-actin molecules are capable of self-assembly, a process that is modulated by binding of myosin (Chen and Reisler, 1991). In myofibrils, the length of the thin filament, and therefore its actin strands, is thought to be controlled by nebulin (Kruger *et al.*, 1991), a filamentous protein that stretches all along the actin filament.

Along each of the two grooves created by the F-actin helix lies a filament of tropomyosin molecules. Tropomyosins are coiled-coil dimers, spanning seven actin residues per molecule, and are joined at each end to another tropomyosin in head-to-tail contact. Each tropomyosin carries a troponin complex consisting of three subunits: troponin C, troponin I, and troponin T. This troponin complex has an elongated structure, with troponins C and I forming a globular head, and troponin T a rodlike tail (Fig. 5). The complex is bound to tro-

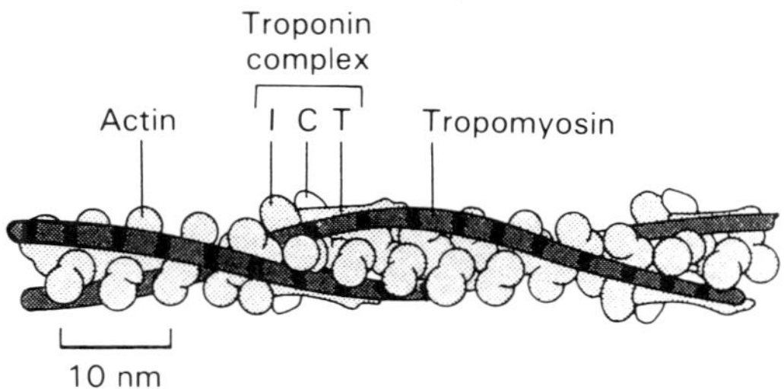

FIG. 5. Model of a segment from a thin filament, indicating the positions of actin, troponin, and tropomyosin subunits. Tropomyosin (itself a dimer, which is not indicated in this drawing) has seven evenly spaced regions of considerable homology, each of which is thought to bind to a separate actin residue. Because every tropomyosin has one troponin complex bound, this results in one troponin complex per seven actin residues. Reprinted from Billeter and Hoppeler (1992); adapted from Alberts *et al.* (1989), with permission.

pomyosin in an antiparallel manner, the NH_2-terminal tail portion of troponin T binding to the head-to-tail contact region of tropomyosin, and troponins C and I binding to tropomyosin, almost 200 Å away (White *et al.*, 1987).

The troponin/tropomyosin complex functions as a Ca^{2+}-sensitive regulatory switch, with troponin C being the Ca^{2+}-binding element. The processes that control the intracellular Ca^{2+} concentration are described in Section II,D. At low intracellular Ca^{2+} concentrations (10^{-8} *M*), two Mg^{2+} ions per troponin C are thought to be bound to the two structural "nonregulatory" binding sites; at high concentrations (10^{-5} *M*), two Ca^{2+} ions additionally bind to the two "regulatory" sites. Under sustained activity (tetanus), eventually all four sites will be occupied by Ca^{2+}. At low Ca^{2+} concentrations, with the two "regulatory" sites not occupied, troponin I can exert its inhibitory effect through tropomyosin onto the actins, which are thus rendered in an inactive conformation, preventing them from activating the ATPase located in the myosin's heads. At high [Ca^{2+}], the two additionally occupied Ca^{2+} binding sites on troponin C induce a conformational change in troponin I that (again through an unknown mechanism via tropomyosin) renders the actins in the activated state, enabling them to activate myosin ATPase (see the following; reviewed in Parmacek and Leiden, 1991). This activation is not just confined to the seven actins connected to one tropomyosin molecule, but rather the whole thin filament cooperatively activates and inactivates as a single unit (Adelstein and Eisenberg, 1980; Brandt *et al.*, 1984; Bertazzon and Tsong, 1990).

3. Z-disc; α-Actinin

In the Z-disc, antipolar thin filaments are anchored by α-actinin-containing bridges that form a stretchable, netlike structure perpendicular to the filaments (Fig. 6; Yamaguchi *et al.*, 1985). The narrow Z-lines of fast fibers contain fewer layers of such bridges than the wide Z-lines of slow fibers (Eisenberg and Kuda, 1976).

4. Cross-Bridge Cycle

Muscle force is generated by the concerted action of many myosin heads interacting with actin in repeated "cross-bridge" cycles of attachment, force generation, and detachment (Huxley, 1957). One such cycle is thought to result in the hydrolysis of one ATP molecule per head. Experiments with isolated myosins immobilized on nitrocellulose-coated glass plates (see earlier) have estimated the force generated by

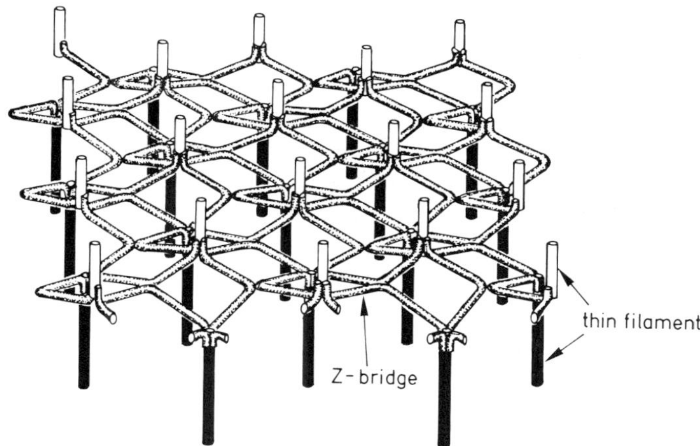

FIG. 6. Model of a "narrow" Z-line. At the Z-line, thin filaments (white and black) are linked by Z-bridges (stippled) in an antiparallel manner. The primary protein of the Z-bridges is α-actinin. The "thick" Z-lines of many fibers have several layers of such Z-bridge networks stacked in this zone. Adapted from Yamaguchi *et al.* (1985) with permission.

one head and cycle to be in the picoNewton range (Kishino and Yanagida, 1988), and the distance moved between 10 and 28 nm (Uyeda *et al.*, 1990).

During the cross-bridge cycle, the myosin head and neck regions exist in at least four conformational stages (Fig. 7; Alberts *et al.*, 1989; Goldman and Brenner, 1987). In State 1, the myosin head does not interact with actin, and has ATP bound in its ATPase site. State 1 is in equilibrium with State 2, in which the ATP has been cleaved by the ATPase, but the resulting ADP and phosphate residue are still retained in the myosin head. Myosin in State 2 can be bound to actin, albeit only weakly, and even in this "weakly bound" state the heads can still rotate (Berger and Thomas, 1991). In State 3, myosin binds actin in the "strongly bound" state and the phosphate is released. The head now undergoes a series of irreversible, poorly understood conformational changes that lead to generation of a force directed to the bound actin, the "power stroke." The resulting State 4 has the myosin heads still tightly bound to actin, with or without the ADP, in the "rigor state." Release of the myosin head from actin in the subsequent return to State 1 requires the binding of a new ATP to the ATPase site of the head.

The transition from State 2 to State 3 is the control step in the cross-bridge cycle. Actin in its activated conformation in the thin filament

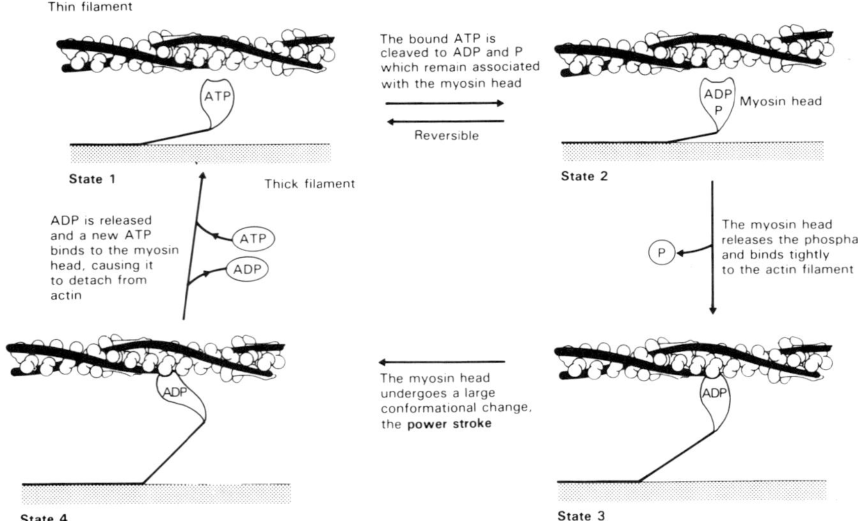

FIG. 7. Model of the cross-bridge cycle. Myosin binds ATP upon detachment from actin, splits the ATP in ADP and phosphate, but retains both molecules in the ATPase site. Only upon interaction with activated actin in the "strongly bound" state is the phosphate released and the myosin undergoes a conformational change, the power stroke. The power stroke results in the exertion of directed force onto the actin and eventually leads to movement. The head is then detached from actin, whereby it releases ADP and binds a new ATP in order to start a new cross-bridge cycle. The different "states" (1–4) correspond to different conformations of the myosin head. Only the transition between State 1 and State 2 is reversible, the others are not. The transition from State 2 to State 3 is the control step in the cross-bridge cycle: actin in activated state greatly facilitates this transition, whereas actin in inactivated state does not. Reprinted from Billeter and Hoppeler (1992); adapted from Alberts *et al.* (1989), with permission.

(two to four Ca^{2+} ions bound to troponin C; see preceding discussion) greatly facilitates transition from State 2 to State 3, whereas actin in inactivated conformation (only two Mg^{2+} ions bound to troponin C) does not; release of phosphate in this case is very slow and cross-bridge cycling is greatly reduced.

In this model, the ATP binding step (release of the myosin head from the actin in the transition from State 4 to State 1) is not the regulatory step in the cross-bridge cycle. Regulation is achieved through the intracellular Ca^{2+} concentration affecting the transition from State 2 to State 3. This Ca^{2+} concentration is controlled by the sarcoplasmic reticulum, the activity of which is ultimately regulated by the action of the motor nerve through control of muscle membrane potential, as described in Section II.

A prerequisite for the functioning of such a model is an immediate, sufficient supply of ATP to the myofibrils that, in the case of muscle, is provided by an energy metabolism adapted to rapid and large fluctuations in energy flow, as described in Section III. In certain types of fatigue, cross-bridge cycling is inhibited by intracellular accumulation of phosphate from the splitting of ATP to ADP and phosphate, and the fall in pH accompanying it (Westerblad *et al.*, 1991).

If a muscle is depleted of ATP it becomes very stiff, because the myosin cannot detach from actin (*rigor mortis*). This is very unlikely to happen *in vivo*, however, because the excitation–contraction coupling process has "safety stops" built in, as described in Section II,F.

In the sarcomere, myosin and actin are arranged in such a manner that the forces generated during cross-bridge cycles are unidirectional, causing the sarcomere to contract as shown in Fig. 2. Extension of contracted sarcomeres, normally done in a relaxed state, has to be performed by forces other than the contractile apparatus proper. Most muscles are, therefore, matched by antagonists that counteract their actions. In many cases, gravity can take over the function of antagonist muscles.

5. Other Myofibrillar Proteins

Although actin and myosin are the most abundant proteins in the sarcomere, many other proteins are necessary for its assembly, maintenance, and function. Figure 8 shows the locations of some of these additional proteins in the sarcomere, and Table I summarizes current hypotheses of their function. Further sarcomeric proteins have been described, but even less is known about their functions.

Titin is one of the largest proteins isolated to date. It is a filamentous protein that contains motifs of the immunoglobulin superfamily group (Labeit *et al.*, 1990), and a single molecule stretches from the Z-line to the M-band (Wang, 1985). Titin forms the elastic filaments that keep the thick filaments centered exactly in the middle between the two Z-lines (Horowits and Podolsky, 1987). Without titin, the thick filaments could pull themselves very close to one side of the sarcomere and away from the other Z-line. They also keep the sarcomere together when it is stretched beyond the point of overlap between thick and thin filaments (Fig. 9; Maruyama, 1986). Titin might also regulate the length of the thick filaments during myofibrillogenesis (Whiting *et al.*, 1989).

The myofibrils of striated muscles are connected to a network of cytoskeletal proteins that links them to the sarcolemma and the nuclear membrane (reviewed in Thornell and Price, 1991). Intermediate

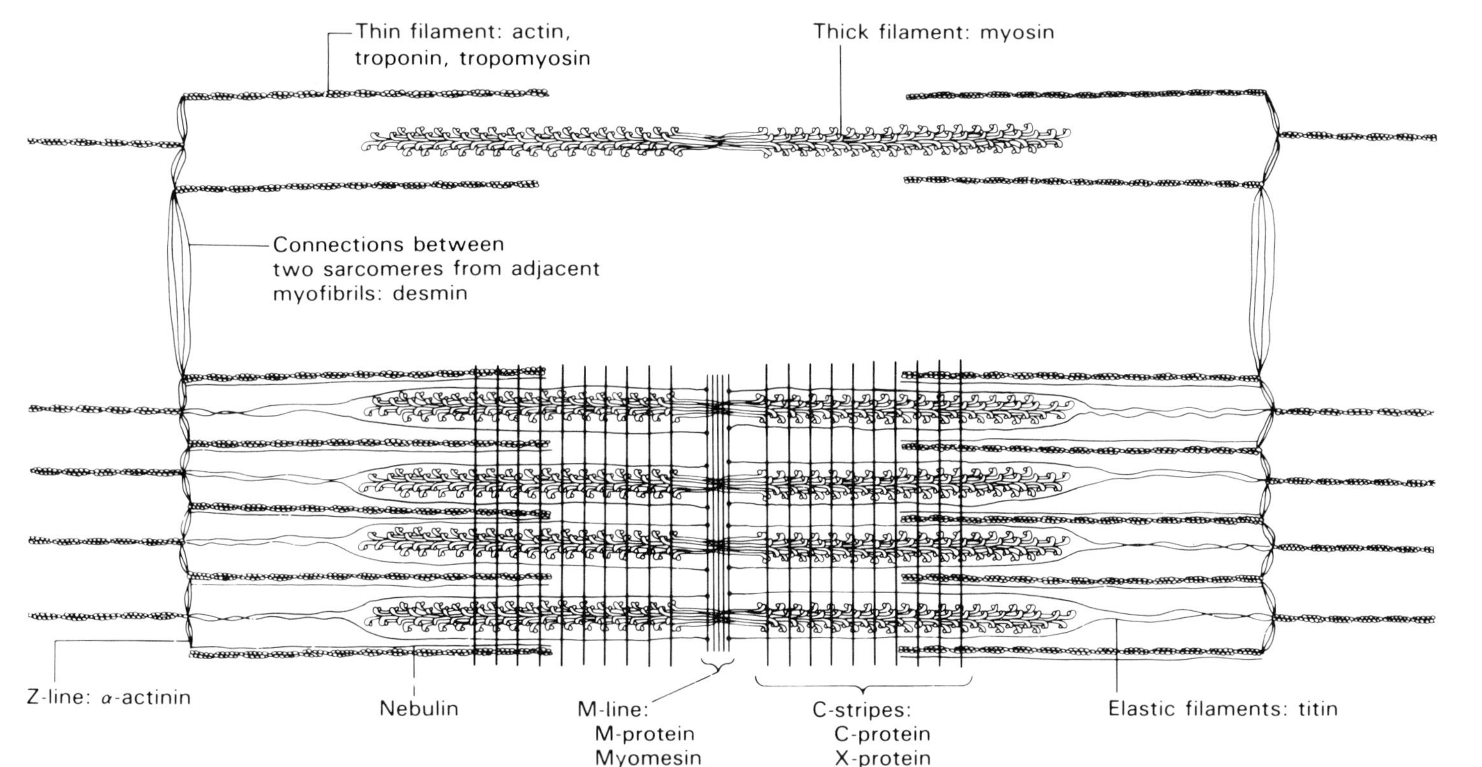

FIG. 8. The location of "auxiliary" proteins in the sarcomere. Table I lists the function (if known) of the proteins represented in this figure. Reprinted from Billeter and Hoppeler (1992), with permission of Blackwell Publications.

TABLE I

SARCOMERIC PROTEINS AND THEIR FUNCTIONS

Element	Protein	Function
Z-line	α-Actinin	Holds thin filaments in place and in register
	Cap Z	Marks the (+) end of the actin core in the thin filament (Casella *et al.*, 1987)
	Calpain	A Ca^{2+}-activated protease (Croall and Demartino, 1991), see also Section II,F
Thin filament	Actin	Core protein of thin filament; interacts with myosin
	Tropomyosin	Transduces the conformational change of troponin to actin
	Troponins	Represent the "switch" that transforms the Ca^{2+} signal into conformational changes of thin filament proteins, leading to cross-bridge cycling
	Nebulin	Along thin filaments; thought to determine the (species specific) length of the sarcomere
Thick filament	Myosin	Core protein of the thick filament; generates mechanical energy (force, movement) from the chemical energy of ATP hydrolysis
C-stripes	C-protein	Thought to hold the thick filaments in a regular array; related in structure to the immunoglobulin superfamily (Epstein and Fischman, 1991)
	H-protein X-protein	Could have a function similar to C-proteins (Bennett *et al.*, 1986)
M-line	M-protein	Hold thick filaments in a regular array; situated in M-line (Grove *et al.*, 1985)
	Myomesin	Forms strong anchoring point for titin
	M-creatine kinase	Provides ATP from creatine phosphate; bound to the M-line (Wallimann *et al.*, 1984) (see also Section III,A)
Elastic filaments	Titin	Keeps the thick filaments in the middle between the two Z-lines during contraction (Wang, 1985); also thought to control the number of myosin molecules contained in the thick filament

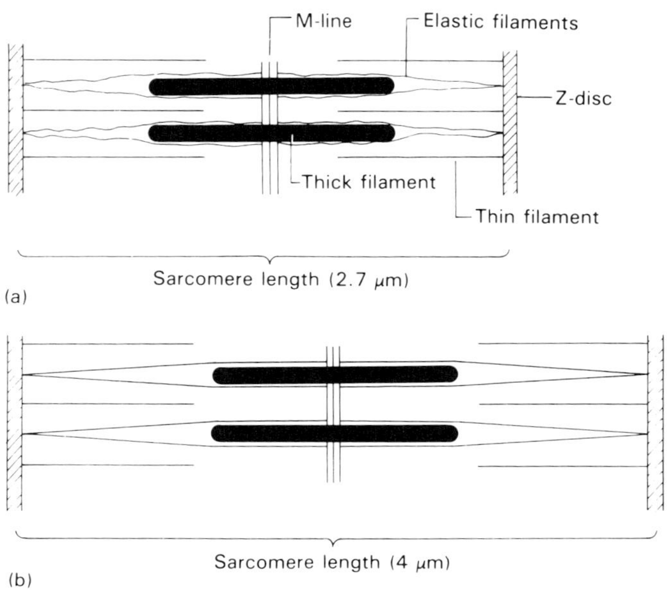

FIG. 9. An illustration of elastic filament function. Elastic filaments link the M-lines of the thick filaments to the Z-lines to keep them in register (a). They are thought to consist of a single molecule of titin. An overstretched sarcomere (b) is still kept together, despite the loss of overlap between thick and thin filaments. Adapted from Horowits and Podolsky (1987) and Maruyama (1986), with permission of Rockefeller University Press and Academic Press.

filaments of the desmin type connect the Z-lines of adjacent myofibrils, keeping them in register and leading to the typical cross-striations. They are also thought to provide a scaffold for the assembly of myofibrils in myogenesis (Epstein and Fischman, 1991).

B. MUSCLE FIBER TYPES

Skeletal muscles perform very different tasks involving a wide range of power outputs, such as holding a human body standing upright for half a day or propelling a cheetah to a hunting speed of 110 km h^{-1}. In mechanical machines, power output is optimized by the use of different gears; muscles achieve their wide range of power output with remarkable efficiency in part from the use of different fiber types.

Individual muscle fibers can differ considerably in their functional and structural properties. Postural work, like holding the body upright, is done with fibers of high endurance, comparatively slow contraction speeds, and preferentially oxidative metabolism. Quick, corrective movements utilize fibers with high contraction speeds and metabolism that relies more on anaerobic glycolysis. This allows a higher energy flow but impedes prolonged activity because it results in the production of lactate and a concomitant decrease in intracellular pH, which affects cross-bridge cycling, as mentioned in Section I,A.

1. The Most Popular Muscle Fiber Type Concept Is Based on Myosin Heavy Chain Isoforms

One chief determinant of the contractile properties of a fiber is the type of myosin heavy chain in its myofibrils (Reiser *et al.,* 1985). Like most other myofibrillar proteins, the myosin heavy chain occurs in many different isoforms (Table II).

The different myosin isoforms gave rise to a "classic" concept of muscle fiber types that is based on the different susceptibilities of their myofibrillar ATPase activities to inactivation by acid or alkali (Brooke and Kaiser, 1970). Examples of fibers typed with this method are shown in Figs. 10, 11, and 12. Staining in this manner demonstrates the myosin heavy chain isoform prevailing in a given fiber (Billeter *et al.,* 1981; Staron, 1991). Additional methods for the detection of different myosin heavy chains in muscle fibers include isoform

TABLE II

METABOLIC PROPERTIES OF MUSCLE FIBER TYPES IN RODENT MUSCLES

Myosin type	Fiber type	Oxidative capacity	Glycolytic capacity
Slow β	I (and ventricle)	+	++
Fast a	IIA	+++	+++
Fast b	IIB	+	+++
Fast x/d[a]	IIX or IID[a]	++	+++
Slow β and fast a	IIC	+(+)	++(+)
Slow tonic	III[b]	NA	NA
Embryonic	emb(IIC)[c]	+	+
Neonatal	neo(IIC)[c]	+	+

[a]IID/X fibers have to date only been found in rodent muscles (Schiaffino *et al.*, 1989; Dix and Eisenberg, 1991).

[b]In rodents, this fiber type is found only in extraocular muscles.

[c]Fibers with a high proportion of embryonic or neonatal myosin heavy chains stain like IIC in the myofibrillar ATPase stain (Jones *et al.*, 1987). A further muscle fiber type specific for masseter muscle, IIM, is found only in carnivores and nonhuman primates (reviewed in Rowlerson, 1990).

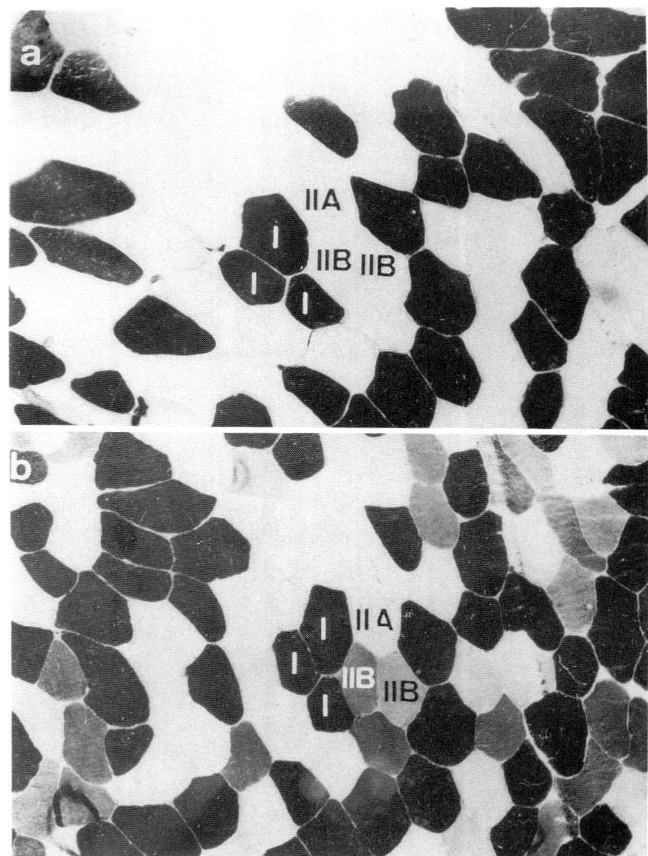

FIG. 10. Fiber types in human *m. vastus lateralis*. (a) Myofibrillar ATPase reaction of a cryostat section (10 μm thickness) after pretreatment at pH 4.3. The ATPase of Type I fibers is active, yielding a strong stain; in Type II fibers it is inactivated, leaving the fiber unstained. (b) Myofibrillar ATPase after pretreatment at pH 4.6. Type IIB fibers stain moderately, Type I fibers strongly, and Type IIA fibers are unstained. (c) Myofibrillar ATPase after pretreatment at pH 10.6. Type II fibers stain strongly, whereas Type I fibers are unstained, yielding a "reverse" picture to (a). (d) SDH (succinate dehydrogenase) reaction, indicating the oxidative capacities of the fibers. Note the slightly higher activity of the Type IIA fiber compared to Type IIB. (e) α-Glycerolphosphate dehydrogenase reaction, indicating the glycolytic capacities of the fibers. Note the variability of Type I fibers and slightly lower reaction in some Type IIA fibers compared to Type IIB. Reprinted from Billeter and Hoppeler (1992) with permission of Blackwell Scientific Publications.

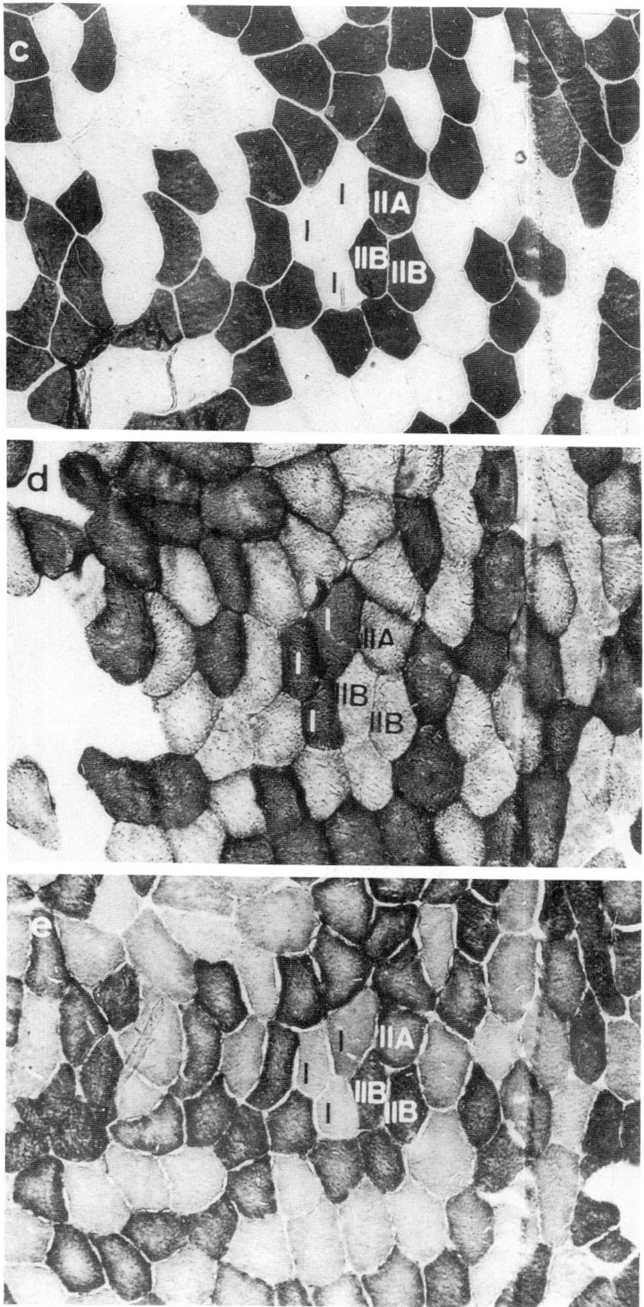

FIG. 10. *(Continued)*

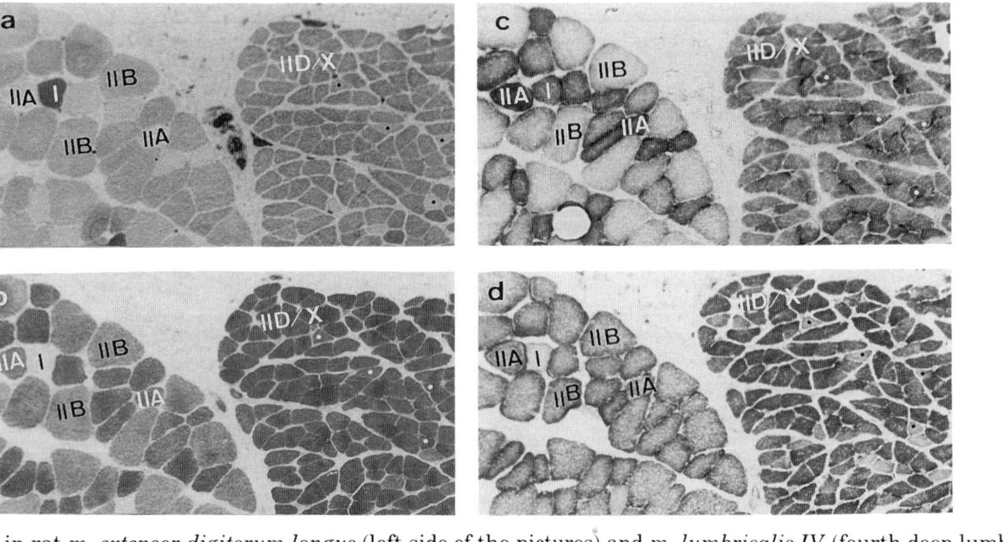

FIG. 11. Fiber types in rat *m. extensor digitorum longus* (left side of the pictures) and *m. lumbricalis IV* (fourth deep lumbrical muscle, right side of the pictures). The photographs were generated and kindly provided to us by A. M. Rowlerson, London. (a) Myofibrillar ATPase after pretreatment at pH 4.65. The few Type I fibers in *m. extensor digitorum longus* stain strongly, Type IIB fibers in *m. extensor digitorum longus* stain moderately, as do the Type IID/X fibers in *m. lumbricalis IV*. Type IIC fibers in *m. lumbricalis IV* (indicated by *) are moderately stained also, but less than Type IID/X, which comprise the majority of fibers in this muscle. Type IIA fibers in *m. extensor digitorum longus* are unstained. (b) Myofibrillar ATPase after pretreatment at pH 10.35. Type I fibers in *m. extensor digitorum longus* stain weakly, and Type IIB fibers stain less than Type IIA, which show very strong staining. Type IID/X in *m. lumbricalis IV* also stain very strongly, and at least some of the Type IIC fibers (*) yield less staining than Type IID/X. (c) SDH (succinate dehydrogenase) reaction, indicating the oxidative capacities of the fibers. Type IIA fibers in *m. extensor digitorum longus* have as high or higher oxidative capacities than Type I, and Type IIB are low oxidative. Type IID/X as well as Type IIC fibers (*) in *m. lumbricalis IV* show high oxidative capacities. (d) α-Glycerolphosphate dehydrogenase reaction, indicating the glycolytic capacities of the fibers. Type IIA and Type IIB fibers show strong staining, Type I are weak, and Type IID/X have high glycolytic capacities. Type IIC fibers (*) are an intermediate between Type I and Type IIA (Billeter *et al.*, 1981; for further information on the fibers in *m. lumbricalis IV*, see Gates *et al.*, 1991).

specific antibodies (Gauthier and Lowey, 1977), gel electrophoretic analysis of the proteins from single fibers (Billeter *et al.*, 1991; Carraro and Catani, 1983), staining for Mg^{2+}-activated ATPase (Mabuchi and Sréter, 1980), and staining for myofibrillar ATPase after preincubation in formaldehyde (Guth and Samaha, 1969) or copper (Matoba *et al.*, 1985).

These methods have led to a refined "myosin-based" fiber typing concept (Tables II and III) that is supported by good, but not perfect, correlations between myosin types and contractile properties (Larsson *et al.*, 1991) and metabolic properties (Engel, 1962), at least within a given species. Other fiber typing schemes rely on combinations of ATPase staining and the metabolic properties of fibers (Peter *et al.*, 1972). These concepts have proven to be more difficult to apply in the exercise sciences because the metabolic properties of muscle fibers can be changed rapidly, whereas the contractile properties are much more stable (Howald *et al.*, 1985).

Biochemical analyses of isolated myosins have shown that the isoforms differ in their ATPase activities (Bárány, 1967). Slow (Type I fiber) myosins cleave ATP at about one-quarter of the rate of fast (Type II fiber) myosins when tested for actomyosin ATPase activity (Maxwell *et al.*, 1982). The myosin ATPase activity correlates well with the speed of shortening of a given muscle (Bárány, 1967), indicating that the myosin heavy chain might be determining the contraction speed. Not only are myosin isoforms the cause for different twitch contraction times, they are also the basis for the different force–velocity relationships of muscle fibers (Lännergren and Hoh, 1984; Ranatunga and Thomas, 1990).

Most of the different myosin heavy chain isoforms of vertebrates described so far are encoded by separate genes. The few primary sequences analyzed to date point to conservation in the head region sequences, with the flexible regions between domains, as well as the "hinge" region, varying somewhat in amino acids. The rod region is more variable than the head (Warrick and Spudich, 1987). *Drosophila* myosins differ from the myosins of vertebrates in that all of their sarcomeric myosins are encoded by a single gene and isoforms are generated by differential splicing (Bernstein *et al.*, 1983).

2. *The Other Myofibrillar Proteins Also Occur in Different Isoforms*

Like the myosins, most of the other myofibrillar proteins described so far occur in isoforms that are expressed differently among the various fiber types. Their expression is not tightly coupled to the histochemical (myosin based) fiber types described in the foregoing, how-

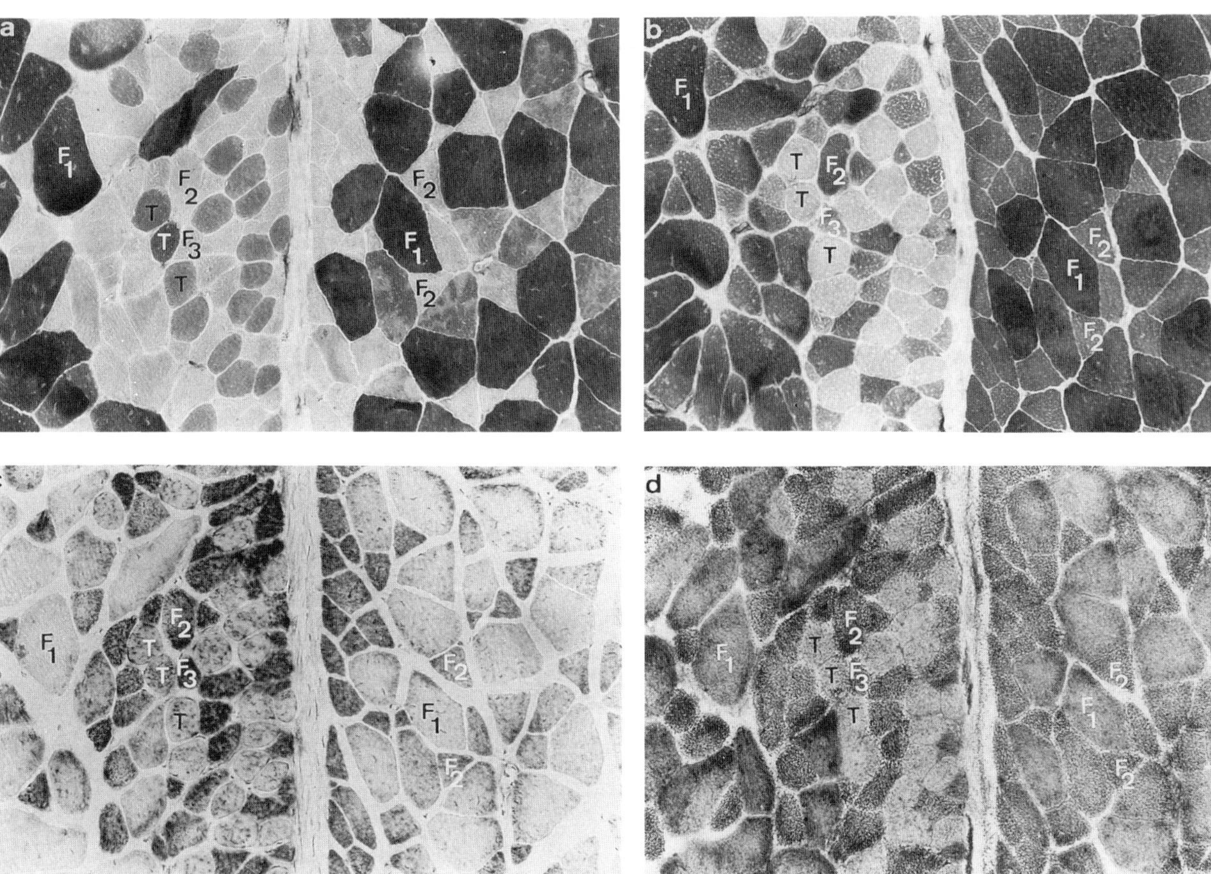

a
F1
F2
T
F3
b
c
d

TABLE III

CONTRACTILE PROPERTIES OF RAT MUSCLES OF DIFFERENT FIBER TYPE COMPOSITIONS[a]

Muscle	Fiber type composition			Time to peak twitch (ms)	V_{max} (length s^{-1})
	% I	% IIA + IIX/D	$ IIB		
m. soleus	77	23[b]	0	34	7.3
m. peroneus longus	9	87	4	17.5	11.9
m. extensor digitorum	6	25	70	14.5	20.6

[a]Data from Ranatunga and Thomas (1990).
[b]IIA + IIC, not IIA + IIX.

ever. In rabbit muscle, for example, a whole variety of combinations of different α- and β-tropomyosins, as well as troponin I and T isoforms, can be found in a given fiber type (Moore and Schachat, 1985).

When assessing the contractile properties of different motor units in a given muscle, a continuum is found (Hennemann, 1980; Burke *et al.*, 1971), rather than the clearly separated categories that might be expected if the myosin type alone determined all aspects of contractility in a given fiber. These variations in the isoform composition of myofibrillar proteins among fibers could, in part, account for such a continuum (Schachat *et al.*, 1985).

Isoforms with fiber-type-specific expression patterns have also been described for other, nonmyofibrillar proteins, such as the Ca^{2+}-ATPase of the sarcoplasmic reticulum (Brandl *et al.*, 1986; Dux *et al.*, 1990), which exist in "fast" and "slow" isoforms. Indeed, the concept of iso-

FIG. 12. Fiber types in *m. iliofibularis* (left) and *m. sartorius* (right) of *Rana*. The photographs were generated and kindly provided to us by A. M. Rowlerson, London. (a) Myofibrillar ATPase after pretreatment at pH 4.75. The large "fast" fibers (F1) react strongly, the slow tonic fibers (T) in *m. iliofibularis* intermediately, and the fast fiber types (Type F2, which occurs in *m. sartorius* as well as *m. iliofibularis*, and Type F3, in *m. iliofibularis*) are unstained. (b) Myofibrillar ATPase after pretreatment at pH 10.6. The tonic fibers (T) are unstained, and Type F1 fibers react the strongest. (c) SDH (succinate dehydrogenase) reaction, indicating the oxidative capacities of the fibers. The large Type F1 fibers, as well as the tonic fibers (T), have low oxidative capacities. Type F2 and Type F3 fibers have higher oxidative capacities. (d) α-Glycerolphosphate dehydrogenase reaction, indicating the glycolytic capacities of the fibers. The tonic fibers (T) show very little staining, whereas Type F2 and Type F3 fibers yield the strongest reactions. The patterns of SDH and α-glycerolphosphate dehydrogenase are very similar, and not reciprocal as in the human muscle shown in Fig. 10. The fiber Types F2 and F3 in *Rana* correspond to the fast fiber types of mammals, and the tonic fibers correspond to the rare tonic fibers in, for example, mouse extraocular muscles.

enzymes originated from the characterization of lactate dehydrogenases in different muscle types (Markert and Ursprung, 1962).

3. Metabolic Properties

The metabolic properties of a muscle fiber (Figs. 10, 11, and 12, and Table II) are primarily determined genetically, but also reflect its pattern of use. In large mammals, such as humans or horses, slow (Type I) fibers have higher oxidative capacities and, therefore, higher mitochondrial volumes and lower glycolytic capacities (Fig. 10). In fast (Type II) fibers, the reverse is true, although the subtype IIA is somewhat more oxidative and less glycolytic than IIB (Hintz *et al.*, 1984). Type IIC fibers are relatively rare and represent fibers intermediate between the Types I and IIA (Billeter *et al.*, 1980).

In rodents (Fig. 11), the situation is different, with Type I fibers displaying only moderate oxidative and low glycolytic capacities, Type IIB fibers showing the low oxidative, high glycolytic pattern, Type IID (also called IIX) fibers having moderately oxidative capacities, and IIA fibers (fast-twitch oxidative glycolytic) having glycolytic capacities similar to IIB fibers, but the highest oxidative capacities of all (Peter *et al.*, 1972).

In anuran amphibians (such as *Rana* and *Xenopus*, Fig. 12; Rowlerson and Spurway, 1988), the fastest (F1) fibers have low oxidative as well as glycolytic capacities, whereas the two other fast fiber types (F2 and F3) have high oxidative and glycolytic capacities, analogous to IID/X and IIA fibers in rodents. However, the true tonic (very slow) fibers have both low oxidative and low glycolytic activities.

4. Individual Muscles Differ in Their Fiber Type Composition, Depending on Their Function

Muscles involved in supporting the skeleton tend to have a high proportion of slow fibers, for example, *m. soleus* in adult rabbits and guinea pigs consists of slow (Type I) fibers only, whereas muscles activated preferentially in fast movements have a higher proportion of Type II fibers, for example, *m. extensor digitorum longus* in the rat, which has more than 95% fast fibers (Table III; Ranatunga and Thomas, 1990). The fiber type compositions of most vertebrate skeletal muscles lie somewhere between these extremes (Garlick *et al.*, 1989). As a general rule, limb muscles close to the bone have a higher percentage of Type I fibers than muscles situated more superficially (Armstrong and Phelps, 1984). This is also the case within many muscles, especially in small mammals, for example, rabbit *m. tibialis anterior* (Pette *et al.*, 1975), and in lower vertebrates (e.g., amphibia; Rowlerson and Spurway, 1988), which show conspicuous fiber type gradients in whole-muscle cross sections.

Allometric studies have revealed that the muscles of small mammals, such as mice, contract distinctly faster than the muscles of large mammals, such as horses. Their myosin ATPase activities parallel the muscles' shortening speeds (Bárány, 1967), and the oxidative capacities and mitochondrial volumes are high in muscles of small mammals and low in larger ones (Mathieu *et al.*, 1981a; Hoppeler *et al.*, 1980). Despite severalfold differences in the actin-activated ATPase activities of their myosins, the (myosin based) fiber type compositions of corresponding muscles in large and small mammals are quite similar. Thus, the ATPase sites are thought to be different in the myosins of large and small mammals. Other structural features of myosin isoforms are conserved, however, so that fast, as well as slow, myosin genes are more closely related among different vertebrates than they are to each other within a given species. This seems also to be the case for the site(s) on the myosin molecule that bring about the differential susceptibility of the histochemical myofibrillar ATPase stain to inhibition by acid or alkali, as shown in Figs. 10, 11, and 12.

The basic pattern of fast/slow fiber type distribution in muscles is generated early in embryonic development. Mononucleated precursor cells migrate from the somites into the limb buds in several waves, where they give rise to multinucleated myotubes, the fates of which differ, depending on the time of migration of their precursor cells. The myotubes eventually mature to adult type fibers, but only after innervation and subsequent use (Rubinstein and Kelly, 1981; Stockdale, 1990; Gunning and Hardeman, 1991).

5. *Plasticity of Muscle Fibers*

Countless experiments have been performed to investigate how, and to what extent, a muscle's fiber type composition can be altered. Such experiments have demonstrated a remarkable plasticity (Pette, 1980, 1990; Salmons and Henriksson, 1981). Continuous electrostimulation, for example, can transform a muscle with Type II fibers only into a muscle consisting entirely of fibers with the characteristics of slow (Type I) fibers, thus dramatically changing its fibers' metabolic as well as contractile properties (reviewed in Pette and Vrbova, 1985; Salmons and Henriksson, 1981). Other approaches used to induce fiber type transformations have included physical training, hypokinesia (including immobilization), denervation and reinnervation, increasing the load by weights or ablation of synergists, and changes in hormonal status.

In adult muscle, one chief determinant of a muscle fiber's characteristics, including its (myosin based) fiber type, seems to be its activity pattern. This is illustrated by the finding that all fibers belonging to an individual motor unit are usually of the same type (Kugelberg, 1973; Larsson *et al.*, 1991), and the activities of key metabolic enzymes in these fibers all fall within a very small range (Nemeth *et al.*, 1979). There are exceptions to this rule, however. In the fourth deep lumbrical muscle of the rat (of which sections are shown in Fig. 11), the small motor units contain Type I, IIC, as well as IIA fibers (Gates *et al.*, 1991). It is possible that these fibers derive from different myoblast lineages in the embryo.

There are additional influences on muscle fiber type determination, such as (unknown) hereditary factors (Simoneau and Bouchard, 1989; Nimmo *et al.*, 1985) or the hormonal status of the animal. Testosterone is thought to be the key factor leading to the sexual dimorphism that has been found in the fiber type composition of specific muscles, such as *m. levator ani* (Carlson *et al.*, 1979), or the jaw-closing muscles in some species (reviewed by Rowlerson, 1990).

Another hormone that influences the fiber type composition of most body muscles is thyroid hormone. Muscles from hypothyroid animals tend to have a higher proportion of Type I fibers than the muscles of euthyroid animals, whereas muscles from hyperthyroid animals have more Type II fibers. Thyroid hormone directly influences the expression of myosin heavy chain genes by interacting with their promotors via the thyroid hormone receptor (Izumo *et al.*, 1986).

II. Control of Contraction: From Nerve Action Potential to Calcium Release

A. Introduction

Many aspects of muscle activation, ranging from the action potential in an α-motoneuron to Ca^{2+} release from the sarcoplasmic reticulum in the muscle fiber, are not generally reviewed in this section, because at least three separate reviews would be required to reach even partial coverage of the recent advances in these fields. Instead, an attempt is made to illustrate evidence that excludes the likelihood of a direct electrical transmission, and the mechanisms that overcome the seemingly insurmountable problems of impedance mismatches. For more detailed information about recent advances in the field, readers are referred to reviews referenced in the following.

In order to enable the motion patterns generated in the central nervous system to become operationally effective, appropriate commands (in this case neuronal action potentials, lasting about 1 to 2 ms) have to be transmitted to the plasma membranes of the skeletal muscle fibers selected for a particular motor task. For information about temporal and spatial summation of muscle responses in motor tasks, readers are referred to a textbook of muscle physiology. For the sake of simplicity, this section addresses the basic mechanisms for a single activation of a muscle fiber. It must be kept in mind that under physiological conditions, the motor units are mostly activated by trains of impulses that lead to mechanical responses in the form of fused tetani.

The necessary steps will be discussed for transformation of an action potential on an α-motoneuron into transient tension development and/or shortening of the appropriate muscle fibers by interaction of the contractile proteins. The latter processes, described in Section I,A, are controlled by the myoplasmic free Ca^{2+} concentration. This mechanism requires that considerable amounts of Ca^{2+} are brought into contact with the contractile proteins in order to allow motion (for review, see Oetliker, 1982). Biochemical experiments and measurements of activation heat lead to an estimate that the amount of Ca^{2+} needed to produce just one twitch in 1 g of muscle is about 200 nmol (Homsher *et al.*, 1972; Smith, 1972; Kobayashi and Sugi, 1980). To promote relaxation, in turn, the same quantity of Ca^{2+} has to be sequestered away from the contractile proteins again (Winegard, 1968, 1970; Hasselbach, 1979; Hasselbach and Oetliker, 1983).

Looking at muscle activation from the point of view of the currents involved, as well as the signals to be transmitted, a model calculation based on quantitative structural data of 1 g of muscle illustrates the quantitative aspects of the problem. Assuming that muscle fibers of 100 μm diameter and 10 cm length have been selected for performing a particular motor task, about 1300 muscle fibers make up 1 cm^3, or 1 g, of muscle tissue. The muscle fibers are contacted by 1300 α-motoneurons via their unmyelinated endings. These neuronal structures form the proximal parts of the respective motor end plates. An example of such neuromuscular junctions is shown in Fig. 13.

The signaling of the motor end plates is in the form of action potentials lasting about 1 to 2 ms, involving a transient change in membrane potential from −70 mV to +20 mV, inside versus outside. The diameters of these axons are in the range of only 1 μm. This is rather small compared with the 30- to 100-μm diameter of the muscle fibers contacted by them.

Unmyelinated nerves and muscle fibers typically produce action cur-

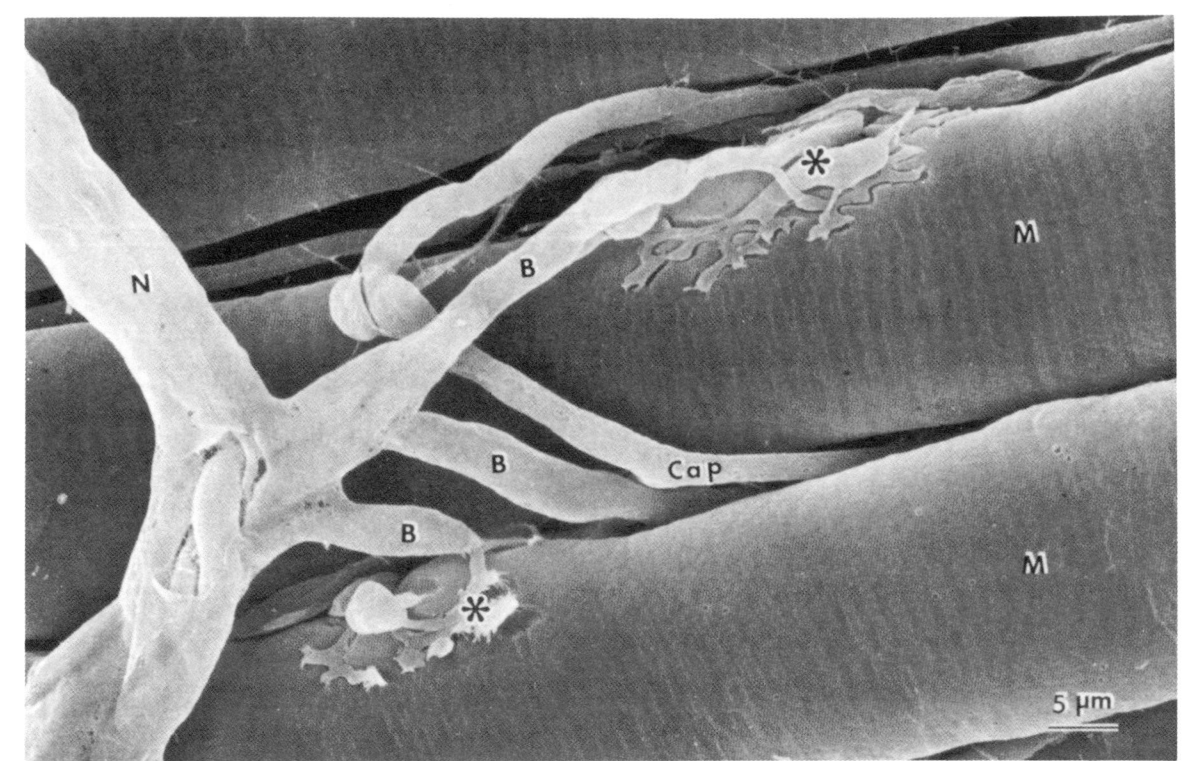

Fig. 13. A scanning electron micrograph of the neuromuscular junction from a Chinese hamster *sternothyroideus* muscle. The nerve (N) forms side branches (B) that terminate on muscle fibers (M), forming neuromuscular junctions, the motor end plates (*); Cap = capillary. Reprinted from Ishikawa *et al.* (1983), with permission of Chapman and Hall.

rents of densities on the order of 1 mA cm^{-2} (Katz, 1966). The current supplied by the 1300 axons would, therefore, amount to 30 μA maximum if the total surface of the axons contacting the muscle fibers is taken into account. The electrical situation is much worse, however, because there is no direct contact between axon and muscle membrane, instead there is a gap of about 500 Å with finite resistance in the extracellular space between the pre- and postsynaptic membranes of the synapses. Thus, the total of the currents reaching the muscle fibers from each of the 1300 axons will amount to only about 10^{-9} A (Katz, 1966).

In our model, a total current of approximately 10^{-9} A at the motor end plates eventually controls a current of about 8 A at the level of the Ca^{2+} release from the sarcoplasmic reticulum (Oetliker, 1982). It is obvious that such a gap of 10 orders of magnitude cannot be bridged by electrical means as far as current is concerned. Using the terminology of an electronic engineer, the expression "severe impedance mismatch" is appropriate. [For a more detailed treatment of these aspects, see the classical book of Katz (1966) from which part of these considerations are drawn, or the chapter by Ginsborg and Jenkinson (1976).]

In the nerve–muscle system, this problem of "impedance mismatch" has been solved by subdivision into three stages:

1. The neuromuscular junction (Fig. 13);
2. The link between surface membrane and t-tubular system (Fig. 15);
3. The transfer of a change in the electrical field across the t-tubular membrane to a release of Ca^{2+} from the sarcoplasmic reticulum at the triadic junction (Figs. 14 and 15).

In each of these three stages, a different mechanism is involved that enables signal transduction and deals with this impedance mismatch, as outlined in the following section.

B. Motor End Plate

The direct action of a current on the order of 10^{-9} A distributed to the 1300 subsynaptic membranes in our model would only induce a depolarization of a few microvolts. This falls far short of the depolarization of 40 mV that is needed to reach (from a resting potential of −90 mV) the potential of −50 mV required to critically depolarize the subsynaptic membranes in order to trigger regenerative a action potential in the extrasynaptic part of the muscle membranes.

This problem is overcome by a synaptic mechanism in which acetylcholine, after release from storage sites in the terminal parts of the

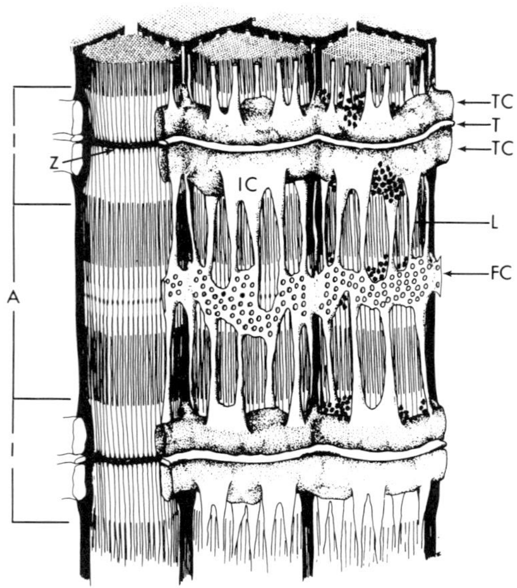

FIG. 14. The arrangement of the t-tubular system and the sarcoplasmic reticulum of a frog twitch fiber. Slightly more than one sarcomere is shown. The sarcoplasmic reticulum forms a three-dimensional network of cisternae and tubules around the myofibrils. Two levels of t-tubular networks (T) are shown; in frog muscle fibers, they are adjacent to the Z-lines, forming the central elements of the triads, which also include two terminal cisternae (TC). The TC connect to longitudinal tubules (L), either directly or through intermediate cisternae (IC). Near the center of the A-band, the longitudinal tubules join the fenestrated collar (FC). In mammalian muscles, two triad structures are found in each sarcomere, with the t-tubules located at the A-band to I-band junction. A schematic of these structures, illustrating the location of the junctional feet, is shown in Fig. 15. I = I-band; A = A-band; Z = Z-line. Reprinted from Peachey and Franzini-Armstrong (1983), with permission of Rockefeller University Press.

motor axons, diffuses to specific receptor sites in the directly opposed postsynaptic membranes (Katz, 1966; Katz and Miledi, 1972). The surface area of the subsynaptic membranes is related to its capacity to generate the current that eventually depolarizes the extrasynaptic region. This surface is greatly enlarged by the presence of invaginations of the plasma membrane, called subsynaptic clefts, where most of the receptors for acetylcholine are located (Heuser and Reese, 1973; Anderson and Cohen, 1974; Bowden and Duchen, 1976, for review). The simultaneous binding of two molecules of acetylcholine brings about the opening of an unspecific transmembrane channel for monovalent cations, allowing about 50,000 Na^+ ions to enter the muscle cell per millisecond (for a detailed review of the molecular processes involved

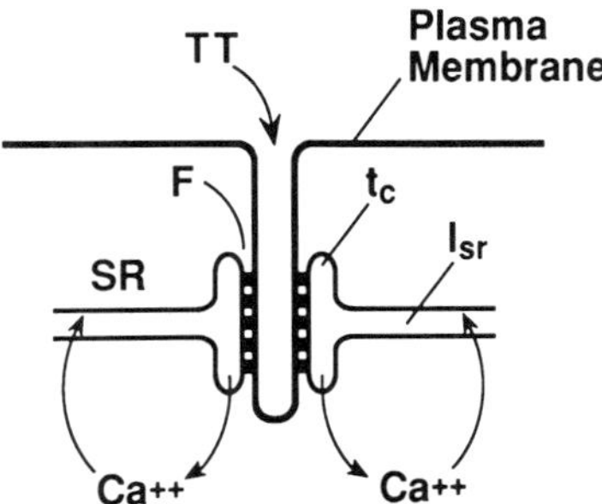

FIG. 15. Illustration of the release of Ca^{2+} in the triad structures and its re-uptake in the longitudinal sarcoplasmic reticulum (lsr). The figure gives a schematic view of the structures involved in the excitation–contraction coupling process; it represents a close-up picture of the t-tubular membranes and the adjacent membranes of the terminal cisternae (tc) with the junctional feet (F) in between. This structure is labeled TC-T-TC in Fig. 14. SR = sarcoplasmic reticulum; TT = transverse tubule. Reprinted from Catterall (1991), with permission of Cell Press.

in activating the nicotinic cholinergic receptors to for membrane-spanning channels, see Steinbach, 1989; Lingle *et al.*, 1992). Such a process takes advantage of the ion gradients established beforehand by the energetic machinery of the much larger and more potent muscle cells.

At rest, the muscle cell exhibits a membrane potential of about −90 mV with respect to the cell's exterior (Hodgkin and Horowicz, 1960). This potential difference is due to ionic gradients between the inside and outside of the cell. They are established and maintained by Na^+/K^+ pumps that expend energy. Under resting conditions, the potential difference is close to the equilibrium potential of the K^+ gradient across the plasma membrane, according to the Goldmann equation (Hodgkin and Katz, 1949). However, the difference is far from the equilibrium potential of the Na^+ gradient, amounting to approximately +60 mV inside relative to outside. This indicates that the muscle membrane is substantially more permeable to K^+ than Na^+. There is a large electrical and chemical driving force for Na^+ ions, whereas K^+ ions tend to move inside with the electrical field and to move outside according to the concentration gradient. Both effects just about cancel out for K^+ at the normal resting potential of −90 mV.

The influence of transmitter release on an action potential propagating through the end plate region of the muscle fiber was evaluated by so-called collision experiments. In these experiments, stimulation of the motor nerve was used to liberate acetylcholine at the motor end plate. A muscle action potential was initiated by direct stimulation of the muscle fiber. Through adjustment of the delay between muscle and nerve stimulations, liberation of acetylcholine could be made to collide with action potentials traveling through the motor end plate region of the muscle fiber at different membrane potentials. The following was found:

1. If transmitter release occurred at a moment when the membrane potential was approximately -15 mV, its influence on actual currents was minimal;
2. If the membrane potential was more positive than -15 mV, a hyperpolarizing effect of transmitter release was seen;
3. If membrane potential was more negative than -15 mV, a depolarizing effect was seen (Del Castillo and Katz, 1954).

From these findings, it was concluded that acetylcholine renders the subsynaptic membrane permeable to roughly the same extent for K^+ as for Na^+.

The arithmetic mean of the two equilibrium potentials for Na^+ (about $+60$ mV) and K^+ (about -100 mV) is about -20 mV. The effect of acetylcholine can, therefore, be described as the opening of an unselective channel for monovalent cations. This brings about a depolarization of appropriate magnitude to induce local depolarizing currents that set up an action potential in the extrasynaptic part of the muscle surface membrane on both sides of the motor end plate (Takeuchi and Takeuchi, 1960). The initiation of the action potentials is due to potential-dependent opening of Na^+-specific channels in the extrasynaptic membrane. These channels are functionally different from those in the subsynaptic part of the muscle fiber, which is the portion of the sarcolemma located below the nerve endings marked by asterisks in Fig. 13.

The Na^+-specific channels, inactivated with time, increase the Na^+ permeability of the membrane to such an extent that it substantially exceeds the K^+ permeability. The delayed increase in K^+ permeability acts synergistically with a decrease in time- and voltage-dependent open probability of the Na^+ channels to bring the membrane potential back to its resting value within a few milliseconds (Hodgkin and Katz, 1949). The action potentials propagate away from the motor end plate in both directions, that is, toward the tendon ends, with a propagation speed of about 2 m s^{-1} in frog muscle (Oetliker and Schümperli, 1982) and 3–4 m s^{-1} in mammalian muscle fibers. In muscle fibers of approximately 500 mm length, the spread of activation takes between 60 and 80 ms from motor end plate to the tendon ends if the motor end plate is located in the middle of the fiber. Innervation of longer muscles by several end plates increases the synchrony of excitation. This would prevent initial passive expansion of not-yet-activated sarcomeres far from the motor end plate at the beginning and at the end of an activation period.

At this stage, the total current produced across the surface membranes of our model muscle cells has reached a magnitude of about 0.4 A g^{-1} muscle, under the assumption that all surface membrane areas are active at the same time. This is not strictly true, but the error introduced is not large and does not influence the proportions of the model calculations that follow for the sarcoplasmic reticulum.

An additional effect of the chemical transmission is a delay, on the order of 1 ms, between the arrival of the action potential at the presynaptic membrane and the onset of the postsynaptic excitatory potential. This delay is due to the diffusion time of the acetylcholine to reach the postsynaptic membrane (Fatt and Katz, 1951; Katz and Miledi, 1965).

The most widely accepted mechanism to explain the release of acetylcholine from the nerve endings at the motor end plate is that the transmitter is released in quantal packages. Release of the packages is due to exocytosis of vesicles contained in the presynaptic nerve endings; this is triggered by a Ca^{2+}-sensitive mechanism that is activated when action potentials reach the nerve endings (Del Castillo and Katz, 1954). This widely accepted hypothesis has been challenged by Dunant and Israel (1985), Israel *et al.* (1979), and Girod *et al.* (1992). They used rapid freezing techniques, combined with simultaneous electrophysiological measurements, in multiple synaptic preparations of the electric organ from *Torpedo marmorata*, a tissue that is homologous to muscle. To compare structural changes with the occurrence of electrical responses, the different preparations were stimulated sequentially in short temporal intervals, while all were frozen at the same time. They found postsynaptic electrical responses in preparations that did not show indications of fusion of the vesicular membranes with the presynaptic nerve membranes. These findings question the functional significance of the vesicles for transmitter release, and support the hypothesis that quantal amounts of transmitter could be released during given open periods of voltage-sensitive channels in the presynaptic membranes.

C. Surface Membrane; T-tubules

To induce Ca^{2+} release from the terminal cisternae of the sarcoplasmic reticulum, the action potentials have to invade the t-tubular system (Huxley and Taylor, 1958; Peachey, 1965; Winegrad, 1968).

This structure that allows excitation to invade the inside of the muscle fiber is illustrated in Figs. 14 and 15. The t-tubular system represents a further possible overload for the action potential (Adrian and Peachey, 1973) because of its much larger surface area (up to six times the surface area of the cylinder in which it is contained; Hodgkin and Nakajima, 1972; Eisenberg and Gage, 1969; Mobley and Eisenberg, 1975). This impedance mismatch is overcome relatively simply by interposing an access resistance at the entrance of the t-tubules. This access resistance (Adrian and Peachey, 1973) prevents too much current from being drawn into the capacitance represented by the t-tubules at the beginning of the action potential. Such an occurrence would reduce the speed of propagation of the action potential along the muscle fiber, or even cause it to fail to propagate along the whole fiber. Because of the restricted extracellular space, the propagation speed of the action potential inside the t-tubules is considerably lower than that along the muscle fiber, only on the order of 7 cm s^{-1} (Gonzalez-Serratos, 1971). However, because of the relatively short distances the action potentials have to travel from the surface membrane to the core of the muscle fiber (about 50 μm in our example), the delays induced by the low propagation velocity are 1 ms at the most.

The significance of the t-tubules for muscular activation was demonstrated by Huxley and Taylor (1958) with their classic local activation experiments under microscopic control. They showed that depolarizing current applied to a muscle fiber via an electrolyte-filled micropipette induced a contraction of the two neighboring half sarcomeres at a certain strength of the current when the pipette was facing an entrance to a t-tubule, whereas the same amount of current was insufficient to elicit contraction when the electrode was displaced from the neighborhood of the t-tubular entrance. For a more complete overview, see the classical synopsis of muscle activation by Huxley (1980).

D. T-tubules; Terminal Cisternae of Sarcoplasmic Reticulum

As mentioned earlier, the action currents due to a single action potential entering the t-tubular system set up a release of Ca^{2+} from the sarcoplasmic reticulum, corresponding to about 8 A g^{-1} muscle tissue in our model if the release lasts for 5 ms (for details of the calculations, see Oetliker, 1982). If we assume that all the t-tubules of the 1 g of muscle in the model undergo an action potential in synchrony and that the surface area of the t-tubules is six times that of the plasma membrane, a total current of 2.4 A is reached at this stage. This is certainly an upper limit for the current carried by the t-tubules, for reasons mentioned earlier in the context of the motor end plate, and because the density of Na^+ channels might be lower in the t-tubules than in the outer plasma membrane.

1. Electrical Considerations

Currents that are established as a result of the t-tubular action potentials are insufficient to substantially alter the potential difference across the sarcoplasmic reticulum membrane. Because the release of Ca^{2+} from the sarcoplasmic reticulum has to be electrically counterbalanced by an uptake of an equivalent quantity of positive charges, the work contained in the arithmetic produce of current × voltage (or current × resistance, corresponding to voltage) × duration cannot, for thermodynamic reasons, exceed the work performed previously by the sarcoplasmic reticulum Ca^{2+} pumps in transporting the Ca^{2+} into the sarcoplasmic reticulum. The work performed by the sarcoplasmic reticulum Ca^{2+} pump to take up 200 nmol of Ca^{2+} can be assessed based on the stoichiometry of the pump, that is, 2 mol Ca^{2+} for the expenditure of 1 mol ATP, and the caloric value of ATP under the prevailing conditions in the myoplasm, yielding approximately 5.7 mJ (Hasselbach and Oetliker, 1983). Solving the equation for resistance (R, in Ω),

$$[\text{current (in A)}]^2 \times \text{time (in s)} \times R(\text{in } \Omega) = \text{work (in J)}$$
$$64\ \text{A}^2 \times 0.005\ \text{s} \times X\Omega = 0.0057\ \text{J}$$

indicates that a maximal theoretical value of $1.9 \times 10^{-2}\ \Omega$ is obtained for the total resistance encountered by the release of Ca^{2+} and the uptake of counter-ions into the sarcoplasmic reticulum for charge compensation. Based on the estimated concentration gradient of Ca^{2+} between the inside of the sarcoplasmic reticulum and the myoplasmic side, approximately three orders of magnitude, a value of $1.1 \times 10^{-2}\ \Omega$ is obtained. This value partially accounts for the fact that not all energy spent by the Ca^{2+} uptake process is available upon release. An estimate of the lumped resistance in the release path in the terminal cisternae can be made from experiments on isolated Ca^{2-} release channels incorporated in planar lipid bilayers. Based on estimates of a single-channel conductance of 10 pS at a free Ca^{2+} concentration of 5 mM, 5×10^{10} channels cm^{-2} facing the t-tubular membrane, and 1700 cm^{-2} of t-tubular surface g^{-1} muscle, a value of $1.2 \times 10^{-3}\ \Omega$ is obtained. This value must be subtracted from the total resistance of $1.1 \times 10^{-2}\ \Omega$ just calculated, yielding about $10^{-2}\ \Omega$ left for the charge-compensating currents. Under the assumption that the total sarcoplasmic reticulum surface of 2.2 m^2 g^{-1} muscle participates in providing compensating current, a maximal specific membrane resistance of the sarcoplasmic reticulum of approximately 200 Ω cm^2 is obtained (for details of the calculations, see Oetliker, 1982, 1989; Hasselbach and Oetliker, 1983).

This value is well above the values for specific membrane resistance calculated from conductance measurements on planar lipid bilayers to which isolated sarcoplasmic reticulum vesicles have been fused beforehand. From these experiments, specific membrane resistances on the order of 3–17 Ω cm^2 were obtained. This indicates that there is a rather large safety factor for current flow related to Ca^{2+} release, and that the resistance in the Ca^{2+} release path is similar to the resistance in the path for the compensating currents. Investigations by Pape *et al.* (1992) indicate that the use of valinomycin as K^+ ionophore does not influence the H^+ content in the myoplasm during Ca^{2+} release. This is a further indication that the permeability for K^+ is such that the very low values for membrane resistance determined *in vitro* can be taken as relevant for the *in vivo* situation. Thus, the buildup of a substantial change in membrane potential across the sarcoplasmic reticulum is very unlikely. The optical signals recorded during activation of skeletal muscle, tentatively interpreted as reflecting changes in membrane potential of the sarcoplasmic reticulum (Bezanilla and Horowicz, 1974; Baylor and Oetliker, 1975; Oetliker *et al.*, 1975), can be interpreted as an indication of changes in myoplasmic Ca^{2+} concentration due to conformational changes of the Ca^{2+}-activated ATPase of the sarcoplasmic reticulum (Oetliker, 1981a,b, 1982, 1989).

This finding also excludes the possibility of direct electrical coupling between the t-tubules and the sarcoplasmic reticulum. The latter is practically contiguous with the rest of the intracellular space in electrical terms. Its resistance is only high for divalent cations, on the order of MΩ (Hasselbach and Oetliker, 1983).

2. *Mechanism of Ca^{2+} Release*

The third and last step in the sequence of events from neuronal excitation to activation of the contractile proteins, Ca^{2+} release, was the last to become even partially understood. The 1970s produced findings that contributed to our understanding of the process, starting with the discovery of specialized structures that span the space between the t-tubular membrane and the adjacent sarcoplasmic reticulum membrane (feet structures; Franzini-Armstrong, 1970), and the discovery by Schneider and Chandler (1973) of what they called charge movement or displacement currents. These displacement currents have magnitudes and characteristics that make them suitable to correspond to conformational changes in the feet structures between the t-tubular membranes and the terminal cisternae of the sarcoplasmic reticulum, as well as being the link between Ca^{2+} release and the change in membrane potential across the t-tubular membranes. These displacement currents have been analyzed in far more detail since then, and different components have been isolated that have particular functional significances (for more information about the recent developments, see Hui and Chandler, 1991).

Information has become available about the structure and function of the proteins involved in Ca^{2+} release (reviewed by Ashley *et al.*, 1991; Catterall, 1991). Currently, the consensus is that it is the involvement of two "binding sites" in the triadic junction that finally enables Ca^{2+} to be released. The DHP (dihydropyridine)-binding site is the voltage sensor. This molecule is closely linked to a structure spanning the region between sarcoplasmic reticulum terminal cisternae and the t-tubular membrane that exhibits binding properties for the alkaloid ryanodine (R-binding site). Figure 16 indicates the structure of these two proteins.

The most widely accepted model today is that the change in the electrical field across the t-tubular membrane induces a conformational change in the DHP-binding site that in turn is transmitted, possibly by structural continuity, to the R-binding site representing the foot structure in the space between the terminal cisternae membrane and the t-tubular walls. Experiments by Tanabe *et al.* (1990) with genetically modified DHP-binding molecules indicate that the loop marked by an asterisk in Fig. 16 is the cytoplasmic domain involved in linking the change in the electric field across the t-tubular membrane to a conformational change in the R-binding site, leading to Ca^{2+} release. This R-binding site forms a pore through the terminal cisternae of the sarcoplasmic reticulum, through which Ca^{2+} can leak out into the myoplasm when the R-binding site is in the appropriate conformation. Taking into account the density of the feet structures (Franzini-Armstrong, 1970) and assuming that each foot functions as a Ca^{2+} channel, the amount of Ca^{2+} released by a single channel is on the order of 300 Ca^{2+} ions ms^{-1}. An ion passage rate of this magnitude is rather small compared to the situation in the acetylcholine-sensitive channels at the level of the neuromuscular junction, where 50,000 Na^{+} ions ms^{-1} enter the muscle cell through a single channel. The proteins forming the Ca^{2+} release channel could be isolated from the heavy fraction of the sarcoplasmic reticulum that contains preferentially triadic junctions, together with terminal cisternae (Smith *et al.*, 1986; Lai *et al.*, 1988; for details and more references, see Catterall, 1991).

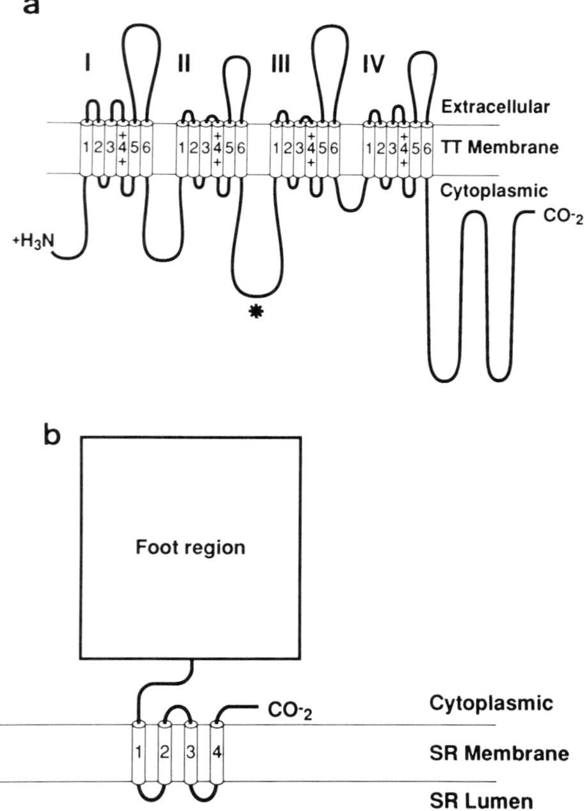

FIG. 16. Illustration of the mechanism by which the voltage-gated Ca^{2+}-channel (a) (dihydropyridine receptor, DHP) and the sarcoplasmic reticulum Ca^{2+}-release channel (b) (ryanodine binding site, R-binding site) are anchored in the membrane of the t-tubules (TT membrane) and the terminal cisternae of the sarcoplasmic reticulum (SR membrane). The loop marked by an asterisk in (a) is thought to interact with the foot region of the ryanodine receptor represented in (b), as described in the text (Tanabe *et al.*, 1990). Reprinted from Catterall (1991), with permission.

E. Relaxation

To stop the cycling of cross-bridges and to enable relaxation, the Ca^{2+} concentration must be lowered to its value before activation (about 10^{-8} *M*). This is achieved by the action of the sarcoplasmic reticulum Ca^{2+} pump.

The sarcoplasmic reticulum is an internal compartment separating about 9% of a muscle fiber's volume from its myofibrillar space (Peachey, 1965; Franzini-Armstrong, 1970; Mobley and Eisenberg,

1975). The sarcoplasmic reticulum is arranged as flat tubes surrounding a set of myofibrils very intimately, such that diffusion distances are minimized, as illustrated in Fig. 14. The surface of the sarcoplasmic reticulum in 1 g of muscle fiber is approximately 2.2 m^2 (Mobley and Eisenberg, 1975), of which 75% is made up by the Ca^{2+}-activated Mg^{2+}-ATPase that is packed there at a density of about 22,000 molecules μm^{-2} (Scales and Inesi, 1976). The concentration of the pumping enzyme is on the order of 100 nmol g^{-1} muscle, or 100 μM. This concentration is similar to the concentration of troponin C in the muscle fiber. A stoichiometric consideration shows that the amount of Ca^{2+} released in a single activation is sufficient to saturate the binding sites on the troponin C as well as those on the Ca^{2+} pumps.

The cooperative nature of thin filament activation (Section I,A.) would even ensure complete activation if not all of its troponin C moieties were able to bind Ca^{2+}. Because the Ca^{2+}-activated ATPase is a relatively slow pump with a turnover rate of about 30 s^{-1} (Hasselbach, 1979; Hasselbach and Oetliker, 1983), there is sufficient time for the half of the Ca^{2+} that was released and bound to the troponin C to activate the cross-bridge cycle before the other half, bound initially to the outside of the sarcoplasmic reticulum, is sequestered. There is also sufficient time for the Ca^{2+} released from troponin C to be taken up into the sarcoplasmic reticulum as well. This process reverses the conformation changes of the regulatory proteins and stops cross-bridge cycling. A molecular model of cross-bridge cycle regulation at the myofilament level is presented in Section I,A. of this paper. The shear number of pump molecules ensures that each pump has to go through only two pumping cycles to sequester the Ca^{2+} released by a single action potential (about 200 nmol g^{-1} muscle, as mentioned earlier; Oetliker, 1982; Hasselbach and Oetliker, 1983).

F. Limits to Performance at the Cellular Level

At first glance one would think that muscle could only contract as long as the availability of energy-rich substrates is sufficient to provide the necessary energy for cross-bridge cycling and tension development. Taking a closer look (Section I,A), one becomes aware that an energy-deficient muscle fiber inevitably comes into the state of rigor. This condition is irreversible *in vivo*, and the fibers are eventually degraded by intracellular proteases that are activated at concentrations of $Ca^{2+} > 1$ mM (Croall and Demartino, 1991).

Looking at muscle from a biochemical point of view, with respect to its available free energy, one realizes that it could become a rather "hazardous" tissue: it contains vast amounts of an efficient ATPase that becomes activated by a rise in the concentration of myoplasmic Ca^{2+}. The myoplasmic Ca^{2+} concentration can only be lowered by energy expenditure. Furthermore, amounts of Ca^{2+} sufficient for complete saturation of the contractile proteins are stored in their immediate neighborhood in the sarcoplasmic reticulum, a few nanometers away, separated only by the sarcoplasmic reticulum membrane. Additionally, the fiber is surrounded by an extracellular medium in which the Ca^{2+} concentration is about five orders of magnitude higher than the resting concentration of Ca^{2+} in the myoplasm. Therefore, a free energy level sufficient to maintain these gradients for Ca^{2+} is a prerequisite for a functional muscle cell.

Experiments of Lüttgau (1965) and Grabowski *et al.* (1972) showed that muscle cells poisoned with CN^- can be fatigued rather quickly by continued stimulation at low frequency (0.5 Hz), in terms of the amplitudes of single twitches. After the single twitches have decreased to 5–10% of their initial amplitude, application of caffeine (which releases Ca^{2+} from the sarcoplasmic reticulum without depolarization of the plasma membrane; Lüttgau and Oetliker, 1968) between 0.5 and 1 m*M* leads to single-twitch amplitudes similar to those before poisoning, or even higher. This response demonstrates that the decreased ATP concentration was not directly responsible for the reduced twitch amplitude, but rather, it was caused by an insufficient release of Ca^{2+}. It was also found that such metabolically poisoned fibers had a resting potential very close to the K^+ equilibrium potential, together with a very low membrane resistance, possibly due to a high K^+ conductivity (Grabowski *et al.*, 1972; Oetliker, 1973). In the meantime, the underlying mechanism for these phenomena became apparent as results from studies of the ATP-dependent gating behavior of plasmalemmal K^+ channels became available (Spruce *et al.*, 1987).

The question of the significance of such a channel is justified, however, its answer must remain speculative. It is possible that it is one of the last measures built in to prevent a disaster for the fiber due to a further reduction of the ATP, an event that would occur if the contractile proteins became activated by an increase in the free myoplasmic Ca^{2+} concentration in a situation in which the ATP content of the fiber was already critically low. In such a scenario, further activation of the contractile apparatus would lead to additional splitting of ATP and lowering of the ATP concentration, with concomitant increases in the concentrations of ADP and Pi. Ultimately, the ATP concentration would be insufficient to maintain the ion gradients. This would lead to an equilibrium of the ionic gradients, eventually bringing the myoplasmic Ca^{2+} concentration to the millimolar range and trapping the contractile proteins in the state of rigor, as described in Section I,A, and activating intracellular Ca^{2+}-dependent proteases as mentioned earlier.

Another hypothesis for the significance of this ATP-dependent K^+ channel is given in Spruce *et al.* (1987): "A possible role for ATP-regulated channels, suggested to us by R. A. Chapman, is that their opening will, by enhancing the efflux of K^+, contribute significantly to vasodilation of skeletal muscle arterioles."

Termination of cross-bridge cycling can only be achieved if the Ca^{2+} released from the sarcoplasmic reticulum is pumped back in again. This requires the availability of a certain minimal amount of free energy contained in the phosphorylation potential. The amount of free energy derived from ATP splitting is determined by the ratio of [ATP]/([ADP] × [Pi]). Under normal circumstances, ADP and Pi are transformed to ATP at very high rates in the mitochondria as long as oxygen (at a sufficient partial pressure) and oxidizable substrates are available, as described in Sections III,C and III,D. If these conditions do not prevail, relaxation will be slowed and a single action potential will induce more ATP splitting, because of the longer time during which the myoplasmic Ca^{2+} concentration is elevated. This further lowers the free energy contained in the ATP splitting reaction and the vicious cycle just described begins, leading after a few revolutions to irreversible exhaustion of the fiber. From NMR measurements of the energy states of muscle, this "point of no return" could be accurately predicted (Dawson *et al.*, 1980).

The safety of the cells requires that muscle cells have a stop built in at a stage preceding the activation of the contractile proteins, ensuring that this dangerous situation is not reached until all production of energy rich substrates has been halted, that is, at the end of the cell's life. At this time, overflow of the muscle cell with Ca^{2+} finally activates intracellular proteases that start the degradation of the contractile apparatus (Croall and Demartino, 1991).

III. Energy Supply for Muscle Contraction

As outlined in Section II,F, an immediate and reliable energy supply, in the form of ATP, is necessary not only for a muscle fiber's function, but even for its very existence. This section is a synopsis of structural conditions that have been identified as playing a role in governing energy and substrate availability in muscle cells.

A. Creatine Kinase/Phosphocreatine System

Skeletal muscle tissue exhibits an extremely wide dynamic range of possible states of energy turnover, as well as the capacity of up-regulating contractile activity almost instantaneously. The ability to respond to a widely fluctuating energy demand is enabled in vertebrate muscle by the accumulation of large quantities of metabolically inactive phosphagen phosphocreatine (PCr). PCr concentrations in muscle range from 20 to 35 m*M*, depending on species and fiber type (Ackerman *et al.*, 1980). The biologically active "energy currency," ATP, is regenerated from PCr by a number of strategically located creatine kinases (CK) at rates that exceed both ATP utilization and ATP replenishment through glycolysis or oxidative phosphorylation (McGilvery, 1973; Wallimann *et al.*, 1984). As a consequence, intramyocellular ATP concentrations can be kept low (2–5 m*M*), barely enough to support a few vigorous contractions. Moreover, the high activity of the creatine kinases prevents a buildup of the ADP concentration, keeping it at very low values of 1–37 μ*M*, depending on the assumptions used for calculating it (Veech *et al.*, 1979; Ackerman *et al.*, 1980). The higher ADP concentrations measured with chemical methods (100–500 μ*M*) are probably due to ATP breakdown or to liberation of actin-bound ADP (Chance *et al.*, 1986). The CK system functions to keep cellular ADP concentrations low. This situation is necessary if ADP concentration is to operate as a regulator of mitochondrial respiration (Chance *et al.*, 1986), and it prevents loss of adenine nucleotides through the adenylate kinase reaction (Fig. 17; Iyengar, 1984). The role of the intracellular compartmentation of creatine kinase isozymes for cellular energy homeostasis has been covered in an excellent review (Wallimann *et al.*, 1992).

In skeletal muscle cells, two subunit "cytosolic" isoforms of CK, Muscle-CK (M-CK) and Brain-CK (B-CK), can be expressed. *In vivo*, M-CK and B-CK subunits combine to give the three possible cytosolic dimeric forms MM-CK, MB-CK, and BB-CK. Although MM-CK is quite specific for adult sarcomeric muscle, the other two isoforms are expressed in skeletal muscle developmentally (Perriard *et al.*, 1978). Mitochondrial creatine kinase (Mi-CK) occurs in dimeric and octameric forms (Wyss *et al.*, 1990). From the two isoforms of Mi-CK, there is one (Mib-CK) that is considered to be specific for striated muscle (Haas and Strauss, 1990).

Wallimann *et al.* (1992) have proposed five main functions for the creatine kinase/phosphocreatine (CK/PCr) system. The first function is generally well accepted, that of a "temporal energy buffer." This

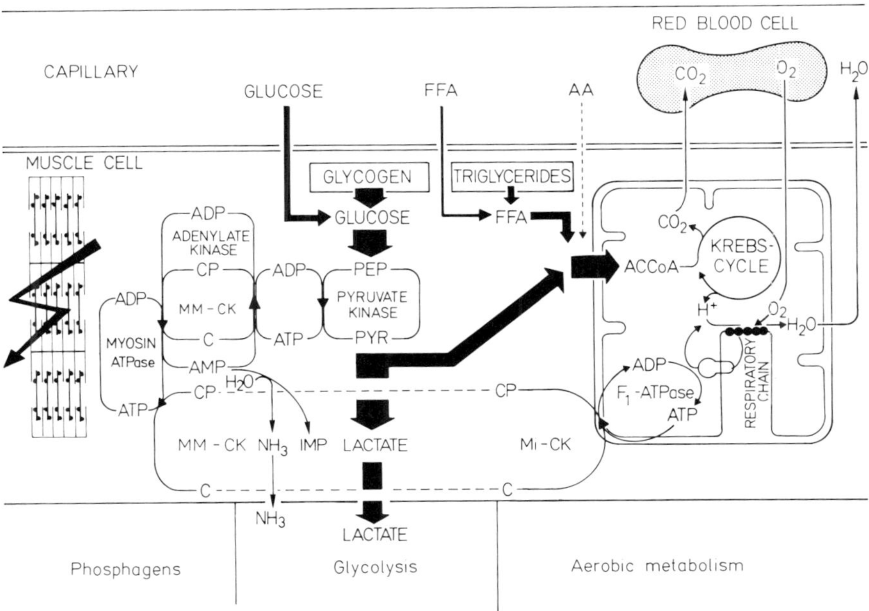

FIG. 17. Schematic of pathways for energy and substrate supply in muscle cells [Mi-CK = mitochondrial form of creatine kinase; MM-CK = M-line bound and cytosolic (soluble) creatine kinase]. The width of lines is an approximation of relative capacities for flux rates. Reproduced with permission of the Company of Biologists Ltd.

function is prevalent in fast-twitch glycolytic muscles that have high PCr concentrations combined with a high relative percentage of cytosolic CK (associated with glycolysis) and a low relative percentage of Mi-CK (Yamashita and Yoshioka, 1991; Ingwall and Fossel, 1983).

The second main function is that of a "spatial energy buffer," for which the term "CP-shuttle" was coined by Bessman and Geiger (1981). The configuration of the CK system is optimized for the phosphate transport function as, for instance, in slow-twitch oxidative muscles (including the heart). In these tissues, the PCr pool size and the relative percentage of cytosolic CK are small, whereas Mi-CK (functionally coupled to oxidative phosphorylation) percentage is large (up to 30% or more of total CK is Mi-CK).

The third function of the CK/PCr system has previously been mentioned, namely, maintaining the low intracellular ADP concentrations necessary for ADP to exert its regulatory role in energy metabolism.

A fourth main function of the CK/PCr system is considered to be that of buffering protons liberated during ATP hydrolysis, while providing Pi at the same time. This function is related to the close structural association of some CK with the ATPases in muscle cells, discussed further in the following. As the CK reaction in the direction of ATP uses both ADP and protons, buildup of these molecular species close to sites of rapid energy consumption is effectively prevented. The increase of the Pi concentration, related to the breakdown of PCr, has the benefit of activating glycolysis and glycogenolysis (Davuluri *et al.*, 1981).

The fifth main function of the CK/PCr system is to provide appropriate ATP/ADP ratios at subcellular sites where ATP is either generated (Mi-CK) or consumed (MM-CK). For this function, the subcellular compartmentation of CK isozymes is of utmost importance (see Wallimann *et al.*, 1992).

1. Structural Association of CK with ATP-Requiring Processes

Depending on the fiber type, 5–10% of the cytosolic MM-CK is bound to the M-line (Fig. 8), where it forms the m4/m4′ m-bridges interlinking neighboring myosin filaments (see Wallimann and Eppenberger, 1985). MM-CK is thus believed to also play a structural role, keeping myosin filaments in register on the I-band level of the sarcomere. Functionally, this M-line-bound MM-CK is capable of regenerating ATP consumed by the actin-activated Mg^{2+}-ATPases at rates exceeding that of the myofibrillar ATPase (Wallimann *et al.*, 1984). Furthermore, specific associations of MM-CK with the sarcoplasmic reticulum have also been demonstrated, indicating a role for CK in ATP-dependent Ca^{2+} uptake, and possibly in Ca^{2+} release (Rossi *et al.*, 1990). A structural and functional association of CK with the Na^+/K^+ pumps of the sarcolemma has also been proposed (see Wallimann *et al.*, 1989).

2. Structural Association of CK with ATP-Providing Processes

Multienzyme complexes are considered by many investigators to be responsible for systemic properties of metabolic pathways that cannot be explained by properties of the individual enzymes that constitute them (see Friedrich and Hajdu, 1987). However, the functional significance of enzyme associations remains contested at present (Brooks and Storey, 1991). The enzymes of the glycolytic pathway are believed to form a multienzyme complex, located in the I-band region of the sarcomere (Maughan and Lord, 1988). At the same location, coeluted with the glycolytic enzyme complex, the major fraction of the "cytosolic" MM-CK is found (Wegmann *et al.*, 1992). Several lines of functional evidence indicate that glycolysis and the CP/PCr system might be metabolically coupled, and are structurally located at the I-band region of the sarcomere (see Wallimann *et al.*, 1992).

The importance of Mi-CK for "channeling" of oxidatively derived energy in muscles is supported by the finding of a severalfold increase in Mi-CK with endurance training or chronic electrical stimulation. These interventions have been shown to decrease overall CK concentration and PCr pool size, compatible with the "shuttle concept" of the CR/PCr system (Schmitt and Pette, 1985; Apple and Rogers, 1986). Mi-CK is located in the mitochondrial intermembrane space, possibly dynamically associated with both membranes and with the adenine nucleotide translocator of the inner mitochondrial membrane (for review, see Wallimann *et al.*, 1992). In this position, Mi-CK is thought to be capable of minimizing the free energy necessary for ATP synthesis, because ATP is immediately transphosphorylated to PCr, thus making ADP available for the ATP/ADP antiport reaction.

B. Glycolysis

If vigorous motor activity is continued for time periods up to a few minutes, glycolysis becomes the dominant metabolic pathway funneling energy into the PCr pool. As discussed in the previous paragraphs, glycolytic enzymes are restricted to the cytoplasm of skeletal muscle fibers, preferentially located in the I-band region of the sarcomere, and possibly associated with the CK/PCr system (Wallimann *et al.*, 1992). A second pathway, involving an adenylate kinase isozyme reaction located in a subcellular compartment (likely actin), has also been reported to be responsible for making glycolysis-derived ATP available to myofibrillar ATPases (Fig. 17; Zeleznikar *et al.*, 1990; Zeleznikar and Goldberg, 1991). The importance of these two proposed pathways for ATP regeneration through glycolysis is difficult to appreciate at present. The issue may be complicated by species or muscle specificity and may, moreover, have to be analyzed in the context of the regulation of adenine nucleotide degradation, which has been shown to occur in a fiber-type-specific manner by Tullson and Terjung (1990).

In any muscle of a given animal species, the importance of glycolysis as a means of providing ATP for muscle contraction depends primarily on the muscle fiber type (fast-twitch glycolytic versus slow-twitch oxidative). In a more general sense, the importance of glycolysis also depends on animal size. In contrast to aerobic capacity (see Section III,C), glycolytic capacity scales positively with animal size (Emmett and Hochachka, 1981), that is, as animal size increases, glycolysis becomes increasingly more important as a means of regenerating ATP for locomotion. Although anaerobic glycolysis could support resting ATP requirements of a shrew for only 13 s, the resting metabolism of an elephant could be supported for several hours (Coulson and Hernandez, 1983). For a broader discussion of the *in vivo* role of glycolysis for vertebrate performance and muscle function, several reviews may be consulted (Hochachka *et al.*, 1988; Coulson, 1987; Connett *et al.*, 1990).

C. Aerobic Metabolism

The high energy turnover states of muscle are anaerobic and non-steady-state. An "oxygen debt" is incurred that must be repaid at a later stage during lower-intensity exercise or after cessation of the exercise (Bangsbo *et al.*, 1990). Under steady-state exercise conditions, oxidative phosphorylation in mitochondria must match the total energy expenditure of contracting muscle cells. Myofibrillar ATPases are responsible for some 70% of the energy consumed. The remainder is used for Ca^{2+} sequestering by the sarcoplasmic reticulum, and for the maintenance of ion gradients across the sarcolemma (Sections II,E and II,F; Rall, 1985). Maximal aerobic metabolism can, on a steady-state basis, support only a fraction of the maximal possible mechanical activity. Thus, it is the respiratory, and not the contractile, machinery that imposes a limit to sustained locomotion.

At the whole-animal level, the state of maximal aerobic metabolism ($\dot{V}_{O_2\,max}$) is characterized by the fact that well over 90% of the oxygen and hence, energy flux, is directed toward working muscle tissue. Under these conditions, the primary problem is related to oxygen transfer to muscle tissue. Skeletal muscle capillaries must deliver oxygen at the rate oxygen is processed by mitochondria in active muscle fibers.

1. *Mitochondria Set the Rate of Oxygen Flow*

We have analyzed the conditions for oxygen transfer in the mammalian respiratory system by combining functional measurements of maximal oxygen flow rates at the whole-animal level with the structural conditions that prevail at each step of the respiratory cascade from lungs to muscle mitochondria (Weibel and Taylor, 1981; Taylor *et al.*, 1987). We found that both the structure and function of mammalian mitochondria are quite invariant among species. An analysis of the maximal oxygen consumption ($\dot{V}_{O_2\,max}$) while running and an estimation of the total mitochondrial volume of the skeletal musculature carried out on eight mammalian species spanning a fivefold range of $\dot{V}_{O_2\,max}$ showed that skeletal muscle mitochondria consumed 4.56 ±

0.61 (mean ± SD) ml O_2 (STPD) min^{-1} cm^{-3} mitochondria at $\dot{V}_{O_2\,max}$ (Hoppeler, 1990). Therefore, the volume fraction of skeletal muscle fiber occupied by mitochondria can be used as a functional descriptor of the skeletal muscle oxidative capacity. This seems to hold true in whole animals (Hoppeler, 1990), single muscles (Hoppeler *et al.*, 1987), and single muscle fibers (van der Laarse *et al.*, 1989).

2. *Oxygen Supply through Capillaries*

Muscle capillaries are anisotropic, running mostly parallel to the muscle fibers with some tortuosity and branching. Several morphometric methods have been developed that allow capillary length density to be estimated in anisotropic tissues, such as skeletal and heart muscle (Mathieu *et al.*, 1981b; Mattfeldt, 1987). The capillary length density, $J_v(c,f)$, is *the* key structural parameter of any capillary network, provided that the mean capillary diameter is known, or known to be constant among the capillary beds being compared. Knowing $J_v(c,f)$ allows the main parameters to be calculated that are critical for capillary exchange function: capillary volume and capillary surface area. Furthermore, capillary length density allows for an approximation of the radius (R) of the "Krogh cylinder."

The value of $J_v(c,f)$ can be determined efficaciously from muscle cross sections and longitudinal sections using a model-based approach that considers the degree of capillary "tortuosity" (Mathieu *et al.*, 1983). The measured capillary tortuosity is primarily dependent on the sarcomere length during muscle fixation (Mathieu-Costello, 1987). Neither animal size, aerobic capacity, hypoxia, nor exercise seems to have a measurable influence on the degree of capillary tortuosity (Mathieu-Costello *et al.*, 1989; Poole and Mathieu-Costello, 1989a,b). Measurements of capillary outer circumference in muscle specimens from a number of mammalian species prepared under standard conditions for electron microscopy indicate that mean capillary perimeter and, hence, capillary cross section vary little among mammalian muscles (Kayar *et al.* 1992). For the purpose of the following discussion, circular capillaries of diameter 4.5 μm are assumed.

3. *Relating Capillaries to Mitochondria*

We have compiled data from heart, *diaphragmus*, *semitendinosus*, and *vastus medialis* muscles from 13 mammalian species to plot capillary length as a function of mitochondrial volume (Fig. 18; Hoppeler and Kayar, 1988). The slope of this relationship is not significantly different from one, indicating that a doubling of mitochondrial volume is matched by a doubling of capillary length. It can be calculated that

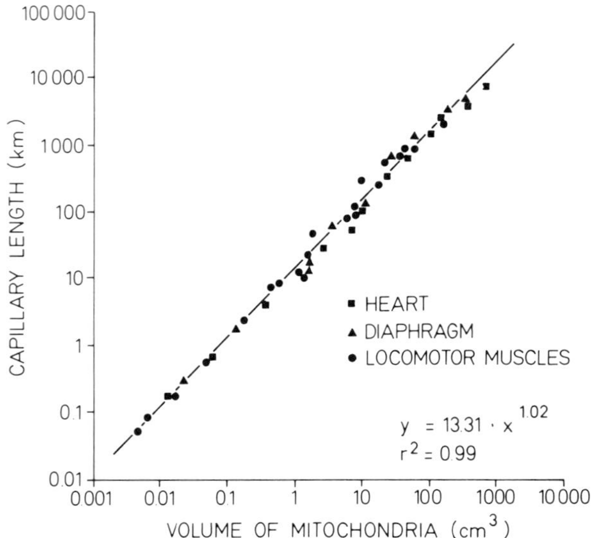

FIG. 18. Species means (13 mammalian species) of capillary length density versus mitochondrial volume density (locomotor muscles include *m. vastus medialis* and *m. semitendinosus*). Adapted from Hoppeler and Kayar (1988) and reproduced with permission of the Company of Biologists Ltd.

for each cubic centimeter of mitochondria [consuming oxygen at 4.5 ml (STPD) min^{-1} and containing approximately 30 m^2 of inner mitochondria membrane (Schwerzmann *et al.*, 1989)], there is some 10 km of capillary length, 1400 cm^3 of capillary surface area, and 0.16 cm^3 of capillary volume.

4. *Capillary Transit Time*

If the assumption is made that all capillaries are perfused during maximal aerobic exercise, we can calculate the average capillary transit time (t_c) as

$$t_c = V_c/\dot{Q},$$

where $\dot{Q}$ is blood flow (in ml s^{-1}, measured with the radioactive microsphere technique) in muscles of animals running at $\dot{V}_{O_2\ max}$ and V_c is the capillary volume (in ml). Kayar *et al.* (1992) showed that skeletal muscle capillary transit times at $\dot{V}_{O_2\ max}$ in 11 vertebrate species ranged from 0.4 s (for a 4-kg fox) to 0.8 s (for a 500-kg horse). When transit times were plotted against blood flows, a hyperbola fitted to the data indicated a limiting value of 0.2 s for the shortest transit times in those muscles with the highest blood flows (Kayar *et al.*, 1992).

5. Hemoglobin Concentration

In a study of adaptive variation comparing "athletic" species, such as dogs and ponies, to "sedentary" species, such as goats and calves, it was found that the athletic animals had both 2.5-fold larger maximal oxygen consumptions and skeletal muscle mitochondrial volumes (Conley *et al.*, 1987). However, the capillary length densities differed by only a factor of 1.7. This difference was complemented by a 1.5-fold higher hematocrit at $\dot{V}_{O_2 \max}$ in the athletic species. Thus, the 2.5-fold greater oxygen delivery in athletic species was achieved in about equal parts by increasing capillary volume and the oxygen transport capacity of the blood.

6. Erythrocyte Spacing

In the theoretical analysis, Federspiel and Popel (1986) have demonstrated that each erythrocyte releases oxygen only over a "zone of influence" extending about one-half the capillary channel width in front of and behind it. Erythrocyte spacing close to these theoretical expectations has been directly observed (Federspiel and Sarelius, 1984). The contribution of the particulate nature of blood to oxygen transport resistance from blood to mitochondria has been estimated to be between 30 and 70% of the total (Federspiel and Popel, 1986). If the higher of the proposed values is correct, one would expect a large drop in oxygen tension (P_{O_2}) between the red blood cells and the muscle cell interior, and a relatively flat P_{O_2} distribution within the muscle fibers (Federspiel, 1986). This pattern of P_{O_2} distribution is also considered to be related to the presence of myoglobin within the myocyte, as well as to the heterogeneous distribution of mitochondria in muscle fibers (see the following).

7. Conditions for Oxygen Diffusion in Muscle Fibers

Intracellular oxygen transport in muscle fibers is greatly assisted by myoglobin, provided three conditions are met (Wittenberg and Wittenberg, 1989):

1. myoglobin must be present in significant concentrations;
2. myoglobin must be partially desaturated;
3. myoglobin must be free to undergo translational diffusion in the sarcoplasma.

Direct observation demonstrates that in heavily exercising dog skeletal muscle, myoglobin saturation is mostly below 50% (Gayeski and Honig, 1986), making it likely that most of the oxygen consumed in these muscle cells is transported by myoglobin-facilitated oxygen diffusion. It may be worth noting that, similar to the distribution of glycolytic enzymes, the concentration of myoglobin seems to be higher in the less crowded I-band than in the A-band (Kawai *et al.*, 1987). This region of the sarcomere also contains a higher volume density of mitochondria (see the following). From these observations, one can speculate that I-band-located mitochondria might benefit from direct oxymyoglobin-mediated oxygen delivery to the respiratory chain, a factor that has been shown to be important in cardiac myocytes (see Wittenberg and Wittenberg, 1989).

8. Distribution of Mitochondria in Muscle Fibers

Mitochondria are not scattered randomly in muscle fibers. In cross sections, it is common to find clusters of subsarcolemmal mitochondria, particularly in regions adjacent to capillaries (Fig. 19). Interfibrillar mitochondria in a variety of muscles have been demonstrated to be distributed in a gradient that is highest near capillaries, and decreases with distance toward the fiber center (James and Meek, 1979; Rakusan and Tomanek, 1986; Kayar *et al.*, 1988a). These observations beg the question of whether mitochondrial location is not related to mitochondrial function. If diffusion distance for oxygen, or any other metabolic substrate, from capillaries to mitochondria were a critical factor to mitochondrial metabolism, then all mitochondria should be located near capillaries. On the other hand, if the distance for exchanging high-energy phosphate compounds between mitochondria and myofibrils were critical to muscle work rate, then mitochondria should be distributed uniformly between myofibrils.

The actual distribution of mitochondria in most muscles is somewhat intermediate to these two cases, suggesting a balance between opposing demands (Kayar and Banchero, 1987; Hoppeler *et al.*, 1991). In muscles trained for endurance exercise, there is an increase in total mitochondrial content of muscle fibers, with a greater relative increase in the subsarcolemmal, compared to the interfibrillar, population of mitochondria (Hoppeler *et al.*, 1973, 1985). This finding must be viewed in the context of speculations that subsarcolemmal and interfibrillar mitochondria might be two biochemically distinct populations (Matlib *et al.*, 1981).

9. Mitochondrial Connectivity

Three-dimensional reconstructions and scanning electron microscopy have shown that muscle mitochondria may actually be quite complex in shape and variable in size (Fig. 20; Bakeeva *et al.*, 1978; Ogata

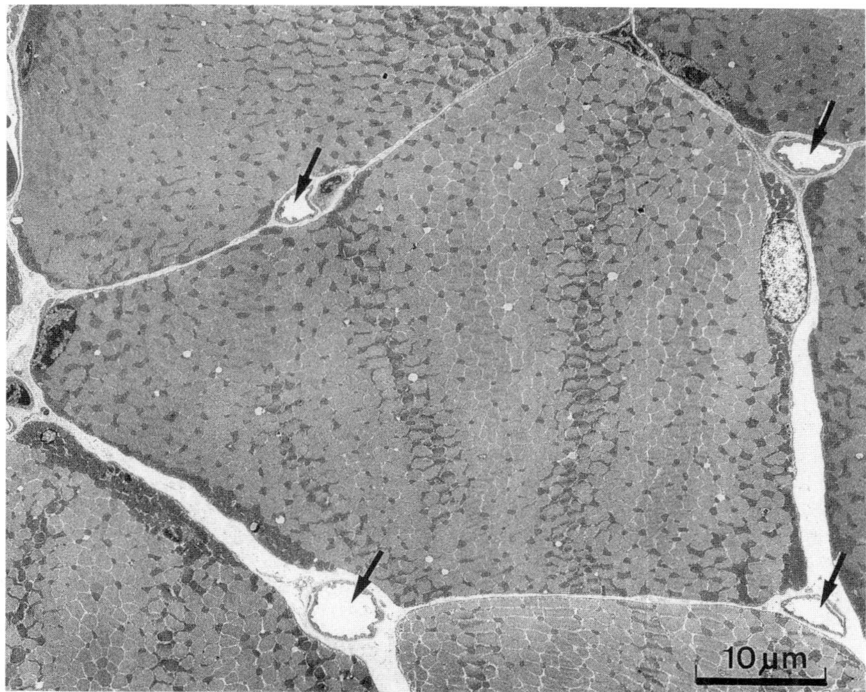

FIG. 19. Cross sections of mammalian skeletal muscle fibers indicating that volume density of mitochondria increases toward the fiber border. Note distinct accumulations of subsarcolemmal mitochondria in the vicinity of capillaries (arrows). Reproduced with permission of the Company of Biologists Ltd.

and Yamasaki, 1985; Kirkwood *et al.*, 1986; Kayar *et al.*, 1988b). It has been proposed that all mitochondrial matter in a muscle fiber might be contiguous, forming a mitochondrial reticulum (Bakeeva *et al.*, 1978). The highly interconnected transverse portions of this mitochondrial network could, by virtue of the high oxygen solubility of their lipid-rich membranes, facilitate oxygen diffusion within muscle fibers (Longmuir, 1981). However, the reconstructions of mitochondria using ultrathin sections suggest that the axis of greater mitochondrial connectivity is longitudinal to the myofibrils (Kayar *et al.*, 1988b) and not oriented transversely, as proposed by Kirkwood *et al.* (1986). Moreover, there is clear evidence of isolated spherical or cylindrical mitochondria in muscles with relatively low mitochondrial volume densities, confirming that mitochondrial connectivity is not essential for mitochondrial function (Kayar *et al.*, 1988b).

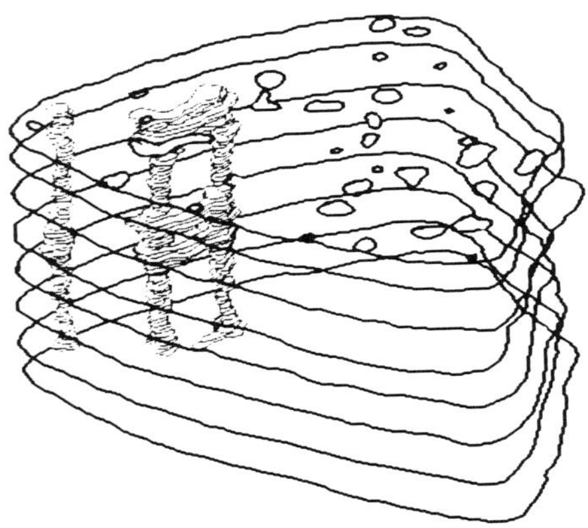

FIG. 20. Reconstruction of two mitochondria in a serially sectioned equine skeletal muscle fiber. Mitochondria are oriented mainly parallel to the main fiber axis, showing transversely oriented portions in the I-band regions. Profiles of additional mitochondria are indicated in the top section. Adapted from Kayar *et al.* (1988b), with permission of Wiley and Sons.

D. Substrate Stores and Transport in Muscle Tissue

Conditions of maximal or near-maximal oxygen uptake can be maintained for minutes only. During long-term locomotion, muscles are therefore limited by substrate, and not by oxygen, availability. Muscle mitochondria are capable of consuming substrates at a faster rate than the circulatory system can supply them and faster than they can be transported into muscle cells. For continued activity, muscles therefore depend on both exogenous (i.e., extramyofiber) and endogenous (i.e., intramyofiber) substrate pools. Substrate depots in the form of glycogen granules and lipid droplets are stored in muscle cells and are readily visible in the electron microscope (Fig. 21).

1. Substrate Stores

During continuous aerobic exercise, the substrate that is used depends on the intensity and duration of the exercise. The highest aerobically sustainable work loads are supported by carbohydrate catabolism (Jones *et al.*, 1980). The contribution of triglycerides is highest at intermediate work intensities (around 50% of $\dot{V}_{O_2\ max}$; Hurley *et al.*, 1986). Human subjects trained for running ultramarathon events are capable of running for 4 h at around 70% of their $\dot{V}_{O_2\ max}$ (Davies and Thompson, 1986).

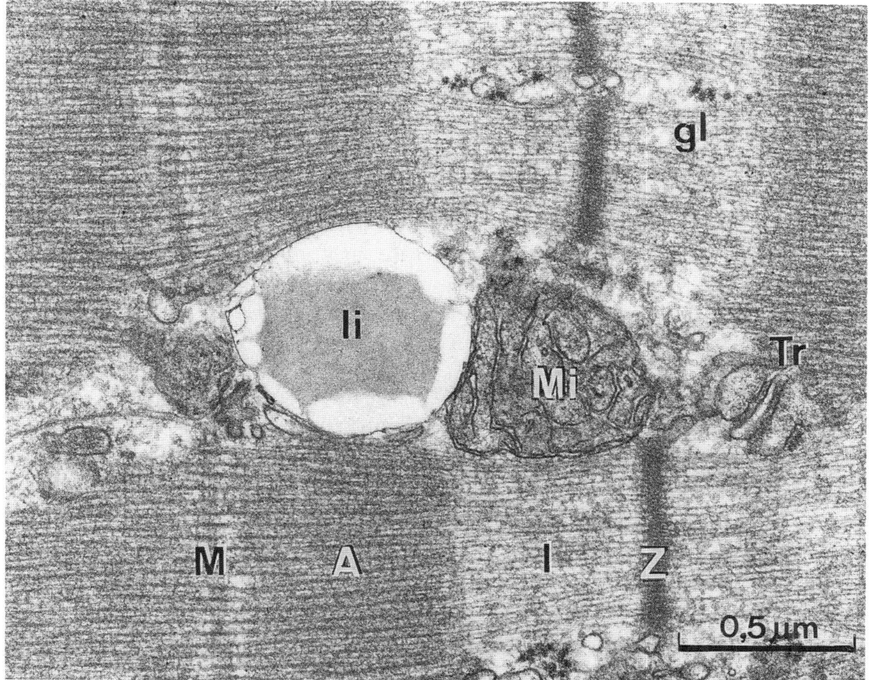

FIG. 21. Longitudinal section through a portion of a skeletal muscle fiber. M = M-band; I = I-band; A = A-band; Z = Z-band; Tr = Triad; gl = glycogen granules; Mi = mitochondrion; li = lipid droplet. Reproduced with permission of the Company of Biologists Ltd.

2. Lipids

Lipid droplets in muscle fibers are usually found adjacent to mitochondria. Morphometry is the method of choice for quantifying these lipid stores because their intracellular location can be positively ascertained. In untrained human muscle, about 0.5% of the fiber volume consists of intracellular lipids. These stores are doubled with endurance exercise training of durations of up to 2 months (Hoppeler *et al.*, 1985). In extreme endurance human athletes, such as professional bicyclists or marathon runners, intracellular lipid stores can make up more than 2% of the fiber volume (Hoppeler, 1986; Staron *et al.*, 1989). Running an ultramarathon (100 km) results in almost complete disappearance of intracellular glycogen and lipid stores (Oberholzer *et al.*, 1976; Kayar *et al.*, 1986).

3. Glycogen

Glycogen can be quantified noninvasively by magnetic resonance, or from muscle samples by biochemical methods. In untrained human *vastus lateralis* muscle, the glycogen concentration is approximately 100 mmol glucose kg^{-1} (wet weight). These glycogen stores can be depleted by 50% in only 6 min of intermittent supramaximal bicycling (112% of $\dot{V}_{O_2 \max}$; Nordheim and Vøllestad, 1990). The ability to mobilize and utilize fat as a fuel is an important mechanism in endurance exercise training because it postpones glycogen depletion and fatigue (Holloszy *et al.*, 1986).

4. Substrate Transport

Sustained performance during submaximal exercise critically depends on delivery of blood-borne fuel to working muscle cells in order to delay the depletion of intracellular glycogen and lipid stores. Two major categories of blood-borne substrates are recognized: plasma glucose derived from hepatic glycolysis or gluconeogenesis, and plasma-free fatty acids (FFA) supplied by adipose tissue triglycerides (Sahlin, 1986). There is considerable, though still somewhat controversial, evidence that lactate can also play an important role as a substrate for mitochondrial respiration in long-term exercise (Brooks, 1986a). In contrast to oxygen, for which over 80% of the arterial content may be extracted during a single passage through the capillary bed in working muscle, the oxidizable fuels lactate, glucose, and free fatty acids are poorly removed from capillaries (Walker *et al.*, 1984). A possible explanation for this phenomenon is that the transport from the capillary lumen into the muscle cell may limit the maximal rate of exogenous fuel supply for each of these substrates.

5. Lactate Turnover

The quantitative contribution of lactate as a fuel source in working skeletal muscle fibers is debated (Hermansen and Vaage, 1977; Brooks and Gaesser, 1980; Brooks, 1986b; Sahlin, 1987). Experiments by Nordheim and Vøllestad (1990) provide evidence that during low-intensity exercise, circulating lactate can be used as a substrate for resynthesis of ATP in Type I fibers, and for glycogen resynthesis in Type II fibers. The transport of lactate into active muscle fibers is proposed to take place both through the capillary bed and directly from glycolytic muscle fibers to adjacent lactate-oxidizing muscle cells. Most of the lactate transport through the cell membrane is believed to be carrier mediated (Moon *et al.*, 1987). Approximately 20% of the total lactate transport may be achieved by direct diffusion of undissociated lactic acid (Juel and Wibrand, 1989).

6. *Glucose Transport*

In muscle tissue, the water-soluble glucose molecule seems to be readily diffusible into the interstitial space through the relatively leaky muscle capillary endothelium. The transport problem for glucose is thought to occur at the sarcolemma. Across the plasma membrane, glucose transport is accomplished by facilitated diffusion via a glucose transporter (Glut 4) belonging to a family of structurally and functionally related glucose transport proteins (Glut 1–5; see Gould and Bell, 1990). Glut 4 is responsible in adipocytes and skeletal muscle tissue for most of the insulin-dependent glucose transport. Insulin-stimulated increase in cellular glucose uptake is mediated primarily by the translocation to the cell membrane of a latent intracellular pool of transporter molecules. In skeletal muscle tissue, contractile activity is capable of increasing the number of glucose transporters on the sarcolemma in the absence of an insulin stimulus (Goodyear *et al.*, 1990). There is a large difference among skeletal muscle fiber types with regard to their insulin- and contraction-stimulated glucose uptake capacities (Henriksen *et al.*, 1990). Exercise training has been shown to increase expression of skeletal muscle Glut 4 protein (Friedman *et al.*, 1990).

7. *Fatty Acid Transport*

Similar to the situation with glucose transport into muscle cells, recent research has shed much light on key aspects of fatty acid transport to and within muscle cells (Fig. 22). Intracellular and circulating triglycerides are believed to contribute about equally to the total lipid oxidized during sustained exercise (Carlson *et al.*, 1971). Possible mechanisms of the cellular control of the two major lipases in heart and red skeletal muscle have been reviewed (Fig. 22; Oscai *et al.*, 1990). It is proposed that fatty acids originating from circulatory chylomicrons and very-low-density lipoproteins are hydrolyzed by a lipoprotein lipase (LPL) that is synthesized in muscle cells and exported to the vascular bed. In contrast, a hormone-sensitive lipase (HSL) is held responsible for intracellular lysis of triglycerides. Consistent with this view is the finding that LPL increases as a consequence of endurance exercise training (Terjung *et al.*, 1983).

The transfer of fatty acids from muscle capillaries to respiring mitochondria is hindered by the poor aqueous solubility of these compounds both in the extracellular space and in the cytosol. Therefore,

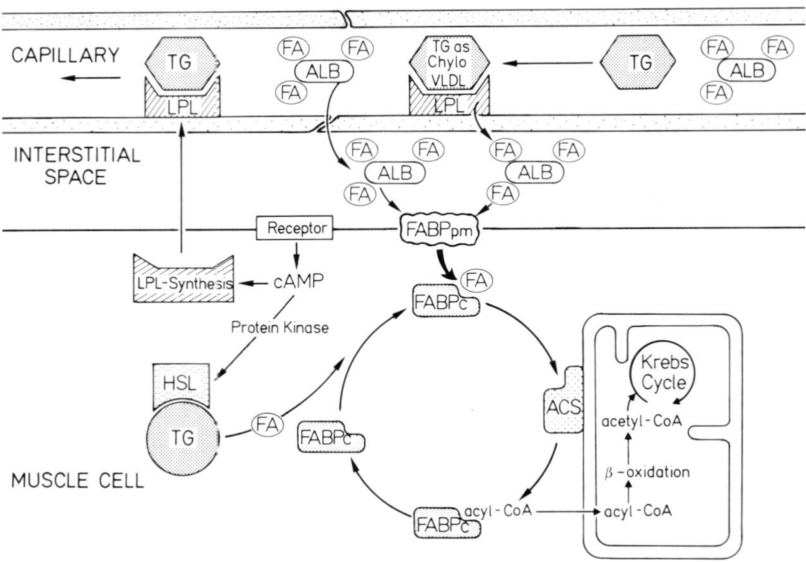

FIG. 22. Schematic of the pathway and regulation of fatty acid supply to muscle mitochondria. ACS = acylcoenzyme A synthetase; ALB = albumin; Chylo = chylomicrons; FA = fatty acid; FABPc = cytoplasmic fatty-acid-binding protein; FABPpm = plasmalemmal fatty-acid-binding protein; LPL = lipoprotein lipase; HSL = hormone-sensitive lipase; TG = triglycerides; VLDL = very-low-density lipoproteins. Adapted from Oscai *et al.* (1990) and Glatz *et al.* (1988), with permission.

various transporter molecules have been implicated in fatty acid movement. Transport through the interstitial space is thought to be albumin mediated (Fig. 22; Rose *et al.*, 1989). In support of this view, it has been shown that the interstitial concentration of albumin is much increased after chronic electrical stimulation of rabbit skeletal muscles (Heilig and Pette, 1988). The transport of fatty acids across the sarcolemma is likely an energy-dependent process involving a fatty acid translocator (Stremmel, 1988a,b; Stremmel *et al.*, 1990).

Intracellular fatty acid transfer has been reviewed extensively (Glatz and van der Vusse, 1989; Matarese *et al.*, 1990; Zachowski and Devaux, 1990). It is recognized that a class of small (14–15 kDa) fatty-acid-binding proteins (FABP) plays a key role in intracellular transport of lipids. FABP are abundant proteins accounting for 3–6% of the total cytosolic protein content in heart and red skeletal muscle. Consistent with this view, FABP have been shown to increase after chronic electrical stimulation in a mixed skeletal muscle (Kaufmann *et al.*, 1989). The major role of FABP seems to be the transcytosolic transport of fatty acids from the sarcolemma to the mitochondria. FABP are thought to act by increasing the effective solubility of fatty acids. The high concentration of FABP and their relatively low affinity for fatty acids suggest that fatty acids are passed from one FABP molecule to the next (see Glatz and van der Vusse, 1989). However, because FABP occur in various types, they may serve many additional functions.

Acknowledgments

The authors are indebted to A. Rowlerson (London) for providing the photographs for Figs. 11 and 12, to L. Tüscher (Bern) for preparing the manuscript, to A. Rowlerson and M. C. Schaub (Zürich) for critical review and suggestions on its first part, and to B. Krieger and K. Babl (Bern) for graphics and photographic support.

References

Ackerman, J. J., Grove, T. H., Wong, G. G., Gadian, D. G., and Radda, G. K. (1980). Mapping of metabolites in whole animals by ^{31}P NMR using surface coils. *Nature (London)* **283,** 167–170.

Adlestein, R. S., and Eisenberg, E. (1980). Regulation and kinetics of the actin–myosin–ATP interaction. *Annu. Rev. Biochem.* **49,** 921–956.

Adrian, R. H., and Peachey, L. D. (1973). Reconstruction of the action potential of frog sartorius muscle. *J. Physiol. (London)* **235,** 103–131.

Alberts, B., Bray, D., Lewis, J., Raff, M., Roberts, K., and Watson, J. D. (1989). "Molecular Biology of the Cell," 2nd ed., pp. 613–625. Garland Publ., New York.

Amos, L. A. (1985). Structure of muscle filaments studied by electron microscopy. *Annu. Rev. Biophys. Chem.* **14,** 291–313.

Anderson, M. J., and Cohen, M. W. (1974). Fluorescent staining of acetylcholine receptors in vertebrate skeletal muscle. *J. Physiol. (London)* **237,** 385–400.

Apple, F. S., and Rogers, M. A. (1986). Mitochondrial creatine kinase activity alterations in skeletal muscle during long-distance running. *J. Appl. Physiol.* **61,** 482–485.

Armstrong, R. B., and Phelps, R. O. (1984). Muscle fiber type composition of the rat hindlimb. *Am. J. Anat.* **171,** 259–272.

Ashley, C. C., Mulligan, I. P., and Lea, T. J. (1991). Ca^{2+} and activation mechanisms in skeletal muscle. *Q. Rev. Biophys.* **24,** 1–73.

Bakeeva, L. E., Chentsov, Y. S., and Skulachev, V. P. (1978). Mitochondrial framework (reticulum mitochondriale) in rat diaphragm muscle. *Biochim. Biophys. Acta* **501,** 349–369.

Bangsbo, J., Gollnick, P. D., Graham, T. E., Juel, C., Kiens, B., Mizuno, M., and Saltin, B. (1990). Anaerobic energy production and O_2 deficit–debt relationship during exhaustive exercise in humans. *J. Physiol. (London)* **422,** 539–559.

Bárány, M. (1967). ATPase activity of myosin correlated with speed of muscle shortening. *J. Gen. Physiol.* **50,** 197–218.

Baylor, S. M., and Oetliker, H. (1975). Birefringence experiments on isolated skeletal muscle fibres suggest a possible signal from the sarcoplasmic reticulum. *Nature (London)* **253,** 97–101.

Beenakkers, A. M. T., van der Horst, D. J., and van Marrewijk, W. J. A. (1984). Insect flight muscle metabolism. *Insect Biochem.* **14,** 243–260.

Bennett, P., Craig, R., Starr, R., and Offer, G. (1986). The ultrastructural location of C-protein, X-protein and H-protein in rabbit muscle. *J. Muscle Res. Cell Motil.* **7,** 550–567.

Berger, C. L., and Thomas, D. D. (1991). Rotational dynamics of actin-bound intermediates in the myosin ATPase cycle. *Biochemistry* **30,** 11036–11045.

Bernstein, X. I., Mogami, K., Donady, J. J., and Emerson, C. P. (1983). *Drosophila* muscle myosin heavy chain encoded by a single gene in a cluster of muscle mutations. *Nature (London)* **302,** 393–398.

Bertazzon, A., and Tsong, T. Y. (1990). Effect of ions and pH on the thermal stability of thin and thick filaments of skeletal muscle: High sensitivity differential scanning calorimetric study. *Biochemistry* **29,** 6447–6452.

Bessman, S. P., and Geiger, P. J. (1981). Transport of energy in muscle: The phosphorylcreatine shuttle. *Science* **211,** 448–452.

Bezanilla, F., and Horowicz, P. (1974). Fluorescence intensity changes associated with contractile activation in frog muscle stained with Nile Blue. *Am. J. Physiol.* **246,** 709–735.

Billeter, R., and Hoppeler, H. (1992). Muscular basis of strength. *In* "Encyclopedia of Sports Medicine" (P. Komi, ed.), pp. 39–63. Blackwell, Oxford.

Billeter, R., Weber, H., Lutz, H., Howald, H., Eppenberger, H. M., and Jenny, E. (1980). Myosin types in human skeletal muscle fibers. *Histochemistry* **65,** 249–279.

Billeter, R., Heizman, C. W., Howald, H., and Jenny, E. (1981). Analysis of myosin light and heavy chain types in single human skeletal muscle fibers. *Eur. J. Biochem.* **116,** 389–395.

Block, B. A. (1991). Evolutionary novelties—How fish have built a heater out of muscle. *Am. Zool.* **31,** 726–742.

Bowden, R. E. M., and Duchen, L. W. (1976). The anatomy and pathology of the neuromuscular junction. *In* "Neuromuscular Junction" (E. Zaimis, ed.), pp. 23–97. Springer-Verlag, Berlin.

Brandl, C. J., Green, M. N., Bozema, K., and MacLennan, D. (1986). Two Ca^{2+} ATPase genes: Homologies and mechanistic implications of deduced amino acid sequences. *Cell (Cambridge, Mass.)* **44,** 597–607.

Brandt, P. W., Diamond, M. S., and Schachat, F. H. (1984). The thin filament of vertebrate skeletal muscle co-operatively activates as a unit. *J. Mol. Biol.* **180,** 379–384.

Brooke, M. H., and Kaiser, K. K. (1970). Muscle fiber types: How many and what kind? *Arch. Neurol. (Chicago)* **23,** 369–379.

Brooks, G. A. (1986a). Lactate production under fully aerobic conditions: The lactate shuttle during rest and exercise. *Fed. Proc., Fed. Am. Soc. Exp. Biol.* **45,** 2924–2929.

Brooks, G. A. (1986b). The lactate shuttle during exercise and recovery. *Med. Sci. Sports Exercise* **18,** 360–368.

Brooks, G. A., and Gaesser, G. A. (1980). End points of lactate and glucose metabolism after exhausting exercise. *J. Appl. Physiol.* **49,** 1057–1069.

Brooks, S. P. J., and Storey, K. B. (1991). Where is the glycolytic complex—A critical evaluation of present data from muscle tissue. *FEBS Lett.* **278,** 135–138.

Bullard, B. (1983). Contractile proteins of insect flight muscle. *Trends Biochem. Sci.* **8,** 68–70.

Burke, R. E., Levine, D. N., Zajac, F. E., Tsairis, P., and Engel, W. K. (1971). Mammalian motor units: Physiological–histological correlation in three types in cat gastrocnemius. *Science* **174,** 709–712.

Carlson, B. M., Herbrychova, A., and Gutmann, E. (1979). Retention of hormonal sensitivity in free grafts of the levator ani muscle. *Exp. Neurol.* **63,** 94–107.

Carlson, L. A., Ekelund, L. G., and Froeberg, S. O. (1971). Concentration of triglycerides, phospholipids and glycogen in skeletal muscle and of free fatty acids and beta-hydroxybutyric acid in blood in man in response to exercise. *Eur. J. Clin. Invest.* **1,** 248–254.

Carraro, U., and Catani, C. (1983). A sensitive SDS–PAGE method separating myosin heavy chain isoforms of rat skeletal muscles reveals the heterogeneous nature of the embryonic myosin. *Biochem. Biophys. Res. Commun.* **116,** 793–802.

Casella, J. F., Craig, S. W., Maack, D. J., and Brown, A. E. (1987). Cap Z(36/32), a barbed end actin-capping protein, is a component of the Z-line of skeletal muscle. *J. Cell. Biol.* **105,** 371–379.

Catterall, A. W. (1991). Excitation–contraction coupling in vertebrate skeletal muscle: A tale of two calcium channels. *Cell (Cambridge, Mass.)* **64,** 871–874.

Chance, B., Leigh, J. S., Kent, J., McCully, K., Nioka, S., Clark, B. J., Maris, J. M., and Graham, T. (1986). Multiple controls of oxidative metabolism in living tissues as studied by phosphorus magnetic resonance. *Proc. Natl. Acad. Sci. U.S.A.* **83,** 9458–9462.

Chen, T., and Reisler, E. (1991). Interactions of myosin subfragment 1 isozymes with G-actin. *Biochemistry* **30,** 4546–4552.

Conley, K. E., Kayar, S. R., Rösler, K., Hoppeler, H., Weibel, E. R., and Taylor, C. R. (1987). Adaptive variation in the mammalian respiratory system in relation to energetic demand. IV. Capillaries and their relationship to oxidative capacity. *Respir. Physiol.* **69,** 47–64.

Connett, R. J., Honig, C. R., Gayeski, T. E. J., and Brooks, G. A. (1990). Defining hypoxia: A systems view of $\dot{V}_{O_2}$ glycolysis, energetics, and intracellular P_{O_2}. *J. Appl. Physiol.* **68,** 833–842.

Cooke, R. (1989). Structure of the myosin head. *Cell Motil. Cytoskel.* **14,** 183–186.

Coulson, R. A. (1987). Aerobic and anaerobic glycolysis in mammals and reptiles *in vivo. Comp. Biochem. Physiol. B.* **87B,** 207–216.

Coulson, R. A., and Hernandez, T. (1983). Alligator metabolism. Studies of chemical reactions *in vivo. Comp. Biochem. Physiol. B.* **74B,** 1–182.

Croall, D. E., and Demartino, G. N. (1991). Calcium-activated neutral protease (calpain) system: Structure, function, and regulation. *Physiol. Rev.* **71,** 813–847.

Davies, C. T. M., and Thompson, M. W. (1986). Physiological responses to prolonged exercise in ultramarathon athletes. *J. Appl. Physiol.* **61,** 611–617.

Davuluri, S. P., Hird, F. J. R., and McLean, R. M. (1981). A re-appraisal of the function and synthesis of phosphoarginine and phosphocreatine in muscle. *Comp. Biochem. Physiol. B.* **69B,** 329–336.

Dawson, M. J., Gadian, D. G., and Wilkie, D. R. (1980). Mechanical relaxation rate and metabolism studied in fatiguing muscle by phosphorous nuclear magnetic resonance. *J. Physiol. (London)* **299,** 465–484.

Del Castillo, J., and Katz, B. (1954). The membrane change produced by the neuromuscular transmitter. *J. Physiol. (London)* **125,** 546–565.

Dix, L., and Eisenberg, B. R. (1991). Expression of a fast myosin heavy chain mRNA in individual rabbit skeletal muscle fibers with intermediate oxidative capacity. *Anat. Rec.* **230,** 52–56.

Dunnant, Y., and Israel, M. (1985). The release of aectylcholine. *Sci. Am.* **252,** 40–48.

Dux, L., Green, H. J., and Pette, D. (1990). Chronic low-frequency stimulation of rabbit fast-twitch muscle induces partial inactivation of the sarcoplasmic reticulum Ca^{2+}-ATPase and changes in its tryptic cleavage. *Eur. J. Biochem.* **192,** 95–100.

Eisenberg, B. R., and Kuda, A. M. (1976). Discrimination between fiber populations in mammalian skeletal muscle using ultrastructural parameters. *J. Ultrastruct. Res.* **54,** 76–88.

Eisenberg, B. R., and Milton, R. L. (1984). Muscle fiber termination at the tendon in the frog sartorius: A stereological study. *Am. J. Anat.* **171,** 273–284.

Eisenberg, R. S., and Gage, P. W. (1969). Ionic conductances of the surface and transverse tubular membranes of frog sartorius fibers. *J. Gen. Physiol.* **53,** 279–297.

Emmett, B., and Hochachka, P. W. (1981). Scaling of oxidative and glycolytic enzymes in mammals. *Respir. Physiol.* **45,** 261–272.

Engel, W. K. (1962). The essentially of histo- and cytochemical studies of skeletal muscle in the investigation of neuromuscular disease. *Neurology* **12,** 778–794.

Epstein, H. F., and Fischman, D. A. (1991). Molecular analysis of protein assembly in muscle development. *Science* **251,** 1039–1044.

Fatt, P., and Katz, B. (1951). An analysis of the end plate potential recorded with an intracellular electrode. *J. Physiol. (London)* **115,** 320–370.

Federspiel, W. J. (1986). A model study of intracellular oxygen gradients in a myoglobin-containing skeletal muscle fiber. *Biophys. J.* **49,** 857–868.

Federspiel, W. J., and Popel, A. S. (1986). A theoretical analysis of the effect of the particulate nature of blood on oxygen release in capillaries. *Microvasc. Res.* **32,** 164–189.

Federspiel, W. J., and Sarelius, I. H. (1984). An examination of the contribution of red cell spacing to the uniformity of oxygen flux at the capillary wall. *Microvasc. Res.* **27,** 273–285.

Franzini-Armstrong, C. (1970). Studies of the triad. Structure of the junction in frog twitch fibres. *J. Cell Biol.* **47,** 488–499.

Friedman, J. E., Sherman, W. M., Reed, M. J., Elton, C. W., and Dohm, G. L. (1990). Exercise training increases glucose transporter protein glut-4 in skeletal muscle of obese zucker (FA/FA) rats. *FEBS Lett.* **268,** 13–16.

Friedrich, P., and Hajdu, J. (1987). The structure, mechanism, control and organization of glycolytic enzymes in relation to their function in muscle—Introduction. *Biochem. Soc. Trans.* **15,** 973–977.

Garlick, P. J., Maltin, C. A., Baillie, A. G. S., Delday, M. I., and Grubb, D. A. (1989). Fiber-type composition of nine rat muscles. II. Relationship to protein turnover. *Am. J. Physiol.* **257,** E828–E832.

Gates, H. J., Ridge, R. A. A., and Rowlerson, A. (1991). Motor units of the fourth deep lumbrical muscle of the adult rat: Isometric contractions and fiber type compositions. *J. Physiol. (London)* **443,** 193–215.

Gauthier, G., and Lowey, S. (1977). Polymorphism of myosin among skeletal muscle fiber types. *J. Cell Biol.* **47,** 760–779.

Gayeski, T. E. J., and Honig, C. R. (1986). O_2 gradients from sarcolemma to cell interior in red muscle at maximal $\dot{V}_{O_2}$. *Am. J. Physiol.* **251,** H789–H799.

Ginsborg, B. L., and Jenkinson, D. H. (1976). Transmission of impulses from nerve to muscle. *In* "Neuromuscular Junction" (E. Zaimis, ed.), pp. 229–364. Springer-Verlag, Berlin.

Girod, R., Eder-Colli, L., Medilanski, J., Dunant, Y., Tabti, N., and Poo, M.-M. (1992). Pulsatile release of acetylcholine by nerve terminals (synaptosomes) isolated from torpedo electric organ. *J. Physiol. (London)* **450,** 325–340.

Glatz, J. F. C., and van der Vusse, G. J. (1989). Intracellular transport of lipids. *Mol. Cell. Biochem.* **88,** 37–44.

Glatz, J. F. C., van der Vusse, G. J., and Veerkamp, J. H. (1988). Fatty acid-binding proteins and their physiological significance. *News Physiol. Sci.* **3,** 41–43.

Goldman, Y. E., and Brenner, B. (1987). Molecular mechanism of muscle contraction. *Annu. Rev. Physiol.* **9,** 629–636.

Gonzalez-Serratos, H. (1971). Inward spread of activation in vertebrate muscle fibres. *J. Physiol. (London)* **212,** 777–799.

Goodyear, L. J., King, P. A., Hirshman, M. F., Thompson, C. M., Horton, E. D., and Horton, E. S. (1990). Contractile activity increases plasma membrane glucose transporters in absence of insulin. *Am. J. Physiol.* **258,** E667–E672.

Gould, G. W., and Bell, G. I. (1990). Facilitative glucose transporters: An expanding family. *Trends Biochem. Sci.* **15,** 18–23.

Grabowski, W., Lobsiger, E. A., and Lüttgau, H. C. (1972). The effect of repetitive stimulation at low frequencies upon the electrical and mechanical activity of single muscle fibres. *Pflüegers Arch.* **334,** 222–239.

Grove, B. K., Cernly, L., Perriard, J. C., and Eppenberger, H. M. (1985). Myomesin and M-protein: Expression of two M-band proteins in pectoral muscle and heart during development. *J. Cell Biol.* **101,** 1413–1421.

Gunning, P., and Hardeman, E. (1991). Multiple mechanisms regulate muscle fiber diversity. *FASEB J.* **5,** 3064–3070.

Guth, L., and Samaha, F. J. (1969). Qualitative difference between actomyosin ATPase of slow and fast mammalian muscle. *Exp. Neurol.* **25,** 138–152.

Haas, R. C., and Strauss, A. W. (1990). Separate nuclear genes encode sarcomere-specific and ubiquitous human mitochondrial creatine kinase isoenzymes. *J. Biol. Chem.* **265,** 6921–6927.

Hammer, J. A. (1991). Novel myosins. *Trends Cell Biol.* **1,** 50–56.

Hasselbach, W. (1979). The sarcoplasmic calcium pump. A model of energy transduction in biological membranes. *Top. Curr. Chem.* **78,** 1–56.

Hasselbach, W., and Oetliker, H. (1983). Energetics and electrogenicity of the sarcoplasmic reticulum calcium pump. *Annu. Rev. Physiol.* **45,** 325–339.

Heilig, A,. and Pette, D. (1988). Albumin in rabbit skeletal muscle: Origin, distribution and regulation by contractile activity. *Eur. J. Biochem.* **171,** 503–508.

Hennemann, E. (1980). *In* "Medical Physiology" (V. B. Mountcastle, ed.), Vol. 1, pp. 674–741. Mosby, St. Louis, MO.

Henriksen, E. J., Bourey, R. E., Rodnick, K. J., Koranyi, L., Permutt, M. A., and Holloszy, J. O. (1990). Glucose transporter protein content and glucose transport capacity in rat skeletal muscles. *Am. J. Physiol.* **259,** E593–E598.

Hermansen, L., and Vaage, O. (1977). Lactate disappearance and glycogen synthesis in human skeletal muscle after maximal exercise. *Am. J. Physiol.* **233,** E422–E429.

Heuser, J. E., and Reese, T. S. (1973). Evidence for recycling of synaptic vesicle membrane during transmitter release at the frog neuromuscular junction. *J. Cell Biol.* **57,** 315–344.

Hintz, C. S., Cpyle, E. F., Kaiser, K. K., Chi, M. M. Y., and Lowry, O. H. (1984). Comparison of muscle fiber typing by quantitative enzyme assays and by myosin ATPase staining. *J. Histochem. Cytochem.* **32,** 655–660.

Hochachka, P. W., Emmett, B., and Suarez, R. K. (1988). Limits and constraints in the scaling of oxidative and glycolytic enzymes in homeotherms. *Can. J. Zool.* **66,** 1128–1138.

Hodgkin, A. L., and Horowicz, P. (1960). The effect of sudden changes in ionic concentration on the membrane potential of single muscle fibres. *J. Physiol. (London)* **153,** 370–385.

Hodgkin, A. L., and Katz, B. (1949). The effect of sodium ions on the electrical constants of giant axon of the squid. *J. Physiol. (London)* **108,** 37–77.

Hodgkin, A. L., and Nakajima, S. (1972). The effect of diameter in the electrical constants of frog skeletal muscle fibres. *J. Physiol. (London)* **221,** 105–120.

Holloszy, J. O., Dalsky, G. P., Nemeth, P. M., and Hurley, B. F. (1986). Utilization of fat as substrate during exercise: Effect of training. *Int. Ser. Sport Sci.* **16,** 183–190.

Homsher, E., Mommaerts, W. F. H. M., Ricchiuti, N. V., and Wallner, A. (1972). Activation heat, activation metabolism and tension-related heat in frog semitendinosus muscle. *J. Physiol. (London)* **220,** 601–625.

Hoppeler, H. (1986). Exercise-induced ultrastructural changes in skeletal muscle. *Int. J. Sports Med.* **7,** 187–204.

Hoppeler, H. (1990). The different relationship of $\dot{V}_{O_2\ max}$ to muscle mitochondria in humans and quadrupedal animals. *Respir. Physiol.* **80,** 137–146.

Hoppeler, H., and Kayar, S. R. (1988). Capillarity and oxidative capacity of muscles. *News Physiol. Sci.* **3,** 113–116.

Hoppeler, H., Lüthi, P., Claassen, H., Weibel, E. R., and Howald, H. (1973). The ultrastructure of the normal human skeletal muscle. A morphometric analysis on untrained men, women, and well-trained orienteers. *Pflüegers Arch.* **344,** 217–232.

Hoppeler, H., Mathieu, O., and Lindstedt, S. L. (1980). Scaling structural parameters of oxygen consumption against $\dot{V}_{O_2\ max}$ and body mass. *In* "Exercise Bioenergetics and Gas Exchange" (P. Cerretelli and B. J. Whipp, eds.), pp. 129–135. Elsevier, Amsterdam.

Hoppeler, H., Howald, H., Conley, K. E., Lindstedt, S. L., Claassen, H., Vock, P., and Weibel, E. R. (1985). Endurance training in humans: Aerobic capacity and structure of skeletal muscle. *J. Appl. Physiol.* **59,** 320–327.

Hoppeler, H., Hudlicka, O., and Uhlmann, E. (1987). Relationship between mitochondria and oxygen consumption in isolated cat muscles. *J. Physiol. (London)* **385,** 661–675.

Hoppeler, H., Mathieu-Costello, O., and Kayar, S. R. (1991). Mitochondria and microvascular design. *In* "The Lung: A Scientific Foundation" (R. G. Crystal, J. B. West, P. J. Barnes, N. S. Cherniack, and E. R. Weibel, eds.), pp. 1467–1477. Raven Press, New York.

Horowits, R., and Podolsky, R. J. (1987). The positional stability of thick filaments in activated skeletal muscle depends on sarcomere length: Evidence for the role of titin filaments. *J. Cell Biol.* **105,** 2217–2223.

Howald, H., Hoppeler, H., Claassen, H., Mathieu, O., and Straub, R. (1985). Influences of endurance training on the ultrastructural composition of the different muscle fiber types in humans. *Pflüegers Arch.* **403,** 369–376.

Hui, C. S., and Chandler, W. K. (1991). Q_β and Q_γ components of intramembranous charge movement in frog cut twitch fibers. *J. Gen. Physiol.* **98,** 429–464.

Hurley, B. F., Nemeth, P. M., Martin, W. H., Hagberg, J. M., Dalsky, G. P., and Holloszy, J. O. (1986). Muscle triglyceride utilization during exercise: Effect of training. *J. Appl. Physiol.* **60,** 562–567.

Huxley, A. F. (1957). Muscle structure and theories of contraction. *Prog. Biophys. Biol. Chem.* **7,** 255–318.

Huxley, A. F., ed. (1980). "Reflections on Muscle." Liverpool Univ. Press, Liverpool.

Huxley, A. F., and Taylor, S. (1958). Local activation of striated muscle fibres. *J. Physiol. (London)* **144,** 436–444.

Huxley, H. E., and Hanson, J. (1954). Changes in the cross-striations of muscle during contraction and stretch and their structural interpretation. *Nature (London)* **173,** 973–976.

Ingwall, J. S., and Fossell, E. T. (1983). Changes in creatine kinase system in the hypertrophied myocardium of the dog and rat. *Cardiovasc. Res.* **7,** 601–617.

Ishikawa, H., Sawada, H., and Yamada, E. (1983). Surface and internal morphology of skeletal muscle. *In* "Handbook of Physiology" (L. D. Peachey, R. H. Adrian, and S. R. Geiger, eds.), Sect. 10, p. 16. Williams & Wilkins, Baltimore, MD.

Israel, M., Dunant, Y., and Manaranche, R. (1979). The present status of the vesicular hypothesis. *Prog. Neurobiol.* **13,** 237–275.

Ito, M., and Hartshorne, D. J. (1990). Phosphorylation of myosin as a regulatory mechanism in smooth muscle. *Prog. Clin. Biol. Res.* **237,** 57–72.

Iyengar, M. R. (1984). Creatine kinase as an intracellular regulator. *J. Muscle Res. Cell Motil.* **5,** 527–534.

Izumo, S., Nadal-Ginard, B., and Madavi, V. (1986). All members of the MHC multigene family respond to thyroid hormone in a highly tissue-specific manner. *Science* **231,** 597–600.

James, N. T., and Meek, G. A. (1979). Stereological analysis of the structure of mitochondria in pigeon skeletal muscle. *Cell Tissue Res.* **202,** 493–503.

Jones, N. L., Heigenhauser, G. J. F., Kuksis, A., Matsos, C. G., Sutton, J. R., and Toews, C. J. (1980). Fat metabolism in heavy exercise. *Clin. Sci.* **59,** 469–478.

Jones, S. P., Ridge, R. M. A. P., and Rowlerson, A. (1987). The non-selective innervation of muscle fibres and mixed composition of motor units in a muscle of neonatal rat. *J. Physiol. (London)* **386,** 377–394.

Juel, C., and Wibrand, F. (1989). Lactatae transport in isolated mouse muscles studied with a tracer technique—Kinetics, stereospecificity, pH dependency and maximal capacity. *Acta Physiol. Scand.* **137,** 33–39.

Kamm, K. E., and Stull, J. T. (1985). The function of myosin light chain kinase phosphorylation in smooth muscle. *Annu. Rev. Pharmacol.* **25,** 593–620.

Katz, B., ed. (1966). "Nerve, Muscle and Synapse." McGraw-Hill, New York.

Katz, B., and Miledi, R. (1965). The measurement of synaptic delay, and the time course of acetylcholine release at the neuromuscular junction. *Proc. R. Soc. London, Ser. B* **161,** 483–495.

Katz, B., and Miledi, R. (1972). The statistical nature of the acetylcholine potential and its molecular components. *J. Physiol. (London)* **224,** 665–699.

Kaufmann, M., Simoneau, J. A., Veerkamp, J. H., and Pette, D. (1989). Electrostimulation-induced increases in fatty acid binding protein and myoglobin in rat fast-twitch muscle and comparison with tissue levels in heart. *FEBS Lett.* **245,** 181–184.

Kawai, H., Nishino, H., Nishida, Y., Masuda, K., and Saito, S. (1987). Localization of myoglobin in human muscle cells by immunoelectron microscopy. *Muscle Nerve* **10,** 144–149.

Kayer, S. R., and Banchero, N. (1987). Volume density and distribution of mitochondria in myocardial growth and hypertrophy. *Respir. Physiol.* **70,** 275–286.

Kayar, S. R., Hoppeler, H., Howald, H., Claassen, H., and Oberholzer, F. (1986). Acute effects of endurance exercise on mitochondrial distribution and skeletal muscle morphology. *Eur. J. Appl. Physiol.* **54,** 578–584.

Kayar, S. R., Hoppeler, H., Essen-Gustavsson, B., and Schwerzmann, K. (1988a). The similarity of mitochondrial distribution in equine skeletal muscles of differing oxidative capacity. *J. Exp. Biol.* **137,** 253–263.

Kayar, S. R., Hoppeler, H., Mermod, L., and Weibel, E. R. (1988b). Mitochondrial size and shape in equine skeletal muscle: A three-dimensional reconstruction study. *Anat. Rec.* **222,** 333–339.

Kayar, S. R., Hoppeler, H., Armstrong, R. B., Laughlin, M. H., Jones, J. H., Longworth, K. E., Conley, K. R., and Taylor, C. R. (1992). Transit time for capillary blood in selected muscles of exercising animals. *Pflüegers Arch.* **421,** 578–584.

Kiehart, D. P. (1990). Molecular genetic dissection of myosin heavy chain function. *Cell (Cambridge, Mass.)* **60,** 347–350.
Kirkwood, S. P., Munn, E. A., and Brooks, G. A. (1986). Mitochondrial reticulum in limb skeletal muscle. *Am. J. Physiol.* **251,** C395–C402.
Kishino, A., and Yanagida, T. (1988). Force measurements by micromanipulation of a single actin filaments by glass needles. *Nature (London)* **334,** 746–747.
Kobayashi, T., and Sugi, H. (1980). Measurement of heat production during the summation of isometric contraction on frog skeletal muscle with an infrared radiometer. *Jpn. J. Physiol.* **30,** 617–629.
Kruger, M., Wright, J., and Wang, K. (1991). Nebulin as a length regulator of thin filaments of vertebrate skeletal muscles—Correlation of thin filament length, nebulin size, and epitope profile. *J. Cell Biol.* **115,** 97–107.
Kugelberg, E. (1973). Histochemical composition, contraction speed and fatiguability of rat soleus motor units. *J. Neurol. Sci.* **20,** 177–198.
Labeit, S., Barlow, D. P., Gautel, M., Gibson, T., Holt, J., Hsieh, C.-L., Francke, U., Leonard, K., Wardale, J., Whiting, A., and Trinick, J. (1990). A regular pattern of two types of 100-residue motif in the sequence of titin. *Nature (London)* **345,** 273–276.
Lai, F. A., Erickson, H. P., Rousseau, E., Liu, Q.-Y., and Meissner, G. (1988). Purification and reconstitution of the calcium release channel from skeletal muscle. *Nature (London)* **331,** 315–319.
Lännergren, J., and Hoh, J. F. Y. (1984). Myosin isoenzymes in single muscle fibres of *Xenopus laevis*: Analysis of five different functional types. *Proc. R. Soc. London, Ser. B.* **222,** 401–408.
Larsson, L., Estroem, L., Lindegren, B., Gorza, L., and Schiaffino, S. (1991). MHC composition and enzyme-histochemical and physiological properties of a novel fast-twitch motor unit type. *Am. J. Physiol.* **261,** C93–C101.
Lingle, C. J., Maconchie, D., and Steinbach, J. H. (1992). Activation of skeletal muscle nicotinic acetylcholine receptors. *J. Membr. Biol.* **126,** 195–217.
Longmuir, I. S. (1981). Channels of oxygen transport from blood to mitochondria. *Adv. Physiol. Sci., Proc. Int. Congr., 28th, 1980,* Vol. 25, pp. 19–22.
Lüttgau, H. C. (1965). The effect of metabolic inhibitors on the fatigue of the action potential in single muscle fibres. *J. Physiol. (London)* **178,** 45–67.
Lüttgau, H. C., and Oetliker, H. (1968). The action of caffeine on the activation of the contractile mechanism in striated muscle fibres. *J. Physiol. (London)* **194,** 51–74.
Mabuchi, K., and Sréter, F. A. (1980). Actomyosin ATPase. II. Fiber typing by histochemical ATPase reaction. *Muscle Nerve* **3,** 223–239.
Markert, C. L., and Ursprung, H. (1962). The ontogeny of isozyme patterns of lactate dehydrogenase in the mouse. *Dev. Biol.* **5,** 363–381.
Marston, S. B. (1989). What is latch? New ideas about tonic contraction in smooth muscle. *J. Muscle Res. Cell Motil.* **10,** 97–100.
Maruyama, K. (1986). Connectin, an elastic filamentous protein of striated muscle. *Int. Rev. Ctyol.* **104,** 81–103.
Matarese, V., Stone, R. L., Waggoner, D. W., and Bernlohr, D. A. (1990). Intracellular fatty acid trafficking and the role of cytosolic lipid binding proteins. *Prog. Lipid Res.* **28,** 245–272.
Mathieu, O., Krauer, R., Hoppeler, H., Gehr, P., Lindstedt, S. L., Alexander, R. McN., Taylor, C. R., and Weibel, E. R. (1981a). Design of the mammalian respiratory system. VII. Scaling mitochondrial volume in skeletal muscle to body mass. *Respir. Physiol.* **44,** 113–128.

Mathieu, O., Cruz-Orive, L. M., Hoppeler, H., and Weibel, E. R. (1981b). Measuring error and sampling variation in stereology: Comparison of the efficiency of various methods for planar image analysis. *J. Microsc. (Oxford)* **121,** 75–88.

Mathieu, O., Cruz-Orive, L. M., Hoppeler, H., and Weibel, E. R. (1983). Estimating length density and quantifying anisotropy in skeletal muscle capillaries. *J. Microsc. (Oxford)* **131,** 131–146.

Mathieu-Costello, O. (1987). Capillary tortuosity and degree of contraction or extension of skeletal muscle. *Microvasc. Res.* **33,** 98–117.

Mathieu-Costello, O., Hoppeler, H., and Weibel, E. R. (1989). Capillary tortuosity in skeletal muscles of mammals depends on muscle contraction. *J. Appl. Physiol.* **66,** 1436–1442.

Matlib, M. A., Rebmann, D., Ashraf, M., Rouslin, W., and Schwartz, A. (1981). Differential activities of putative subsarcolemmal and interfibrillar mitochondria from cardiac muscle. *J. Mol. Cell. Cardiol.* **13,** 163–170.

Matoba, H., Allen, J. R., Bayly, W. M., Oakley, C. R., and Gollnick, P. D. (1985). Comparison of fiber types in skeletal muscle from ten animal species based on sensitivity of the myofibrillar actomyosin ATPase to acid or copper. *Histochemistry* **82,** 185–183.

Mattfeldt, T. (1987). Estimation of length, surface, and number of anisotropic objects: A review of models and design-based approaches. *Acta Stereol.* **6,** 537–548.

Maughan, D., and Lord, C. (1988). Protein diffusivities in skinned frog skeletal muscle fibers. *Adv. Exp. Med. Biol.* **226,** 75–84.

Maxwell, L. C., Faulkner, J. A., and Murphy, R. A. (1982). Relationship among fibre type, myosin ATPase activity and contractile properties. *Histochem. J.* **14,** 981–997.

McGilvery, R. W. (1973). The use of fuels for muscular work. *In* "Metabolic Adaptation to Prolonged Physical Exercise" (H. Howald and J. R. Poortmans, eds.), pp. 12–26. Birkheuser, Basel.

McLachlan, A. D., and Karn, J. (1982). Periodic charge distributions in the myosin rod amino acid sequence match cross-bridge spacings in muscle. *Nature (London)* **299,** 226–231.

Metzger, J. M., Greaser, M. L., and Moss, R. L. (1989). Variations in cross-bridge attachment rate and tension with phosphorylation of myosin in mammalian skinned skeletal muscle fibers. Implications for twitch potentiation in intact muscle. *J. Gen. Physiol.* **93,** 855–883.

Mobley, B. A., and Eisenberg, B. R. (1975). Sizes of components in frog skeletal muscle measured by methods of stereology. *J. Gen. Physiol.* **66,** 31–45.

Moon, T. W., Brill, R. W., Hochachka, P. W., and Weber, J.-M. (1987). L-(+)-Lactate translocation into the red blood cells of the skipjack tuna (*Katsuwonus pelamis*). *Can. J. Zool.* **65,** 2570–2573.

Moore, G. E., and Schachat, F. H. (1985). Molecular heterogeneity of histochemical fibre types: A comparison of fast fibres. *J. Muscle Res. Cell Motil.* **6,** 513–524.

Moore, R. L., and Stull, J. T. (1984). Myosin light chain phosphorylation in fast and slow skeletal muscles *in situ*. *Am. J. Physiol.* **247,** C462–C471.

Nemeth, P., Pette, D., and Vrbova, G. (1979). Malate dehydrogenase activity indicating metabolic homogeneity of single fibres of the motor unit. *J. Physiol. (London)* **301,** 73P–74P.

Nimmo, M. A., Wilson, R. H., and Snow, D. H. (1985). The inheritance of skeletal muscle fibre composition in mice. *Comp. Biochem. Physiol. A* **81A,** 109–116.

Nordheim, K., and Vøllestad, N. K. (1990). Glycogen and lactate metabolism during low-intensity exercise in man. *Acta Physiol. Scand.* **139,** 475–484.

Oberholzer, F., Claassen, H., Moesch, H. and Howald, H. (1976). Ultrastrukturelle, biochemische und energetische Analyse einer extremen Dauerleistung (100 Km-Lauf). *Schweiz. Z. Sportmed.* **24,** 71–98.

Oetliker, H. (1973). The influence of fatigue on membrane conductance in frog skeletal muscle. *Experientia* **29,** 747.

Oetliker, H. (1981a). Does indodicarbocyanine fluorescence reflect membrane potential of the sarcoplasmic reticulum in skeletal muscle? *Adv. Physiol. Sci., Proc. Int. Congr., 28th, 1980,* Vol. 5, pp. 345–361.

Oetliker, H. (1981b). Divalent cation concentration-dependent fluorescence of isolated sarcoplasmic reticulum vesicles stained with indodicarbocyanine. *J. Physiol. (London)* **318,** 11–12.

Oetliker, H. (1982). An appraisal of the evidence for a sarcoplasmic reticulum membrane potential and its relation to calcium release in skeletal muscle. *J. Muscle Res. Cell Motil.* **3,** 247–272.

Oetliker, H. (1989). Energetical considerations related to calcium release from the sarcoplasmic reticulum in skeletal muscle. *Biomed. Biochim. Acta* **48,** 313–318.

Oetliker, H., and Schümperli, R. A. (1982). Influence of sarcomere length, tonicity, and external sodium concentration on conduction velocity in frog muscle fibres. *J. Physiol. (London)* **332,** 203–221.

Oetliker, H., Baylor, S. M., and Chandler, W. K. (1975). Simultaneous changes in fluorescence and optical retardation in single muscle fibres during activity. *Nature (London)* **257,** 693–696.

Ogata, T., and Yamasaki, Y. (1985). Scanning electron-microscopic studies on the three-dimensional structure of mitochondria in the mammalian red, white and intermediate muscle fibers. *Cell Tissue Res.* **241,** 251–256.

Oscai, L. B., Essig, D. A., and Palmer, W. K. (1990). Lipase regulation of muscle triglyceride hydrolysis. *J. Appl. Physiol.* **69,** 1571–1577.

Pape, P. C., Konishi, M., and Baylor, S. M. (1992). Valinomycin and excitation contraction coupling in skeletal muscle fibres of the frog. *J. Physiol. (London)* **449,** 219–235.

Parmacek, M. S., and Leiden, J. M. (1991). Focus on molecular biology—Structure, function, and regulation of troponin-C. *Circulation* **84,** 991–1003.

Peachey, L. D. (1965). The sarcoplasmic reticulum and transverse tubules of the frog's sartorius. *J. Cell Biol.* **25,** 209–231.

Peachey, L. D., and Franzini-Armstrong, C. (1983). Structure and function of membrane systems of skeletal muscle cells. *In* "Handbook of Physiology" (L. D. Peachey, R. H. Adrian, and S. R. Geiger, eds.), Sect. 10, p. 35. Williams & Wilkins, Baltimore, MD.

Perriard, J. C., Caravatti, M., Perriard, E. R., and Eppenberger, H. M. (1978). Quantitation of creatine kinase isoenzyme transition in differentiating chicken embryonic breast muscle and myogenic cell cultures by immunoadsorption. *Arch. Biochem. Biophys.* **191,** 90–100.

Peter, J. B., Barnard, R. J., Edgerton, V. R., Gillespie, C. A., and Stempel, K. E. (1972). Metabolic profiles of three fiber types of skeletal muscle in guinea pigs and rabbits. *Biochemistry* **11,** 2627–2633.

Pette, D. (1980). "Plasticity of Muscle." de Gruyter, Berlin.

Pette, D. (1990). "The Dynamic State of Muscle Fibers." de Gruyter, Berlin.

Pette, D., and Vrbova, G. (1985). Neural control of phenotypic expression in mammalian muscle fibers. *Muscle Nerve* **8,** 676–689.

Pette, D., Ramurez, B. U., Wüller, W., Simon, R., Exner, G. U., and Hildebrand, R. (1975). Influence of intermittent long-term stimulation on contractile, histochemical and metabolic properties of fibre populations in fast and slow rabbit muscles. *Pflüegers Arch.* **361,** 1–7.

Poole, D. C., and Mathieu-Costello, O. (1989a). Skeletal muscle capillary geometry: Adaptation to chronic hypoxia. *Respir. Physiol.* **77,** 21–30.

Poole, D. C., and Mathieu-Costello, O. (1989b). Capillary tortuosity in rat soleus muscle is not affected by endurance training. *Am. J. Physiol.* **256,** H1110–H1116.

Pringle, J. W. S. (1976). The muscles and sense organs involved in insect flight. *In* "Insect Flight" (R. C. Rayney, ed.), pp. 3–15. Blackwell, Oxford.

Rakusan, K., and Tomanek, R. J. (1986). Distribution of mitochondria in normal and hypertrophic myocytes from the rat heart. *J. Mol. Cell. Cardiol.* **18,** 299–305.

Rall, J. A. (1985). Energetic aspects of skeletal muscle contraction: Implication of fiber types. *Exercise Sport Sci. Rev.* **13,** 33–74.

Ranatunga, K. W., and Thomas, P. E. (1990). Correlation between shortening velocity, force–velocity relation and histochemical fibre-type composition in rat muscles. *J. Muscle Res. Cell Motil.* **11,** 240–250.

Reiser, P. J., Moss, R. L., Giulian, G. G., and Greaser, M. L. (1985). Shortening velocity in single fibers from adult rabbit soleus muscles is correlated with myosin heavy chain composition. *J. Biol. Chem.* **260,** 9077–9080.

Rose, H., Hennecke, T., and Kammermeier, H. (1989). Is fatty acid uptake in cardiomyocytes determined by physicochemical fatty acid partition between albumin and membranes? *Mol. Cell. Biochem.* **88,** 31–36.

Rossi, A. M., Eppenberger, H. M., Volpe, P., Cotrufo, R., and Wallimann, T. (1990). Muscle-type MM creatine kinase is specifically bound to sarcoplasmic reticulum and can support Ca^{2+} uptake and regulate local ATP/ADP ratios. *J. Biol. Chem.* **265,** 5258–5266.

Rowlerson, A. M. (1990). Specialization of mammalian jaw muscles: Fibre type compositions and the distribution of muscle spindles. *In* "Neurophysiology of the Jaws and Teeth" (A. Taylor, ed.), pp. 1–51. Macmillan, New York.

Rowlerson, A. M., and Spurway, N. C. (1988). Histochemical and immunohistochemical properties of skeletal muscle fibres from *Rana* and *Xenopus*. *Histochem. J.* **20,** 657–673.

Rubinstein, N. A., and Kelly, A. M. (1981). Development of muscle fiber specialization in the rat hindlimb. *J. Cell Biol.* **90,** 128–144.

Sahlin, K. (1986). Metabolic changes limiting muscle performance. *Int. Ser. Sport Sci.* **16,** 323–343.

Sahlin, K. (1987). Lactate production cannot be measured with tracer techniques. *Am. J. Physiol.* **252,** E439–E440.

Salmons, S., and Henriksson, J. (1981). The adaptive response of skeletal muscle to increased use. *Muscle Nerve* **4,** 94–105.

Scales, D., and Inesi, C. (1976). Assembly of the ATPase protein in sarcoplasmic reticulum membranes. *Biophys. J.* **16,** 735–751.

Schachat, F. H., Bronson, D. D., and McDonald, O. B. (1985). Heterogeneity of contractile proteins. A continuum of troponin–tropomyosin expression in mammalian skeletal muscle. *J. Biol. Chem.* **260,** 1108–1113.

Schiaffino, S., Gorza, L., Sartore, S., Saggin, L., Ausoni, S., Vianello, M., Gundersen, K., and Lomo, T. (1989). Three myosin heavy chain isoforms in type 2 skeletal muscle fibres. *J. Muscle Res. Cell Motil.* **10,** 197–205.

Schmitt, T., and Pette, D. (1985). Increased mitochondrial creatine kinase in chronically stimulated fast-twitch rabbit muscle. *FEBS Lett.* **188,** 341–344.

Schneider, M. F., and Chandler, W. K. (1973). Voltage dependent charge movement in skeletal muscle: A possible step in excitation–contraction coupling. *Nature (London)* **242,** 244–246.

Scholey, J. (1991). Motor proteins in cell division. *Trends Cell Biol.* **1,** 122–129.

Schwerzmann, K., Hoppeler, H., Kayar, S. R., and Weibel, E. R. (1989). Oxidative capacity of muscle and mitochondria: Correlation of physiological, biochemical, and morphometric characteristics. *Proc. Natl. Acad. Sci. U.S.A.* **86,** 1583–1587.

Simoneau, J.-A., and Bouchard, C. (1989). Human variation in skeletal muscle fiber-type proportion and enzyme activities. *Am. J. Physiol.* **257,** E567–E572.

Smith, I. C. H. (1972). Energetics of activation in frog and toad muscle. *J. Physiol. (London)* **220,** 583–599.

Smith, J. S., Coronado, R., and Meissner, G. (1986). Single channel measurements of the calcium release channel from skeletal muscle sarcoplasmic reticulum. Activation by Ca^{2+} and ATP and modulation by Mg^{2+}. *J. Gen. Physiol.* **88,** 573–588.

Spruce, A. E., Standen, N. B., and Stanfield, P. R. (1987). Studies of the unitary properties of adenosine-5′-triphosphate-regulated potassium channels of frog skeletal muscle. *J. Physiol. (London)* **382,** 213–236.

Staron, R. S. (1991). Correlation between myofibrillar ATPase activity and myosin heavy chain composition in single human muscle fibers. *Histochemistry* **96,** 21–24.

Staron, R. S., Hidida, R. S., Murray, T. F., and Hagerman, N. T. (1989). Lipid depletion and repletion in skeletal muscle following a marathon. *J. Neurol. Sci.* **94,** 29–40.

Steinbach, J. H. (1989). Structural and functional diversity in vertebrate skeletal muscle nicotinic acetylcholine receptors. *Annu. Rev. Physiol.* **51,** 353–365.

Stockdale, F. E. (1990). Muscle fiber development and origins of fiber diversity. *In* "The Dynamic State of Muscle Fibers" (D. Petty, ed.), pp. 121–125. de Gruyter, Berlin.

Stremmel, W. (1988a). Fatty acid uptake by isolated rat heart myocytes represents a carrier-mediated transport process. *J. Clin. Invest.* **81,** 844–852.

Stremmel, W. (1988b). Uptake of fatty acids by jejunal mucosal cells is mediated by a fatty acid binding membrane protein. *J. Clin. Invest.* **82,** 2001–2010.

Stremmel, W., Diede, H. E., Rodilla-Sala, E., Vyska, K., Schrader, M., Fitscher, B., and Passarella, S. (1990). The membrane fatty acid-binding protein is not identical to mitochondrial glutamic oxaloacetic transaminase (mGOT). *Mol. Cell. Biochem.* **98,** 191–199.

Takeuchi, A., and Takeuchi, N. (1960). On the permeability of end plate membrane during the action of transmitter. *J. Physiol. (London)* **154,** 52–67.

Tanabe, T., Beam, K. G., Adams, A. B., Niidome, T., and Numa, S. (1990). Regions of the skeletal muscle dihydropyridine receptor critical for excitation–contraction coupling. *Nature (London)* **346,** 567–569.

Taylor, C. R., Karas, R. H., Weibel, E. R., and Hoppeler, H. (1987). Adaptive variation in the mammalian respiratory system in relation to energetic demand. *Respir. Physiol.* **69,** 1–127.

Taylor, E. W. (1979). Mechanism of actomyosin ATPase and the problem of muscle contraction. *CRC Crit. Rev. Biochem.* **6,** 103–164.

Terjung, R. L., Mackie, B. G., Dudley, G. A., and Kaciuba-Uscilko, H. (1983). Influence of exercise on chylomicron triacylglycerol metabolism: Plasma turnover and muscle uptake. *Med. Sci. Sports Exercise* **15,** 340–347.

Thornell, L. E., and Price, M. G. (1991). The cytoskeleton in muscle cells in relation to function. *Biochem. Soc. Trans.* **19,** 1116–1120.

Tullson, P. C., and Terjung, R. L. (1990). Adenine nucleotide degradation in striated muscle. *Int. J. Sports Med.* **11,** S47–S55.

Uyeda, T. Q. P., Kron, S. J., and Spudich, J. (1990). Myosin step size. Estimation from slow sliding movement of actin over low densities of heavy meromyosin. *J. Mol. Biol.* **214,** 699–710.

Vale, R. D., and Goldstein, L. S. B. (1990). One motor, many tails: An expanding repertoire of force-generating enzymes. *Cell (Cambridge, Mass.)* **60,** 883–885.

Vallee, R. (1991). Cytoplasmic dynein: Advances in microtubule-based motility. *Trends Cell Biol.* **1,** 25–29.

van der Laarse, W. J., Diegenbach, P. C., and Elzinga, G. (1989). Maximal rate of oxygen consumption and quantitative histochemistry of succinate dehydrogenase in single muscle fibres of *Xenopus laevis*. *J. Muscle Res. Cell Motil.* **10,** 221–228.

Veech, R. L., Lawson, J. W., Cornell, N. W., and Krebs, H. A. (1979). Cytosolic phosphorylation potential. *J. Biol. Chem.* **254,** 6538–6547.

Vibert, P., and Cohen, C. (1988). Domains, motions and regulation in the myosin head. *J. Muscle Res. Cell Motil.* **9,** 296–305.

Walker, P. M., Mickle, D. A. G., Tanner, W. R., Harding, R., and Romaschin, A. D. (1984). Decreased uptake of exogenous substrates following graded muscle stimulation. *Am. J. Physiol.* **246,** H690–H695.

Wallimann, T., and Eppenberger, H. M. (1985). Localization and function of M-line-bound creatine kinase. M-band model and creatine phosphate shuttle. *Cell Muscle Motil.* **6,** 239–285.

Wallimann, T., Schloesser, T., and Eppenberger, H. M. (1984). Function of M-line-bound creatine kinase as intramyofibrillar ATP regenerator at the receiving end of the phosphorylcreatine shuttle in muscle. *J. Biol. Chem.* **259,** 5238–5246.

Wallimann, T., Schnyder, T., Schlegel, J., Wyss, M., Wegmann, G., Rossi, A. M., Hemmer, W., Eppenberger, H. M., and Quest, A. F. (1989). Subcellular compartmentation of creatine kinase isoenzymes, regulation of CK and octameric structure of mitochondrial CK: Important aspects of the phosphoryl–creatine circuit. *Prog. Clin. Biol. Res.* **315,** 159–176.

Wallimann, T., Wyss, M., Brdiczka, D., Nicolay, K., and Eppenberger, H. M. (1992). Intracellular compartmentation, structure and function of creatine kinase isoenzymes in tissues with high and fluctuating energy demands: The "phosphocreatine circuit" for cellular energy homeostasis. *Biochem. J.* **281,** 21–40.

Wang, K. (1985). Sarcomere-associated cytoskeletal lattices in striated muscle: A review hypothesis. *Cell Muscle Motil.* **6,** 315–369.

Warrick, H. M., and Spudich, J. A. (1987). Myosin structure and function in cell motility. *Annu. Rev. Cell Biol.* **3,** 379–421.

Warwick, R., and Williams, P. L., eds. (1973). "Grays Anatomy," 35th Br. ed., p. 481. Saunders, Philadelphia.

Wegmann, G., Zanolla, E., Eppenberger, H. M., and Wallimann, T.. (1992). *In situ* compartmentation of creatine kinase in sarcomeric muscle: The actomyosin overlap zone as a molecular sieve. *J. Muscle Res. Cell Motil.* **13,** 420–435.

Weibel, E. R., and Taylor, C. R. (1981). Design of the mammalian respiratory system. *Respir. Physiol.* **44,** 1–164.

Westerblad, H., Lee, J. A., Lannergren, J., and Allen, D. G. (1991). Cellular mechanisms of fatigue in skeletal muscle. *Am. J. Physiol.* **261,** C195–C209.

White, S. P., Cohen, C., and Phillips, G. N. (1987). Structure of co-crystals of tropomyosin and troponin. *Nature (London)* **325,** 826–828.

Whiting, A., Wardale, J., and Trinick, J. (1989). Does titin regulate the length of muscle thick filaments? *J. Mol. Biol.* **205,** 263–268.

Winegrad, S. (1968). Intracellular calcium movements of frog skeletal muscle during recovery from tetanus. *J. Gen. Physiol.* **51,** 65–83.

Winegrad, S. (1970). The intracellular site of calcium activation of contraction in frog skeletal muscle. *J. Gen. Physiol.* **55,** 77–88.

Wittenberg, B. A., and Wittenberg, J. B. (1989). Transport of oxygen in muscle. *Annu. Rev. Physiol.* **51,** 857–878.

Wyss, M., Schlegel, J., James, P., Eppenberger, H. M., and Wallimann, T. (1990). Mitochondrial creatine kinase from chicken brain. Purification, biphysical characterization, and generation of heterodimeric and heterooctameric molecules with subunits of other creatine kinase isoenzymes. *J. Biol. Chem.* **265,** 15900–15908.

Yamaguchi, M., Izumimoto, M., Robson, R. M., and Stromer, M. H. (1985). Fine structure of wide and narrow vertebrate muscle Z-lines. *J. Mol. Biol.* **184,** 621–644.

Yamashita, K., and Yoshioka, T. (1991). Profiles of creatine kinase isoenzyme compositions in single muscle fibres of different types. *J. Muscle Res. Cell Motil.* **12,** 37–44.

Zachowski, A., and Devaux, P. F. (1990). Transmembrane movements of lipids. *Experientia* **46,** 644–655.

Zeleznikar, R. J., and Goldberg, N. D. (1991). Kinetics and compartmentation of energy metabolism in intact skeletal muscle determined from ^{18}O labeling of metabolite phosphoryls. *J. Biol. Chem.* **266,** 15110–15119.

Zeleznikar, R. J., Heyman, R. A., Graeff, R. M., Walseth, T. F., Dawis, S. M., Butz, E. A., and Goldberg, N. D. (1990). Evidence for compartmentalized adenylate kinase catalysis serving a high energy phosphoryl transfer function in rat skeletal muscle. *J. Biol. Chem.* **265,** 300–311.

Zenker, W., Snobl, D., and Boetschi, R. (1990). Multifocal innervation and muscle length. A morphological study on the role of myo-myonal junctions, fiber branching and multiple innervation in muscles of different size and shape. *Anat. Embryol.* **182,** 273–283.

ADVANCES IN VETERINARY SCIENCE AND COMPARATIVE MEDICINE, VOL. 38A

The Mechanical Design of the Muscular System

LAWRENCE C. ROME

Department of Biology, University of Pennsylvania, Philadelphia, Pennsylvania 19104

I. Introduction

One of the most fascinating areas of physiology is the study of how parameters of a given system are fine-tuned to provide optimal performance under a variety of conditions. This paper will examine how the mechanical properties of the muscular system are designed to power locomotion. As we shall see, the muscular system provides an exceptional model to examine the principles of physiological design. First, the performance at the level of the whole-animal (locomotion) can be easily studied and is intuitively important (it makes sense for animals to move rapidly and/or efficiently). Second, because of the regular geometry of the muscle, integration from the level of the cross-bridge to whole animal movements is relatively straightforward. Third, locomotor performance shows tremendous dynamic range both within a given animal and between species (a given animal can locomote over a wide range of speeds, and different species use very different types of locomotion). On one hand, this provides the system with many important problems to solve and, on the other hand, provides the investigator with many interesting opportunities to test theories of muscle design. Fourth, there are many components that make up the muscular system, and some show tremendous variation, both within a given animal (e.g., different fiber types) and between species.

In particular, maximum velocity of shortening (V_{max}), rate of relaxation, rate of activation, mechanical gearing of the fibers, and myofilament lengths are known to show considerable variation. Some of these parameters vary in different fiber types, change during development, change in response to physiological stress (e.g., exercise training, thermal adaptation), and vary in different species. Much of the field of muscle physiology focuses on *how* this variation is achieved at the molecular level. However, in this paper, we will be concerned with *why* this variation occurs in the first place. What are the rules (design constraints) by which evolution sets particular values for these parameters in a given animal?

A final, and seemingly unique, feature of muscle is that muscle function depends not only on what proteins are present, but also on the manner in which they are used. For instance, the force, power, and efficiency of muscle are not fixed numbers, but rather they depend on the ratio of the velocity at which the muscle is shortening (V) to V_{max}, or V/V_{max} (illustrated in Fig. 9). This last feature has two important consequences. First, unlike most biological systems, where the flux of a reaction can be increased by adding more enzyme, in muscle, the right protein must be added to enable an animal to make movements efficiently. For instance, increasing the amount of slow muscle will not enable an animal to locomote rapidly. Second, there are curves that define performance (the sarcomere length–tension curve, force–velocity curve, etc.), and there are optimal places on these curves for the muscle to work. Thus, whereas in most biological systems, qualitative a posteriori reasons are usually given for the value of a given parameter, in the muscular system, one can generate hypotheses a priori (e.g., the system is designed to conform to the constraints described by the curves), and then test them.

II. Strategy for Determining the Mechanical Design of the Muscular System

A. Components of the System

Most aspects of the physiology of muscle contraction are fairly well understood at each level of organization. However, very little is understood about how function is integrated up and down levels of organization. It is well known that there are many different modes of locomotion among vertebrates and also a great variety of different locomotor abilities within an animal. Intuitively, one might expect that some of the components must vary to permit this wide range of locomotor capabilities. As with any biological system, some components are kept constant while others vary.

It is useful to distinguish between two types of variation that can occur. The most common manner in which the rate of a function is varied is by having more or less of a given component. For instance, the maximum rate of oxygen consumption ($\dot{V}_{O_2\ max}$) of tissues increases in direct proportion to the volume of mitochondria present (see papers in this Volume by Conley and Billeter *et al.*, and by Lindstedt and Thomas in Volume 38B). Mitochondria in large and small mammals appear to be similar (i.e., they are highly conserved), thus the high mass-specific tissue $\dot{V}_{O_2\ max}$ observed in small mammals is achieved by adding more mitochondria. Because the rate is increased by altering the quantity of the component, we call this *quantitative variation.*

The other manner by which the rate of a function is increased is by switching from an enzyme with a low kinetic rate to one with a higher kinetic rate. Thus, with the same quantity of enzyme, the new enzyme provides a higher reaction rate. We call this *qualitative variation*, as there is a qualitative change in the proteins involved.

Although there is important quantitative variation in muscular systems, as we will see, qualitative variation is the most striking and sets the limits for the design of the system. Thus, our initial strategy is to briefly examine the components of the muscular system at various levels of organization, and to determine which vary and which remain constant among animals.

1. Molecular Level

Force is generated in muscle by myosin heads binding to actin. In normal contraction, the cross-bridges go through cyclical attachment and detachment while splitting ATP. Muscle shortening is caused by the myosin heads pulling the actin thin filaments past the myosin thick filaments. A high maximum velocity of shortening (V_{max}) is thought to be associated with a fast cross-bridge detachment rate (Huxley, 1957), and this cross-bridge step is thought to be set primarily by the type of myosin heavy chain and, to a lesser extent, the myosin light chain in the muscle fiber (Sweeney *et al.*, 1988). V_{max} varies a 1000-fold between the slowest smooth muscle and the fastest skeletal muscle, principally owing to alterations of the myosin protein (qualitative variation). Because of its impact on mechanics and its great variability, V_{max} is a very important design parameter. Actin, on the other hand, seems to be generally conserved among vertebrates. Nevertheless, there can be changes in the length of actin thin filaments that are important for the sarcomere length (SL)–tension relationship. For example, mammals have longer thin filaments than amphibians or fish, thus their SL–tension relationship is shifted to longer SL (see Sosnicki *et al.*, 1991; Rome and Sosnicki, 1991).

Activation and relaxation of muscle involve the properties of the thin filament regulatory proteins [troponin, (Tn) and tropomyosin] as well as the sarcoplasmic reticulum (SR). The rate of activation and relaxation can be influenced by the kinetics of Ca^{2+} binding to and release from troponin. The rate of activation and relaxation can also be influenced by the rate of Ca^{2+} release and uptake by the SR. There is quantitative variation in the densities of Ca^{2+} release sites and pumps among different muscles and animals. There is also qualitative variation in Ca^{2+} pumps and troponin C that influences the specific Ca^{2+} pumping rate and the kinetics of Ca^{2+} release from Tn, respectively.

2. Cellular Level

At the level of the cell, there is large quantitative variation in the overall density of the SR (Appelt *et al.*, 1991). Slow fibers have a low density of SR, whereas fast fibers have a high density of SR. This difference underlies, to some extent, the fast-twitch and slow-twitch capabilities of the fibers. There can also be quantitative variation in the density of metabolic machinery, such as mitochondrial volume and activity of glycolytic enzymes. This variation is discussed in other papers.

3. *Organ Level*

The mechanical effect a given muscle fiber has on the locomotion of an animal depends a great deal on the gearing of the muscle fiber. The gear ratio of the fiber (which we have defined as Δ body movement/ ΔSL; see Rome and Sosnicki, 1991) depends on both the orientation of the muscle fibers within the muscle (whether parallel, pennate, or helical) and the positioning of the origin and insertion of the muscle (i.e., the moment a muscle makes around a particular joint).

Finally, the nervous system plays an extremely important role in the functioning of the muscular system. Motor neurons innervate muscle fibers of the same type to form a motor unit (Henneman *et al.*, 1965). The function of the muscular system is controlled by the nervous system (by recruitment), but the muscle can only act within the bounds set by the functional parameters of the fibers mentioned in the preceding. What the nervous system can do, however, is to control which types of muscle fibers are active at any one time, and it is the active fibers that define the muscular system. For instance, the nervous system can recruit different muscle fibers with different V_{max} to power the different movements for which the fibers are functionally suited. Hence, the nervous system endows the muscular system with the ability to instantly alter its functional capabilities.

B. Approach

Although we know what the important components of the muscular system are and how they vary, we do not know how the system is designed. To evaluate how the muscular system is designed requires us to understand how the various components are integrated into the system. More specifically, given the variation that occurs in each component, we must ask by what rules are the values of these parameters chosen?

Although there have been many excellent studies on isolated muscle and many others on animal locomotion, there have been very few studies in which the appropriate measures have been made on both the isolated muscle and the whole animal. In this paper we restrict our attention to this last, but rare, type of experiment, in the belief that this these studies will improve our insight into muscle design.

Over the past nine years, my laboratory has performed both isolated-muscle and whole-animal experiments in an effort to elucidate:

1. Whether there are identifiable rules by which values for various components of the muscular system are set to enable an animal to perform the wide range of activities of which it is capable;
2. Whether or not these rules hold for the great variety of locomotor behaviors in which vertebrates engage.

We have used two complementary intellectual approaches to address these questions. First we asked: Based on our knowledge of muscle function, how would we design the system? This approach used by itself is flawed, because there is no way to know what the design goals are. The second approach is to take advantage of the fact that animals are already designed, and by examining how muscle is used *in vivo*, try to discern what pattern was followed. This approach by itself is very difficult if one does not know what to examine. Combining these approaches has been very effective, that is, use Approach 1 to guess what the design rules might be, then use Approach 2 to test whether or not this really is the case.

We have also used both integrative and comparative approaches in our research. First, to elucidate the rules by which the system is designed, it has been necessary to perform experiments on the subcellular, cellular, organ, and whole-animal levels of organization. This integrative approach has only been possible because of the extremely orderly, almost crystalline structure of muscle. By knowing both the macroscopic and microscopic structure of muscle, we can calculate from gross body movements how the thick and thin filaments move past one another, which is the level at which force generation takes place. Second, it is difficult to find suitable animal models on which we can perform experiments at all levels of organization, and thus we have used a comparative approach to find the best model for attacking each particular problem.

C. What Are the Rules?

From cell physiology, we may anticipate that there are rules (Fig. 1) that are followed when an animal muscular system is designed. During steady activation, the force that muscle generates depends on the amount of filament overlap between myosin and actin, or more precisely, the number of myosin cross-bridges that can interact with actin sites (Gordon *et al.*, 1966). It would seem sensible for animals to vary

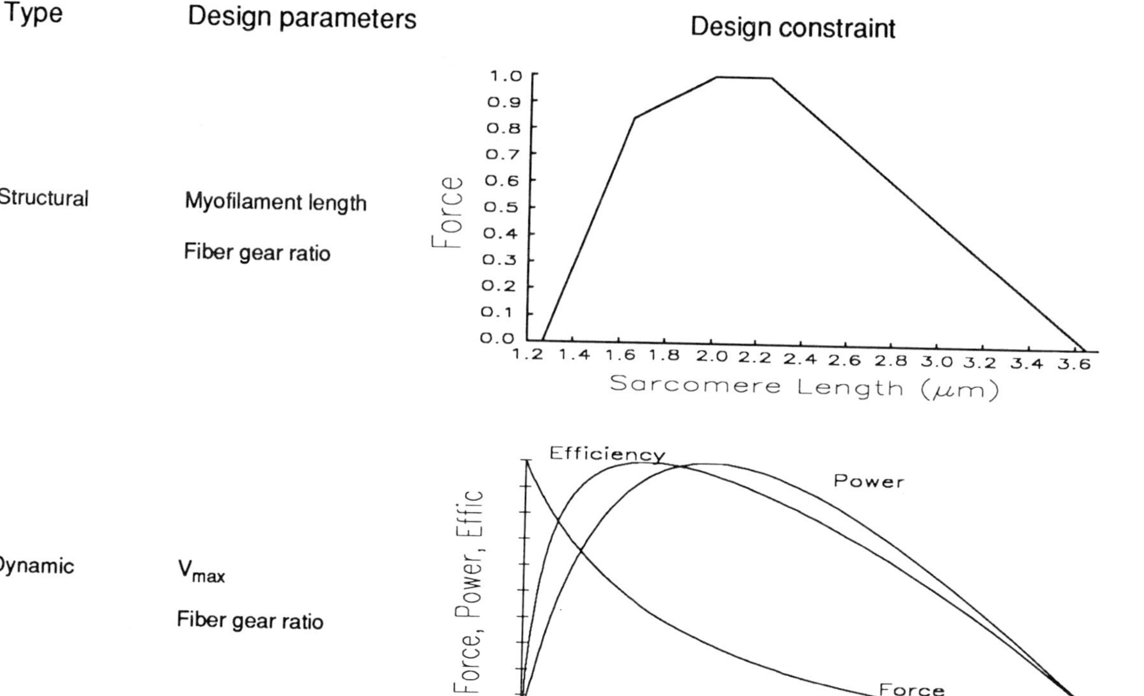

FIG. 1. For two design considerations, the important design parameters (muscle properties that can be varied during evolution) and the *potential* design constraints are shown (system values that are kept constant).

the gear ratios of their muscle fibers and their myofilament lengths so that no matter what movements the animal makes, the muscle would operate at optimal myofilament overlap (i.e., where the muscle generates near-maximal force). As such, gear ratios and myofilament lengths can be viewed as the *design parameters* (those components that can be varied during evolution). Myofilament overlap can be viewed as a *design constraint* or *design goal* (i.e., what has to remain constant or the rule by which the variation in parameters is adjusted). As both design parameters are anatomical features of the muscle, one being at the organ level and the other at the molecular level, these factors can be viewed as a *structural* design consideration.

There is also a dynamic design consideration that accounts for the fact that muscle shortens during locomotion. The force that muscle generates is a function of V/V_{max}, where V is the velocity of shortening. More importantly, the mechanical power that a muscle generates and the efficiency with which it generates the mechanical power are functions of V/V_{max} as well. Again we might anticipate that the muscular system would be designed in such a way that no matter what movement the animal makes, the muscle fibers operate over a range of V/V_{max} values (0.15–0.40) for which the fibers generate maximal power at maximum efficiency. Thus, the design parameters V_{max} and fiber gear ratio are varied in such a way that they operate under the design constraint of V/V_{max}.

It is important to emphasize that myofilament overlap and V/V_{max} are *potential* constraints and are derived exclusively from experiments on *isolated* muscle. We needed to determine whether animals actually use their muscles over this narrow range of values during their full range of locomotion, a determination that had never been made. Although it would seem reasonable that the muscle is used in this fashion, the following considerations show that it does not have to be. First, isolated muscle can be used over most of its SL range and its full range of V/V_{max} without damage. Second, cells are subservient to the goals of the animals, not the other way around. Survival of the animal may require the cells to work under conditions in which cell physiology suggests they may not be most effective.

D. Different Locomotor Behaviors May Require Different Muscle Designs

Different types of animals face very different problems during locomotion, thus their muscles must perform different activities. This is an important issue, because the potential design constraints just listed formally refer only to shortening muscle, whereas other types of muscle contractions (e.g., isometric or lengthening) may not necessarily follow these constraints.

The types of muscle contractions involved in different locomotor behaviors can be easily categorized based on the medium in which the animal is locomoting, and whether its movements are cyclical or singular. Swimming and flying differ from terrestrial locomotion in locomotor mechanics and in types of muscle contractions. The fundamental difference between them is the amount of drag the animals face from the medium. During swimming and flying, muscles must generate substantial mechanical power to overcome drag (force of drag × velocity of movement). Because only shortening muscle can generate mechanical power, the foregoing constraints are likely to be important.

During terrestrial locomotion, however, the drag from air is inconsequential at most running speeds. Work, however, is required to decelerate the center of mass and limbs and subsequently reaccelerate them. This is done by muscles first performing negative work (i.e., muscle lengthening during activation to decelerate the animal) and subsequently performing positive work to reaccelerate the animal. During lengthening contractions, the muscle and tendons may store energy elastically and then return it to the animal in the propulsive phase. Although the shortening phase of the contraction should come under the constraints listed, the lengthening and isometric parts of the contraction might not. Thus, the design of the muscle might be more influenced by other constraints, yet the lack of information on the mechanics and energetics of lengthening contractions makes it difficult to postulate what these constraints might be.

One exception to the preceding description of muscle behavior during terrestrial locomotion is "one-shot" locomotion. One-shot locomotion is exemplified by the frog's jump. As opposed to steady locomotion, in a jump there is a rapid acceleration of the animal's mass (increase in kinetic energy), hence power must be produced by the muscle. Again, because only shortening muscle generates mechanical power, muscle shortening is essential for this form of locomotion.

Because the potential design constraints mentioned refer formally only to shortening muscle, our research to test these design constraints has focused on animals that primarily use shortening contractions, that is, fish and frogs.

III. Design Constraint 1: Myofilament Overlap

The simplest design constraint, and one that researchers have considered for many years, is myofilament overlap (the SL–tension relationship). Because of the sliding filament structure of muscle, muscle generates maximum force over a fairly narrow range of SL (Gordon *et al.*, 1966). Researchers have generally assumed that the muscle is used only over those SL, but there has been relatively little evidence to justify this assumption. If myofilament overlap is a design constraint, it raises an interesting question: How do animals produce such a wide range of movements if their fibers always operate at optimal myofilament overlap?

A. Fish Studies

Our recent experiments on carp provide the most extensive study of these issues. To swim, a fish must bend its backbone. By a combination of high-speed motion pictures and anatomical approaches, we have found that at low swimming speeds, the red muscle (Fig. 2), which

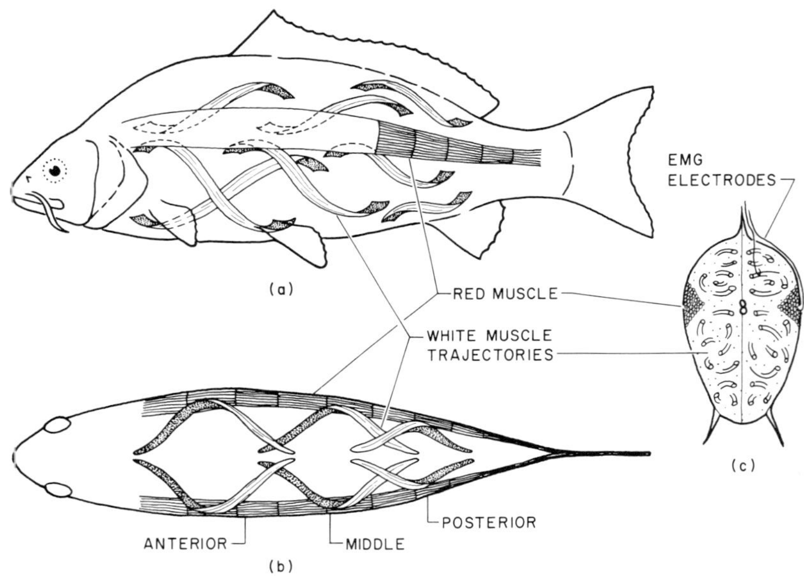

FIG. 2. Longitudinal view (a), dorsal view (b), and cross-sectional view (c) of carp. The red muscle is a thin sheet of muscle just under the skin, extending to a depth of only 10% of the distance to the backbone (the cross section of the red muscle is exaggerated for illustrative purposes). Because the red fibers run parallel to the body axis, SL excursion depends on both curvature of spine and distance from the spine. The trajectories of the white muscle fibers shown in a and b are based on Alexander's (1969) description. The white fibers lie closer to the median plane than the red ones, and they run helically rather than parallel to the long axis of the body. Consequently, they shorten by only about one-fourth as much as the red fibers for a given curvature change of the body (see text). Placement of electromyography (EMG) electrodes used to determine the activity of the red and white muscles are shown in c. Reproduced from Rome *et al.* (1988).

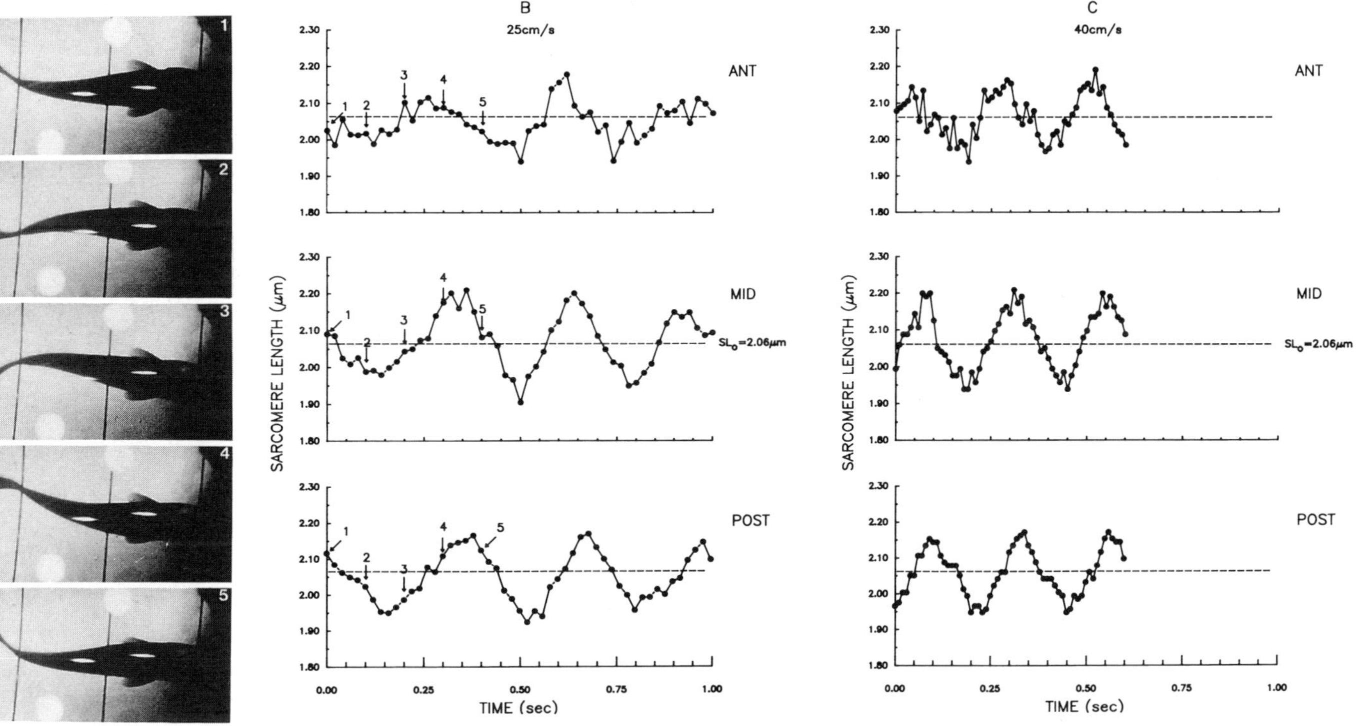

FIG. 3. Typical record showing changes in sarcomere length in anterior, middle, and posterior positions of a fish swimming at 25 and 40 cm s^{-1}. Five frames (A) separated by 0.1 s are shown; numbers on the photographs correspond to data points in the 25 cm s^{-1} graphs (B). For comparison, the corresponding sarcomere length versus time graph is shown for the same fish swimming at 40 cm s^{-1} (C). Note the reduction in noise and increased sarcomere length excursion in the anterior sections of the fish at the higher speed. Sarcomere length excursion (average difference between shortest and longest lengths measured in a sequence) was calculated from the amplitude of the graph. Muscle velocity was calculated from the slope of the graph. Reproduced from Rome *et al.* (1990a).

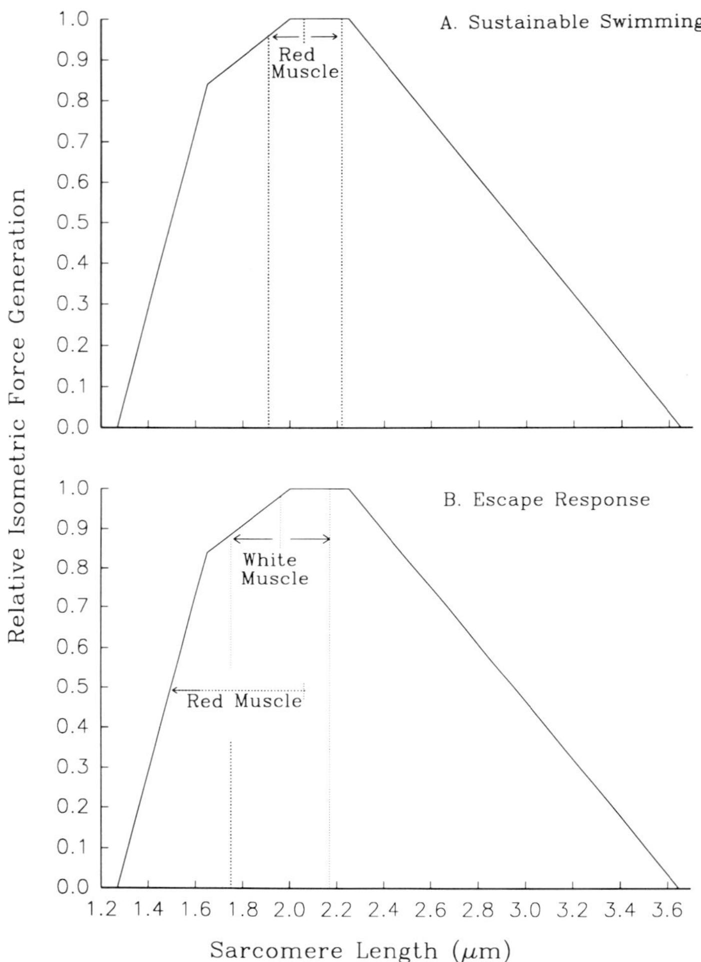

FIG. 4. Design constraint 1: myofilament overlap. During all movements, muscle is used at nearly optimal myofilament overlap. During steady swimming (A), carp uses red muscle over a SL of 1.91–2.23 μm, where no less than 96% maximal tension is generated. If the red muscle were to power the more extreme escape response (B), it would have to shorten to 1.4 μm, a length at which it generates little tension and can be damaged. Instead, the white muscle with its fourfold higher gear ratio is used. In the posterior region of the fish, the white muscle shortens to only 1.75 μm, where at least 85% maximal tension is generated. In the rest of the fish the excursion is smaller and the force higher. Reproduced from Rome and Sosnicki (1991).

powers this movement, undergoes cyclical SL excursions between 1.89 and 2.25 μm, centered around a sarcomere length of 2.07 μm (Fig. 3; Rome *et al.*, 1990a). Further, we determined from electron microscopy that thick and thin filament lengths of the red (1.52 and 0.96 μm) and white (1.56 and 0.99 μm) muscle in carp are similar to those in frogs (Sosnicki *et al.*, 1991). Using the frog SL–tension relationship, the red muscle was shown to be operating over a range of SL where no less than 96% maximal tension is generated (Fig. 4; Rome and Sosnicki, 1991).

We then examined the most extreme movement that carp make, the escape response (pictured in Fig. 5). This response involves a far greater curvature of the backbone than does steady swimming. *If* the red muscle were powering this movement, it would have to shorten to a SL of 1.4 μm, at which low forces are generated and even irreversible damage can occur (Fig. 5B; Rome *et al.*, 1988; Rome and Sosnicki, 1991). Instead, it is the white muscle that performs the movement, because the white muscle has a different orientation than the red. The red muscle fibers run parallel to the long axis of the fish (Fig. 2), just beneath the skin. In contrast, the white muscle fibers run in a helical orientation with respect to the long axis of the fish. Alexander (1969) predicted that the helical orientation would endow the white fibers with two features that are crucial to their function.

First, the helical pattern would enable the white fibers to be at the same SL irrespective of their distance from the backbone. If the fibers ran parallel to the long axis, those far from the backbone would shorten to much smaller SL (and at a much faster V) than those close to the backbone. This would make it impossible for all the white fibers to work at optimal myofilament overlap and optimal V/V_{max}. Second, this helical orientation would endow the white muscle with the higher gear ratio necessary to power extreme movements.

We have experimentally verified both of Alexander's predictions (Rome and Sosnicki, 1991). Figure 6 shows cross sections of bent fish, which illustrate two important points. First, the SL in the white muscle is identical both close to and far from the backbone, so that the white fibers work as a unit. Second, comparing the left and right sides of the fish, the SL difference (which is equivalent to the SL excursion) for the white muscle is much smaller than that for the red. Hence, the white muscle can produce a given backbone curvature with a much smaller SL excursion than would be required for the red, and hence it has a higher gear ratio (in fish, defined as Δ backbone curvature/ΔSL).

Therefore, for a given backbone curvature, the white muscle undergoes only about one-fourth the SL excursion of the red (i.e., it has a

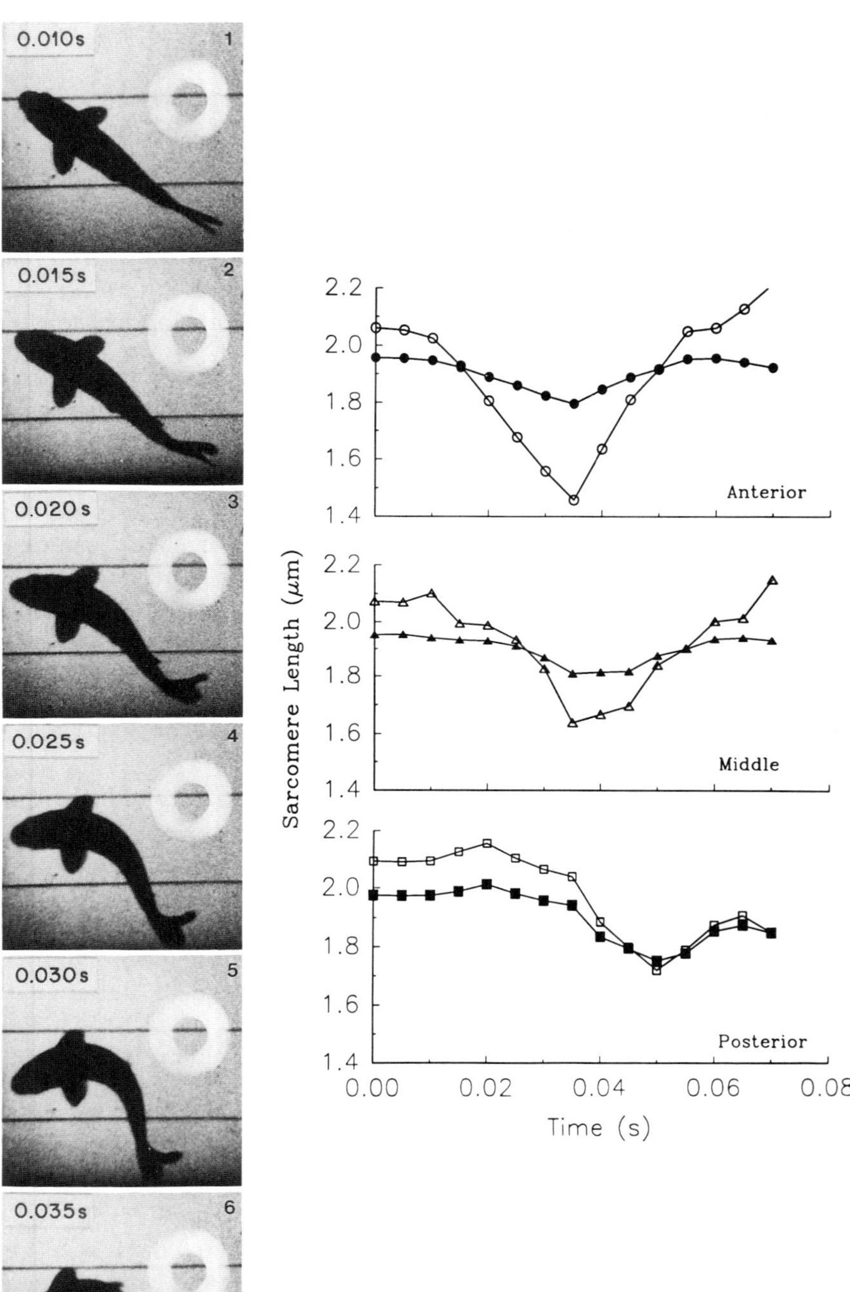

0.010s
1
0.015s
2
0.020s
3
0.025s
4
0.030s
5
0.035s
6
Sarcomere Length (μm)
Anterior
Middle
Posterior
2.2
2.0
1.8
1.6
1.4
0.00
0.02
0.04
0.06
0.08
Time (s)

four-fold higher gear ratio). To power this most extreme movement of fish, in the worst case (posterior), the white muscle must shorten to a SL of 1.75 μm (Fig. 5, posterior). At this SL, the muscle generates about 85% maximal force. However, most of the volume of the white muscle (middle and anterior sections) does not shorten as much, and therefore generates even more force (95% for anterior and 92% for middle). When the white muscle is used in less extreme movements (i.e., during fast swimming), the curvature of the backbone is not nearly as great, and the white muscle generates nearly maximal force (Rome and Sosnicki, 1991).

As we have shown, the myofilament overlap is never far from its optimal value, even during the most extreme movements. Therefore, it appears that animals are designed in such a way that *no matter what the movement*, the muscles used generate nearly optimal forces. As such, myofilament overlap can be considered a design constraint (i.e., a part of the system that is kept constant). Given the movements that animals need to make, animals adjust two design parameters, fiber orientation and myofilament lengths, as they evolve so that the muscle fibers being used always operate at near-maximal myofilament overlap and force generation (see Fig. 16).

B. Frog Studies

Because frog jumping requires a one-shot movement rather than cyclical contractions as in fish, frogs were used as a model to determine whether the myofilament overlap design constraint is unique for cyclical locomotion or is more general. Lutz and Rome (1994) have addressed the question of where on the SL–tension curve fibers operate during a maximal jump on the semimembraneous muscle (SM) of *Rana pipiens*.

During a maximal jump, a frog accelerates from a stationary crouched position to high vertical and forward velocities in under 100 ms (Calow and Alexander, 1973; Hirano and Rome, 1984). To produce

---

Fig. 5. The startle response of a carp. A resting carp received a 100-ms, 150-Hz sound pulse through an underwater speaker in the aquarium about 30 cm from the fish. The response was filmed at 200 frame s^{-1}, and six consecutive frames (separated by 5 ms) are numbered and shown on the left. The SL excursions of the white muscle in the anterior, middle, and posterior positions are shown on the right. The SL excursion of the white muscle is greatest in the posterior, because here the backbone undergoes its largest curvature. Note that the red fibers *do not* actually shorten to the SL shown because they can not shorten fast enough. Adapted from Rome *et al.* (1988) and Rome and Sosnicki (1991).

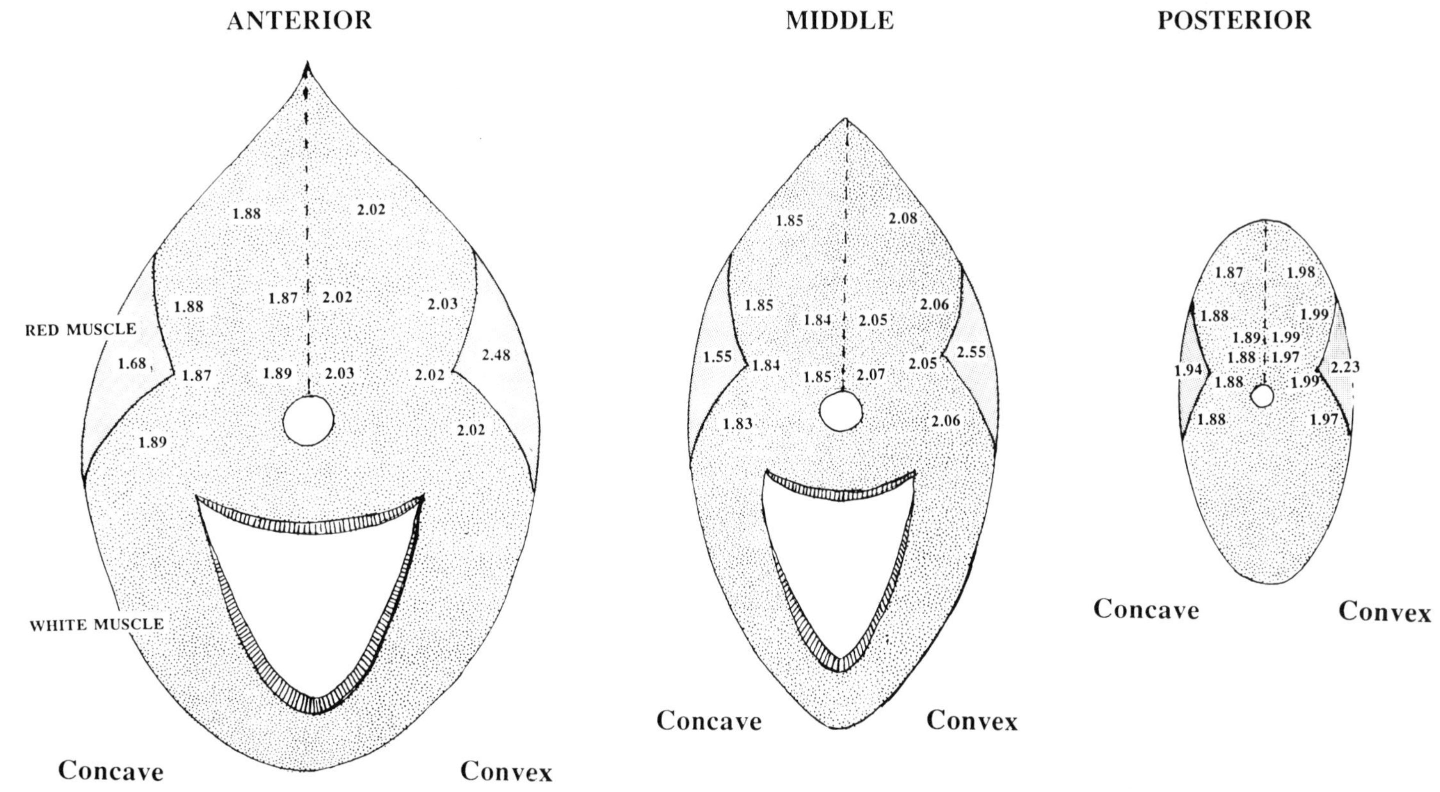

FIG. 6. SL distribution in cross sections from the three positions along the craniocaudal axis of the fish. The cross sections of the anterior, middle, and posterior regions of a bent fish are shown, as well as the SL at different points in the section. Unlike the red muscle, the white muscle SL excursion did not depend on distance from the backbone. There was little variation in SL of the white muscle in a given section, and there was no consistent pattern to the variation, suggesting that white muscle operates as a unit. At all three positions, however, the SL of the red muscle is very different from that of the white. Reproduced from Rome and Sosnicki (1991).

these rapid increases in potential and kinetic energy, the muscles of the frog must generate high mechanical powers. Ultimately, the distance frogs travel is directly proportional to power production (Hirano and Rome, 1984).

For the muscles of the frog to generate their maximum mechanical power, they must (1) operate over the plateau of the sarcomere length (SL)-tension curve where maximum force is generated, (2) operate at the optimal V/V_{max} where maximum power is generated, and (3) be maximally activated.

By a combination of high speed motion pictures, anatomical techniques, and EMGs, Lutz and Rome determined the SL changes and stimulation pattern the SM undergoes during frog jumping (Fig. 7A,B). Because frog muscle is more tractable than that of fish, tension, and active SL could be measured directly. This showed that the standard frog SL-tension curve (Gordon *et al.*, 1966) fit the SM data, and hence it was used to represent the SL-tension curve of the SM (Fig. 8A). During jumping, the SM operates mostly over the plateau where maximum force is generated. This suggests that during evolution the muscle origin, insertion, and angle of pennation have been adjusted to give the appropriate gear ratio (Δ body movement/Δ SL) so that fibers shorten mainly over the plateau during jumping.

The fact that the SM shortened somewhat beyond the plateau of the SL-tension curve deserves some attention. First, it never produced less than 90% of the force generation at optimal SL. In addition, during jumps, force and power fell rapidly just prior to takeoff, so that very little power was generated at the shorter SL's; most of the power was generated over the plateau of the SL-tension relationship. Nonetheless, because the total mechanical energy change during a frog jump depends on the amount of work (force $\times$ Δlength) performed by the muscles, shortening on to the ascending limb of the curve may increase overall work performed (by increasing Δlength [strain] while incurring only a small loss of force).

C. Other Species

Data from several other species support our contention that myofilament length is an important design constraint in muscular systems. Dimery (1985) calculated that in galloping rabbits fibers undergo SL excursions of 1 μm (1.7 to 2.7 μm), over which range there would be no less than 80% maximal tension generated (N.B.: Mammalian fibers have longer thin filaments than do frogs and fish, hence their SL–tension relationship is shifted to longer SL.) Cutts (1986), using simi-

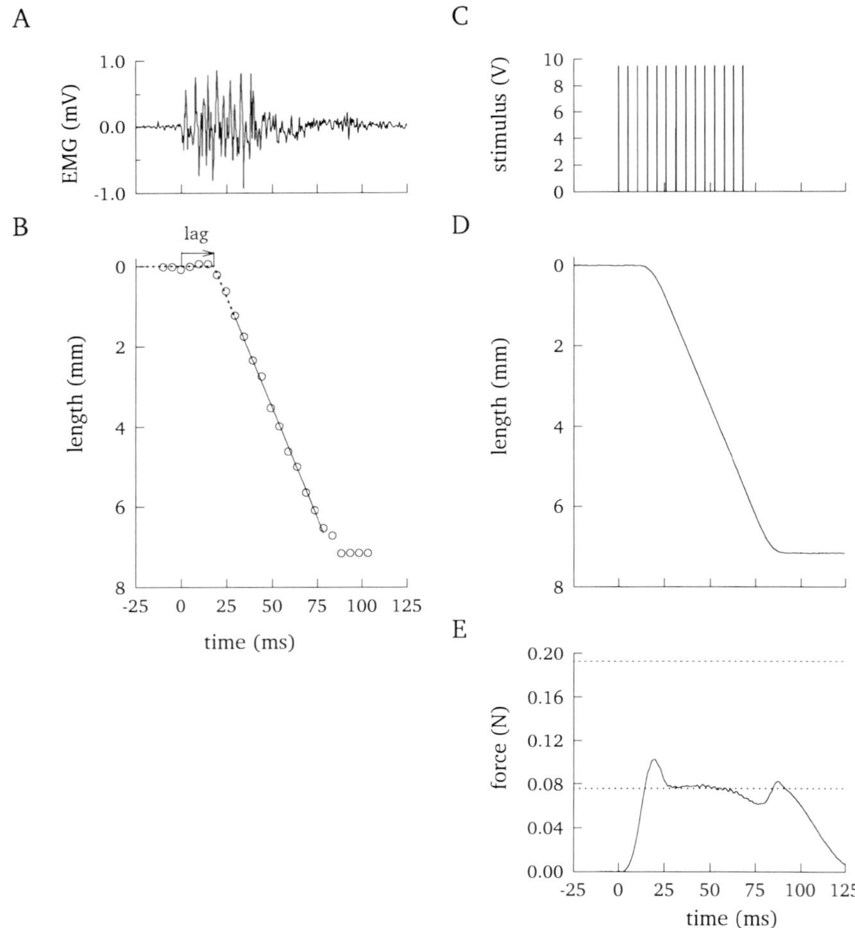

FIG. 7. Muscle function during jumping in frogs. Parts A and B, show the length change and stimulation pattern the SM undergoes during a maximal jump. Parts C and D, show an isolated muscle bundle driven through the *in vivo* length change and stimulation pattern, and E shows the resulting force production of the muscle. Isolated muscle bundles were stimulated at either 200 or 120 pps, but this had only a minor effect on the results. The stimulation duration was determined form the EMG. The phase of the stimulus with respect to the length change was determined in the following fashion. The initiation of shortening was determined by extrapolating the constant velocity portion of the length record back to zero length (B). The lag betewen the stimulus and shortening was defined as the time between the onset of the EMG and the initiation of shortening. Because the early portion of the length record was curved during jumping, digital smoothing was used to obtain the correct shape of the computer generated length change (D). The dashed line (E) is isometric force and the dotted live is the steady-state force generated by the same muscle at the same V during a force–velocity experiment (see Fig. 2). The jump shown in A and B is the longest measured (distance = 0.8m, V = 3.78 ML s^{-1}), and was reproduced in C–E. Reproduced with permission from Lutz and Rome, 1994.

A

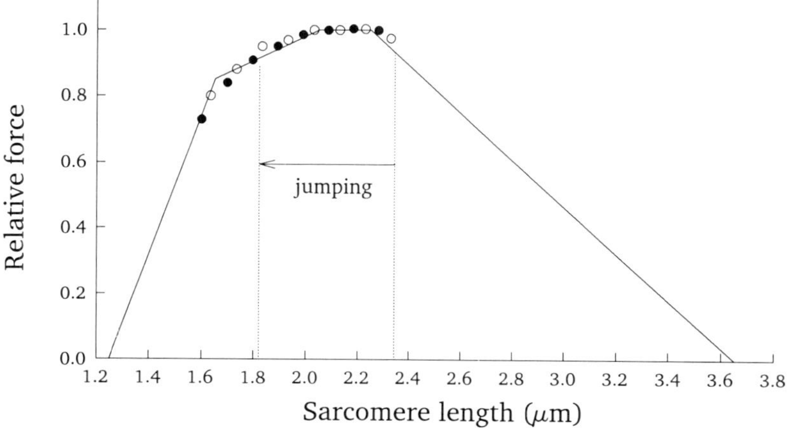

B

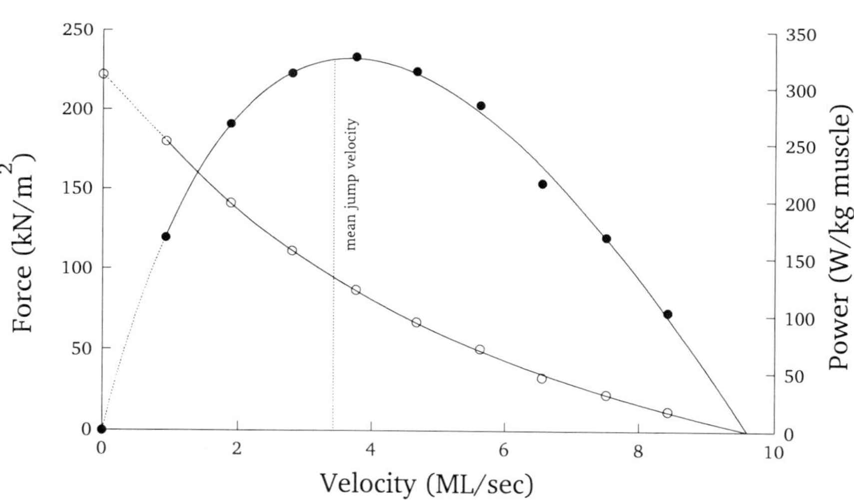

FIG. 8. Where does the SM muscle operate on its SL-tension and force–velocity curves? (A) the open symbols represent the results from one experiment, the solid symbols represent the results from the other. The SL-tension relationship was not studied at SL > 2.35, because frogs do not use these during jumping and because of well known experimental problems associated with "fixed-end" contractions at long SL's (Gordon *et al.*, 1966). The muscle shortens from an SL of 2.34 to about 1.83 μm at take off. Note that during jumps, force and power fell rapidly prior to take off (data not shown), so that the most power was generated at somewhat longer SL. Even at 1.83μ, however, the muscle still generated over 90% tension. (B) shows a typical force–velocity and power–velocity curve. The power curve was simply calculated from the force-velocity fit. At the *V* used during jumping, the muscle operates over the portion of the power curve where at least 99% of maximum power is generated. Reproduced with permission from Lutz and Rome, 1994.

lar techniques to Dimery (1985), concluded that bird muscles operate over about 0.5-μm excursions (1.7 to 2.2 μm). Both results appear to be centered around the plateau region of their respective SL–tension curves.

IV. Design Constraint 2: V/V_{max}

A. Why Is V/V_{MAX} Important?

V_{max} varies greatly in muscle, therefore, we have concentrated our focus on this design parameter and on V/V_{max}. Differences in V_{max} have been found in muscles from different animals, in different muscles in a given animal, in a given muscle at different temperatures, and recently, with the advent of single-muscle fiber techniques, in different muscle fiber types within the same muscle. To a first approximation, fibers with different V_{max} generate the same maximum isometric force per cross section and have the same maximum efficiency, whereas the maximum power generated and rate of ATP splitting in a fiber with high V_{max} are considerably greater than in a fiber with low V_{max} (Fig. 9; *N.B.*: One ATP is split for each cross-bridge going through the attachment–detachment cycle).

However, from Hill's work (Hill, 1938, 1964) we know that a muscle fiber's mechanical properties (force generation and power production) and energetic properties (ATP utilization and efficiency) are not simply a function of the fiber's V_{max}. They also depend on V/V_{max}. According to the Huxley 1957 model, the mechanical and energetic properties of the fiber depend on V/V_{max} as well.

Figure 9 shows the force, power output, rate of energy utilization, and efficiency of muscle fibers with two different V_{max} values. For a given velocity of shortening, the force and mechanical power per cross-sectional area can be considerably higher in the fiber with a high V_{max} (at V_1 they are quite similar, whereas at V_2 they are quite different; Figs. 9A and 9B). Therefore, to generate a given force or power, an animal need use fewer fibers with high V_{max} than fibers with low V_{max}. It would thus seem advantageous to only have muscle fibers with high V_{max}. However, there is an energetic price to be paid for a high V_{max}. In Fig. 9C, the rate of ATP utilization in the fiber with high V_{max} is considerably greater than in the fiber with low V_{max} at all velocities of shortening. Therefore, there appears to be an adaptive balance between the mechanics and energetics of contractions. This balance can be best seen in Fig. 9D. The fibers with low V_{max} are more efficient at

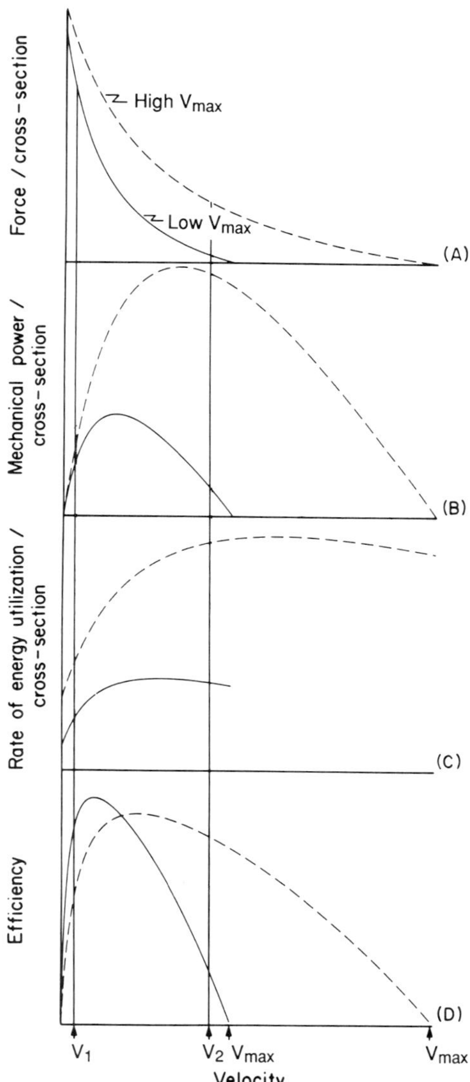

FIG. 9. Relative force, power, rate of energy utilization, and efficiency (A–D) as a function of relative shortening velocity for a muscle with a high V_{max} (dashed curves) and a muscle with a low V_{max} (solid curves). Respective V_{max} values are shown on the velocity axis. V_1 and V_2 are arbitrarily chosen examples of low and high shortening velocities. Values for curves are derived from heat, oxygen, and mechanics measurements on frog muscle (Hill, 1938, 1964; Rome and Kushmerick, 1983).

low V (e.g., V_1). At higher velocities (*e.g.*, V_2), however, the fibers with high V_{max} are more efficient. Thus, to produce both slow movements and fast movements efficiently, the animal should use the fibers for which V_{max} is matched to the V at which it needs to shorten.

If V/V_{max} is a design constraint, we would predict that a given muscle fiber type would be used over a range of V/V_{max} of about 0.15 to 0.40, over which efficiency and power output are maximized. As a result, the two design parameters, V_{max} and gear ratio, should be set so that during body movement V falls within this range of V/V_{max}. If this in fact happens, how do animals produce such a wide range of speeds?

When we made these predictions, V/V_{max} had never been measured. To determine V/V_{max}, one must determine:

1. Which muscle fibers are active during a given activity;
2. The V at which the fibers shorten;
3. The V_{max} of the different fiber types.

It was the necessity of making the first of these measurements that led us to work on fish. Because of the anatomical separation of the different muscle fiber types in fish (Fig. 2), it is possible to monitor activities of different fiber types with electromyography (EMG). For instance, we know from EMG that at slow speeds only the red fibers are active, and at fast speeds the white muscle fibers are recruited as well (Fig. 10). Thus, by knowing the speed and temperature at which the fish is swimming, we know which muscle fiber types are actually being used.

This anatomical separation of muscle fibers also helps in other regards. First, it is relatively easy to isolate bundles of fibers that are *pure* in fiber type on which to measure V_{max} and other mechanical properties. This is a great aid, because in most other animals the only way to obtain a pure fiber type is to dissect single fibers. With the exception of frogs, fragility of the cell membrane necessitates using "skinned" single fibers, in which the cell membrane has been removed or permeabilized. Skinned fiber experiments are not optimal for determining functional ability during locomotion because it is not possible to examine the kinetics of activation and relaxation. Second, as illustrated with regard to myofilament overlap, the anatomical separation of fibers enables us to define two separate gear ratios, one for the red muscle and one for the white, rather than a continuum as in other animals. Thus, V can be measured from the slope of the SL–time graph (Figs. 3 and 5).

To test the importance of V/V_{max} as a design constraint, we have

---

FIG. 10. Electromyograms from red (R) and white (W) muscle of carp at 10° and 20°C. Reproduced from Rome *et al.* (1984).

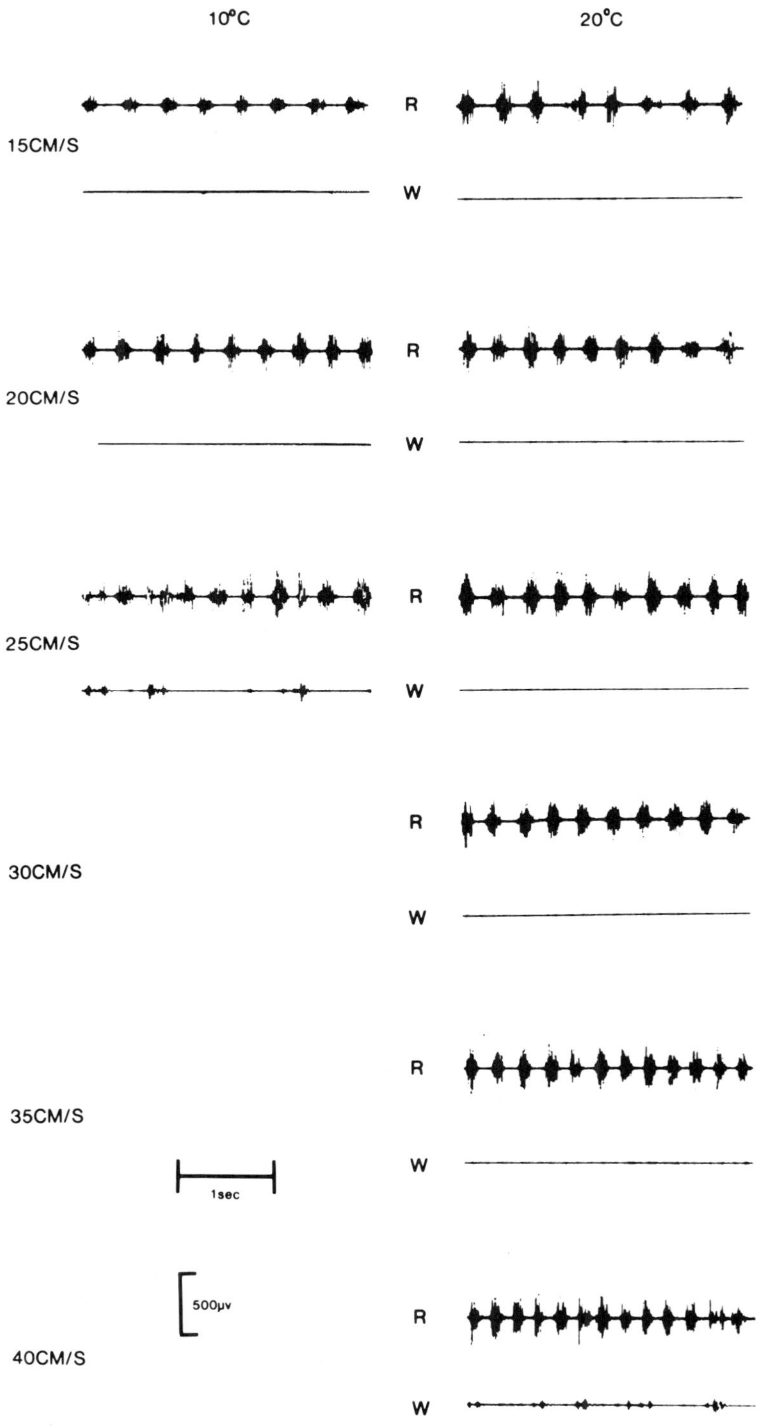

10°C

20°C

15CM/S

R

W

20CM/S

R

W

25CM/S

R

W

30CM/S

R

W

35CM/S

R

W

1sec

500μv

40CM/S

R

W

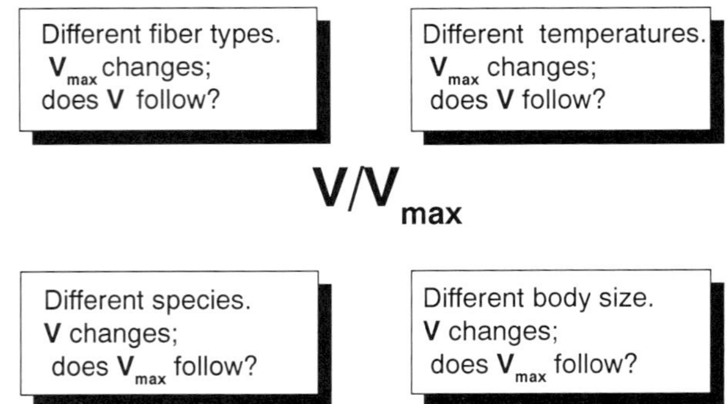

FIG. 11. The importance of V/V_{max} was examined in four different situations. In each case there was an anticipated change in either the numerator (V) or denominator (V_{max}), and we tested to see if a concomitant change occurred in the other parameter so that V/V_{max} remained constant.

examined this parameter in five situations (four of which are shown in Fig. 11). Each situation was chosen to change either the numerator, V, or the denominator, V_{max}, to see if there was a concomitant change in the other parameter to maintain constant V/V_{max}.

B. Different Fiber Types in Carp

The first question we asked is why do animals have different fiber types (i.e., different V_{max} within the same animal)? Are the faster fibers used for faster movement (higher V) so that they operate at the same V/V_{max}?

As illustrated in Fig. 12, we found that the V_{max} of carp red muscle was 4.65 muscle length s^{-1} (ML s^{-1}), and the V_{max} of white muscle was 2.5 times higher, 12.8 ML s^{-1} (Rome *et al.*, 1988). During steady swimming, the red muscle is used over velocity ranges of about 0.7 to 1.5 ML s^{-1} (Fig. 12A, shaded part of the curve; Rome *et al.*, 1990a). This corresponds to V/V_{max} of 0.17–0.36, which is where maximal power is generated. At higher swimming speeds (higher V) the fish recruit white muscle because the mechanical power output of the red muscle actually decreases.

It is clear from Fig. 12A that the red muscle cannot possibly power the escape response. To power the escape response the red muscle would have to shorten at 20 ML s^{-1}, which it clearly can not do, as this is four times its V_{max}. But even if the white muscle, with its higher V_{max}, were placed in the same orientation as the red (i.e., same gear ratio), it could not power the escape response either, because its V_{max} is only about 13 ML s^{-1}. However, because of its fourfold higher gear ratio in its normal orientation, the white muscle needs to shorten at

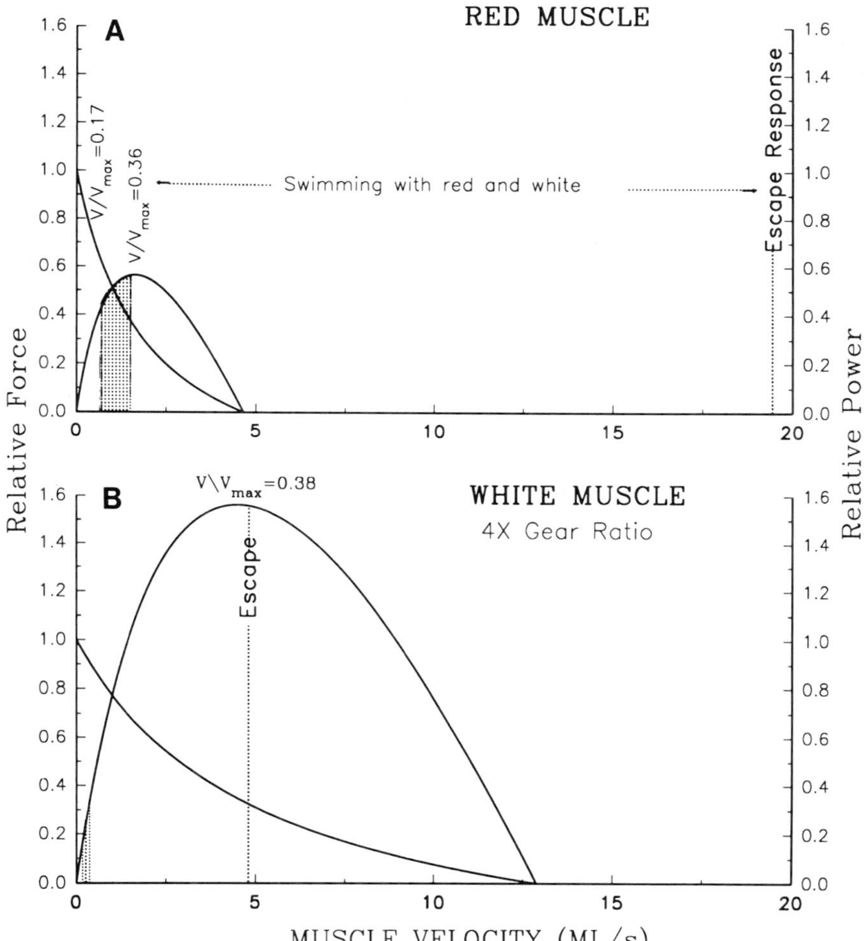

FIG. 12. Design constraint 2: V/V_{max}. During slow and fast movements the *active* fibers always shorten at V/V_{max} of 0.17–0.38, at which maximum power and efficiency are generated. During steady swimming (red muscle), the fibers are used at V/V_{max} of 0.17–0.36 (A). The red fibers cannot power the escape response because they would have to shorten at 20 ML s^{-1}, or four times their V_{max}. The escape response is powered by the white muscle, which only needs to shorten at 5 ML s^{-1} (V/V_{max} = 0.38) because of its fourfold higher gear ratio (B). The white muscle would not be well suited to power slow-swimming movements, as it would have to shorten at a V/V_{max} of 0.01–0.03, where power and efficiency are low. Thus, fast movements are obtained with fibers with high V_{max} and large gear ratio.

only 5 ML s^{-1} to power the escape response (Fig. 12B), corresponding to V/V_{max} of about 0.38, which is the value at which white muscle generates maximal power. Because all white muscle (both close to and far from the backbone) shortens the same amount, and thus with the same V and V/V_{max}, all the white muscle will generate maximal power. Therefore, the helical orientation of the fibers that makes the white muscle work as a unit is an important adaptation for maintaining an optimal V/V_{max}.

Red and white muscle forms a system with two gears that power very different movements. The red muscle powers slow movements, whereas the white muscle powers very fast movements, however, each works at the appropriate V/V_{max}. In terms of backbone curvature, the white muscle can produce 10-fold faster movements (2.5-fold higher V_{max} × 4-fold higher gear ratio). It is important to realize that the effectiveness of the white muscle gear depends on the product of its gear ratio and V_{max}. For instance, if the red muscle had the gear ratio of the white, it could not produce the escape response, but neither could the white muscle if it had the gear ratio of the red. The white muscle requires both the correct V_{max} and the correct gear ratio to produce very rapid movements.

As discussed earlier, the red and white muscles have a maximum V/V_{max} at which they can be used. However, there also appears to be a minimum V/V_{max} at which they are used, and this parameter is also an important determinant of muscle function. The importance of this fact is demonstrated by considering that if the white muscle produces fast movements so effectively, why doesn't the fish have only one fiber type and let the white muscle power the slow swimming movements as well? The white muscle certainly could power slow swimming, but it is not used for this because its high V_{max} and fourfold higher gear ratio would make its V/V_{max} at slow swimming speeds too low (i.e., 0.01–0.03, values for which the muscle is nearly isometric and efficiency is nearly 0; shaded portion of Fig. 12B). At slightly faster swimming speeds, when the white muscle starts to be recruited to augment the power of the red, the same problem exists. If the fish were to continue swimming steadily, the V of the white muscle would be only slightly faster than the shaded portion in Fig. 12B, and the white muscle would still be shortening with too low a V/V_{max} to generate power efficiently. Under these circumstances, however, the fish employs "burst and coast swimming," in which it makes rapid tail beats (i.e., short duty cycle) with the white muscle to obtain a high V and utilize optimal V/V_{max}.

At very slow swimming speeds (i.e., those below steady swimming, to the left of the shaded portion in Fig. 12A), carp also use the "burst and coast" pattern, but with recruitment of the red muscle. This pattern allows the fish's red muscle to make the normal sarcomere excursion in less time (shorter duty cycle) than the tail beat period predicted for steady swimming. As a result, the V and V/V_{max} will be higher, resulting in normal efficiencies. Thus, carp appear to adopt a "burst and coast" pattern of swimming (red at very slow speeds, white at fast speeds) when the muscle fiber type being used would have too small V/V_{max} to generate power with high efficiency if the fish were swimming steadily (Rome *et al.*, 1990a).

In summary, given the constraints at both high V/V_{max} and low V/V_{max}, to achieve a full repertoire of movement animals must have different fiber types with different V_{max} and different gear ratios. Therefore, V/V_{max} does appear to be an important design constraint in muscular systems (i.e., muscle operates over a narrow portion of the curve, 0.17–0.38, over which maximal power is generated at high efficiencies).

C. Different Muscle Temperatures in Carp

Raising the temperature of a muscle increases its V_{max}. If V/V_{max} is an important design constraint, then an animal should use its muscle over a higher range of V at a higher temperature. Indeed, the V_{max} of carp red muscle is 1.6-fold greater at 20° than at 10°C (Fig. 13; Rome and Sosnicki, 1990). We found that at a given swimming speed, the V at which the muscle shortens is independent of temperature (Rome *et al.*, 1990a, 1992a). However, carp at 20°C could swim to speeds of 45 cm s^{-1} using red muscle exclusively, whereas, at 10°C they could only swim to speeds of 30 cm s^{-1} with red muscle. The corresponding V were 1.6-fold higher at 20°C than at 10°C (2.04 versus 1.28 ML s^{-1}, respectively), resulting in the same V/V_{max} (0.36) at the respective maximum speed at both temperatures (Fig. 13, shaded portions). In addition, the 10°C fish could swim steadily at lower speeds than at 20°C, corresponding to a 1.6-fold lower V, and resulting in the same V/V_{max} at the lowest speed. Thus, at both 10° and 20°C, carp use their red muscle over the same range of V/V_{max} (0.17–0.36). Conversely, V/V_{max} seems to set the operating range of the red muscle at different temperatures and, hence, the range of swimming speeds. As such, it is a powerful determinant of locomotor behavior.

D. Fast-Swimming Species versus Slow-Swimming Species

At a given temperature, the marine scup can swim twice as fast using its red muscle as the freshwater carp (80 versus 45 cm s^{-1} at 20°C; Rome *et al.*, 1992a). We predicted that the scup's maximum V

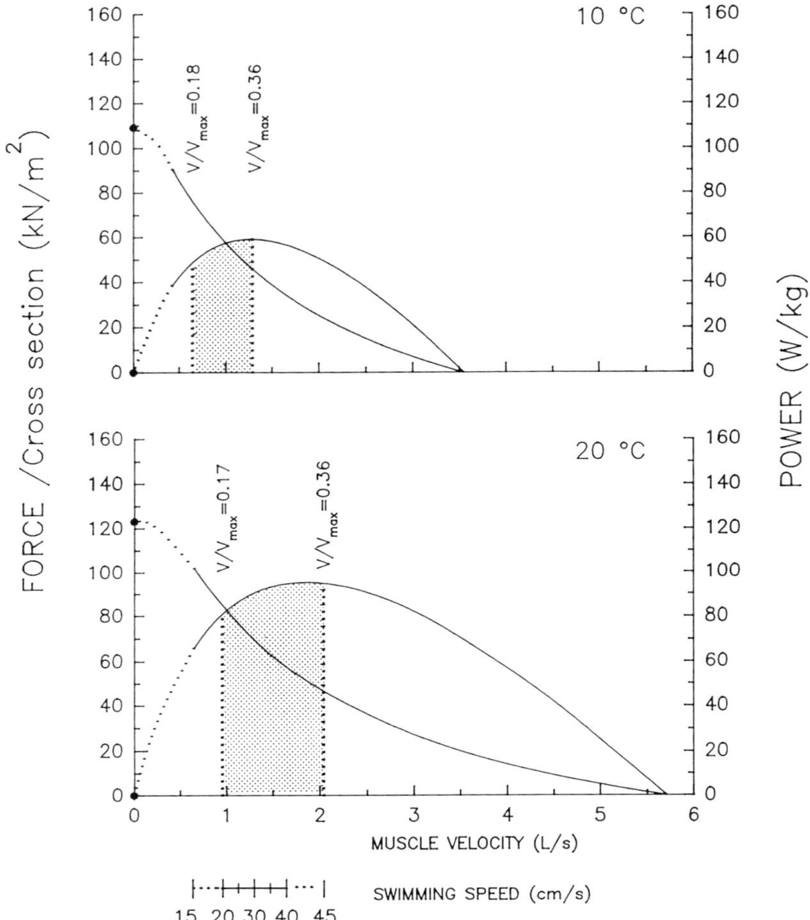

FIG. 13. Influence of temperature on mechanical properties of carp red muscle and on their use during swimming. This figure shows the average force–velocity and power–velocity curves of carp red muscle at 10° and 20°C, based on the results of Rome and Sosnicki (1990). Because muscle shortening velocity during steady swimming was independent of temperature, the swimming speed axis has been placed on the graph as well. Thus, during steady swimming, the curves provide the power, force, and muscle shortening velocity as a function of swimming speed. The shaded regions represent V during steady swimming with red muscle. The dotted vertical lines at each temperature represent transition swimming speeds. At slower swimming speeds than that of the leftmost line, the carp used "burst and coast" swimming with red muscle. At higher swimming speeds than that of the rightmost line, white muscle is recruited and the carp used "burst and coast" swimming. For each temperature, V/V_{max} at the transition points is given. Note that V/V_{max} over which the red muscle is used is the same at both temperatures. Reproduced from Rome *et al.* (1990a)

(i.e., the V while swimming at 80 cm s^{-1}) would be twice as high as the carp's maximum V (*i.e.*, while swimming at 45 cm s^{-1}). If maintaining V/V_{max} were important in muscle design, then we expected that the V_{max} of scup would also be twice as high as that of the carp.

Remarkably, because scup employ a less undulatory style of swimming than carp (*i.e.*, smaller backbone curvature and smaller SL excursion), at their respective maximal swimming speeds using red muscle, the V at which the muscles shorten are equal (about 2.04 ML s^{-1}; Rome *et al.*, 1992a). This finding led us to change our prediction, so that we now predicted that carp and scup should have the same V_{max}. This was exactly what we found (Rome *et al.*, 1992b). Hence, both the fast fish and the slow fish use their red muscle over the same V/V_{max}, even though it corresponds to much higher swimming speed in scup (Fig. 14).

Two additional points should be noted from these findings. First, fast-swimming fish do not necessarily need fast-contracting muscle. Second, scup should not necessarily be viewed as being better designed than carp because they can swim faster. Carp can swim steadily at much lower speeds than scup (Fig. 14), an ability that may be important to their niche.

E. Frog Jumping

During jumping, the muscles of frogs also work at the appropriate V/V_{max} (0.33) for maximum power generation. Lutz and Rome (1994) found that the V_{max} of the SM was 10.35 ML s^{-1} and that the V at which maximum power was generated was 3.44 ML s^{-1} (Fig. 8B) which matches closely the average V during jumping (3.43 ML s^{-1}; Fig. 7B). Although the SM is composed of the two fastest-twitch amphibian fibers [amphibian type 1 and type 2 (Smith and Ovalle, 1973)], there is relatively little difference (~20%) in V_{max} between these fiber types (Edman *et al.*, 1988; Lannergren, 1987). These results suggest that the V_{max} of both fiber types have been adjusted during evolution such that during jumping, the muscle operates at the appropriate V/V_{max} for maximal power generation.

F. Scaling of V/V_{max} with Body Mass in Mammals

The scaling of the mechanics and energetics of locomotion with body size is of considerable interest (see paper by Taylor in this Volume and by Lindstedt and Thomas in Volume 38B). Unfortunately, the mammalian muscular system is not well suited for determining V/V_{max}, as

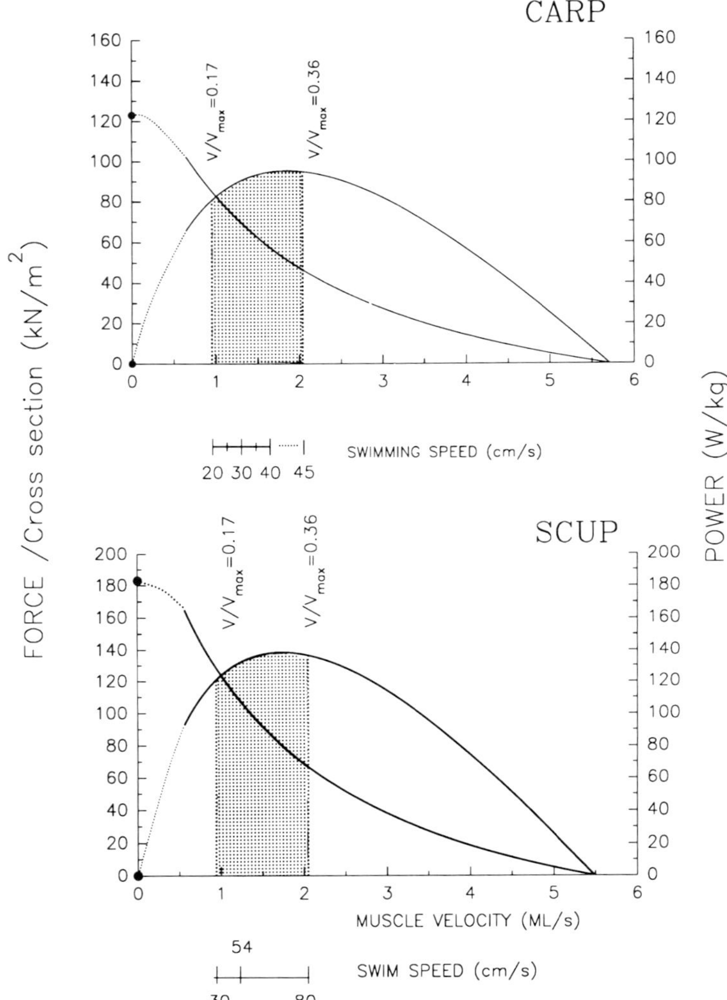

FIG. 14. V/V_{max} of the fast-swimming scup and slow-swimming carp. The force–velocity and power–velocity curves are shown for the red muscle of both species at 20°C. Despite the fact that scup can swim to 80 cm s^{-1} with the red muscle, whereas carp can only swim to about 45 cm s^{-1}, V_{max} of both species are nearly the same. At the maximal swimming speed at which the red muscle is used (80 cm s^{-1} in scup and 45 cm s^{-1} in carp), it shortens at the same V (about 2.04 ML s^{-1}). Hence, both species use their red muscle over the same range of V/V_{max}, but this occurs at higher swimming speeds in the scup. Constructed with data from Rome and Sosnicki (1990) and Rome *et al.* (1990a, 1992a,b).

it lacks the anatomical separation of different fiber types that we have used to advantage in the fish. Furthermore, the muscles in mammals perform a variety of activities in addition to shortening (e.g., active lengthening and remaining isometric; Goslow *et al.*, 1981) and thus these muscles might be designed with a different set of constraints. Nonetheless, the constraint on shortening muscle to have high efficiency and power output is likely to be the most stringent, and hence V/V_{max} would be expected to be important in this muscle as well.

Stride frequency of mammals at physiologically equivalent speeds scales in proportion to $M_b^{-0.15}$ (Fig. 15A; Heglund and Taylor, 1988), where M_b is body mass. As SL excursion seems to be independent of M_b, V should scale as $M_b^{-0.15}$ as well. As a result, V in large mammals should be much lower than in small ones, and to maintain constant V/V_{max} large mammals should have much lower V_{max}.

When we sought data in the literature to test this hypothesis, we found that V_{max} of large mammals had never been measured, largely for technical reasons. Therefore, we developed single-fiber histochemistry techniques and combined them with myosin light and heavy chain electrophoretic techniques to enable us to differentiate the three major fiber types (Sosnicki *et al.*, 1989). We then used single skinned fiber mechanics techniques, and measured V_{max} of three fiber types within a single mammalian muscle (Rome *et al.*, 1990b). We found that within the horse *soleus*, mean V_{max} varied more than 10-fold between fiber types, and there was no overlap in V_{max} values between different fiber types. The mean V_{max} were 0.33, 1.33, and 3.20 ML s^{-1}, respectively, for the slow oxidative (SO), fast oxidative (FOG), and fast glycolytic (FG) fibers. This wide range of V_{max} provides the horse with tremendous flexibility, presumably to adjust its recruited fiber population to a wide range of motor activities.

By comparing V_{max} measured on the horse (M_b = 450 kg) to those previously measured on the rat (M_b = 0.38 kg) and on the rabbit (M_b = 5 kg) by Moss and co-workers (Reiser *et al.*, 1985; Eddinger *et al.*, 1986; Moss, 1986), we examined scaling of V_{max} over a 1200-fold size range (Rome *et al.*, 1990b). Figure 15B shows that there is a relatively small scaling effect on the Type IIB (FG) fibers ($V_{max} = 5.01\ M_b^{-0.07}$) and a larger scaling effect on the Type I (SO) fibers ($V_{max} = 0.99\ M_b^{-0.18}$). Because of the lack of information on V_{max} of Type IIA (FOG) fibers in the smaller species, we cannot calculate its scaling exponent.

This difference in scaling exponents between slow and fast fibers may reflect the different biological constraints put on the use of different fiber types. Slow oxidative fibers are used continuously during locomotion, so their V_{max} should be matched to V to yield the appropri-

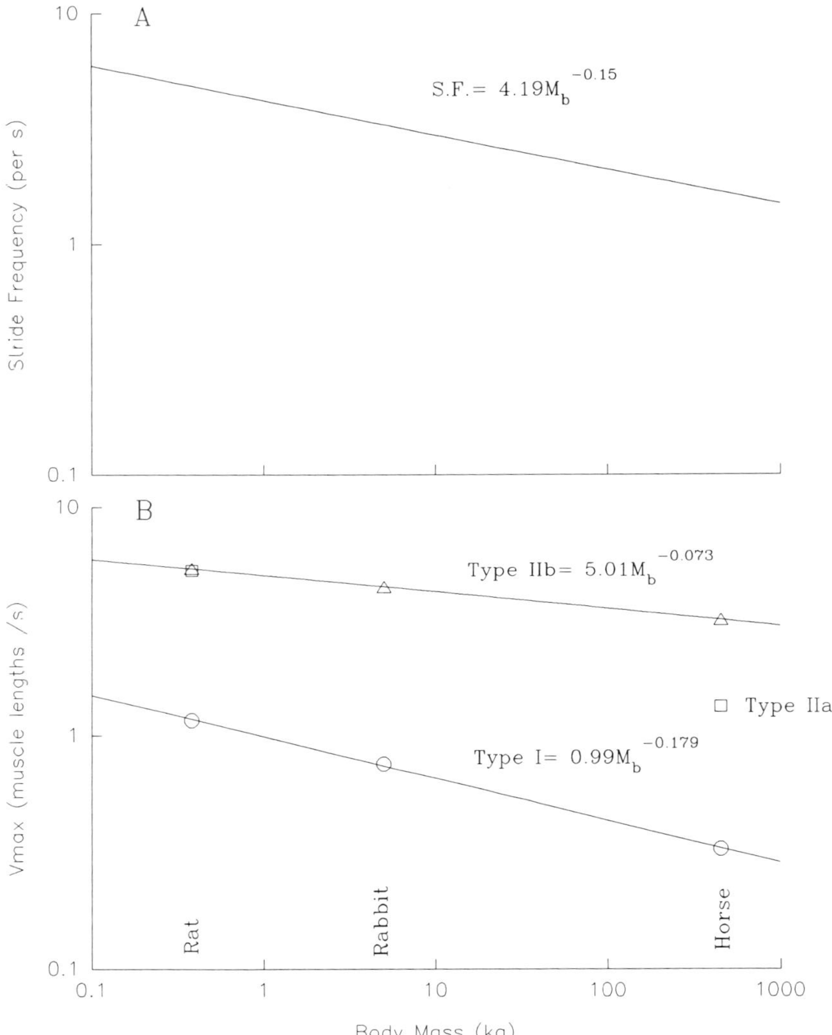

FIG. 15. Scaling of stride frequency and V_{max} with body size. (A) Stride frequency at the trot-gallop transition scales as $M_b^{-0.15}$. Graph is drawn from data in Heglund and Taylor (1988). (B) V_{max}, determined in skinned fibers at 15°C using the slack test, is plotted as a function of adult body mass. When plotted on log scales, there is a linear ($r^2 = 0.99$) relationship for both Type IIB and I fibers. The rat data are from Eddinger *et al.* (1986) and the rabbit data are from Reiser *et al.* (1985) and Moss (1986). The rat IIA data are from Metzger and Moss (1987). Reproduced from Rome *et al.* (1990b).

ate V/V_{max} for maximal efficiency. The scaling exponent of -0.18 for SO fibers is similar to the exponent of -0.15 for stride frequency (Fig. 15; Heglund and Taylor, 1988) and V, suggesting that the SO fibers of large and small animals operate at the same V/V_{max} and at similar efficiencies.

In contrast, FG fibers did not scale appropriately for maintaining constant V/V_{max}; instead, the FG fibers have higher V_{max} in large animals than needed. However, FG fibers are only used infrequently, for example, when the horse is jumping or running at maximal speed. In these circumstances, the horse's muscular system may have sacrificed efficiency for increased mechanical power. Because the mechanical power output is higher in fibers with high V_{max} than in fibers with low V_{max}, a smaller scaling exponent for the FG fibers may be beneficial.

Altringham and Young's (1991) study on the effect of body size on power output during work loops of mammalian diaphragm muscle in mammals ranging from mice to rabbits further supports the hypothesis that muscle that is used repetitively scales appropriately for V/V_{max}. They found that the frequency at which optimal power output was achieved decreased with increasing body size and scaled as $M_b^{-0.16}$, nearly identical to the value we found for V_{max} in slow fibers ($M_b^{-0.18}$). This similarity suggests that the twitch properties, which are included in work loop measurements, and V_{max} scale with the same exponent. Because ventilatory frequency scales similarly to stride frequency ($M_b^{-0.15}$), it appears that these muscles are perfectly scaled for their function.

The necessity that V_{max} change with body size (to maintain constant V/V_{max}) appears to have a number of mechanical and energetic consequences. Taylor and colleagues (1980) found that the energetics of locomotion scales as $M_b^{-0.30}$ (see paper by Taylor in this Volume and by Lindstedt and Thomas in Volume 38B), a finding they attributed to the increased cost of generating force in small animals. Rome (1992) has extended Taylor's hypothesis by assuming that the cost of generating force is proportional to V_{max}, and that at physiologically equivalent speeds, large and small mammals are recruiting the same muscle fiber types. This extended hypothesis quantitatively explains the observed locomotor energetics both at absolute speeds and at physiologically equivalent speeds, as well as a number of the mechanical findings.

G. Summary of Steady-State Design Constraints

In five different situations that might lead to changes in V/V_{max} (Fig. 11), evolution has acted to maintain constant V/V_{max}. In the three

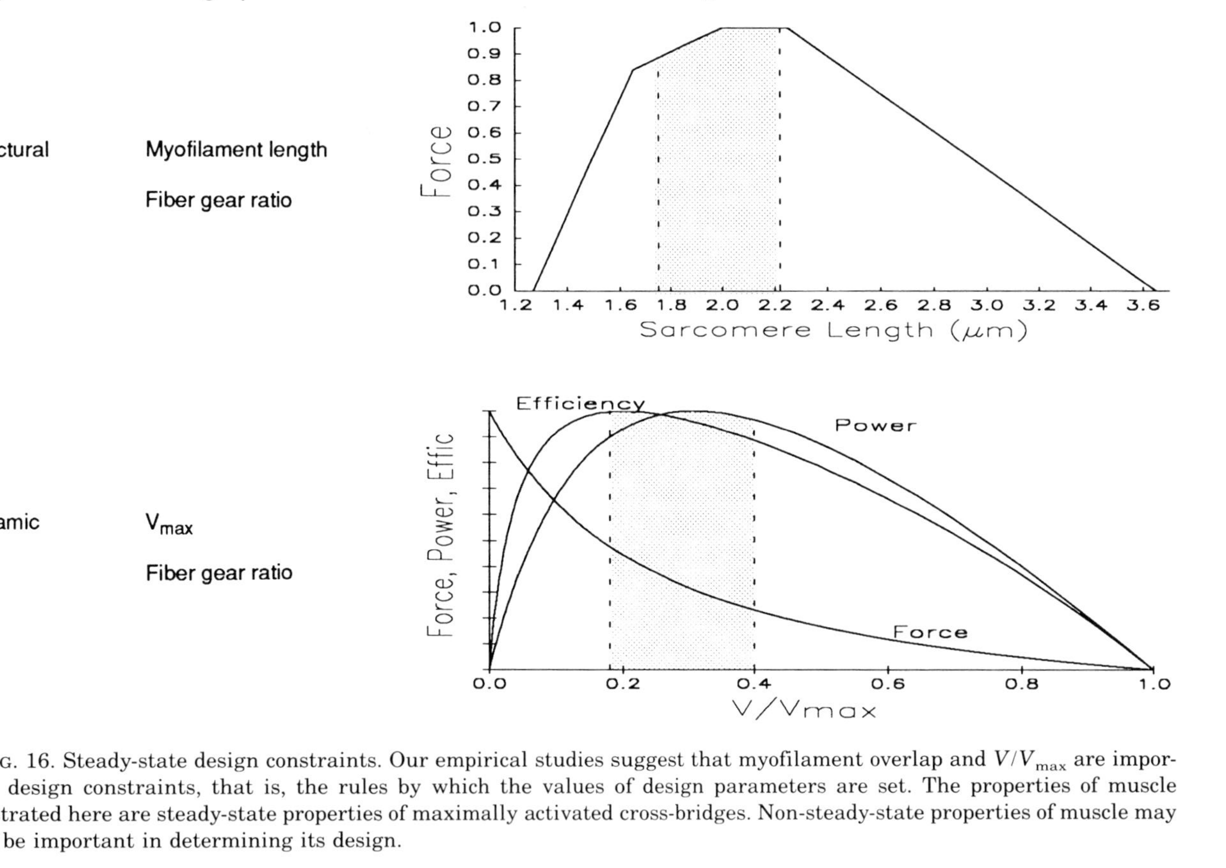

FIG. 16. Steady-state design constraints. Our empirical studies suggest that myofilament overlap and V/V_{max} are important design constraints, that is, the rules by which the values of design parameters are set. The properties of muscle illustrated here are steady-state properties of maximally activated cross-bridges. Non-steady-state properties of muscle may also be important in determining its design.

fish experiments and in the frog experiment, V/V_{max} was found to fall between 0.17 and 0.38, values for which power and efficiency are maximized. These findings demonstrate that V/V_{max} is an important design constraint. The way that animals produce a wide range of movements is by using fibers with different V_{max} and with different gear ratios. In the case of mammals, it appears that V_{max} scales appropriately with body size to maintain constant V/V_{max}, although we do not know precisely what this value is.

In conclusion, it appears that animals use their muscles over a narrow range of myofilament overlaps and over a narrow range of V/V_{max}, where muscle generates maximum force and maximum power with optimal efficiency (Fig. 16). Three design parameters (gear ratio, V_{max}, and myofilament lengths) appear to have been adjusted during evolution so as to conform with these design constraints no matter what movement is made. Hence, these constraints appear to be two of the rules that govern the design of muscular systems.

Before proceeding, it should be recognized that these constraints apply formally only to shortening muscle. Their apparent generality may be explained by the following considerations. First, fish, for instance, must generate positive mechanical work to overcome the force of drag, and *only* shortening muscle can produce positive mechanical work. Second, even though some of the fish's musculature may produce lengthening contractions (van Leeuwen *et al.*, 1990; Williams *et al.*, 1989), the design constraints on power generating muscle are likely to be more stringent than those on lengthening muscle (i.e., lengthening muscle generates high forces while requiring little energy, and in many cases in nature is simply replaced by tendon). Thus, it appears that muscle is designed primarily for shortening.

V. Non-Steady-State Mechanical Properties of Muscle

A. How Well Do Steady-State Mechanics of Maximally Activated Cross-Bridges Describe Muscle Mechanics during Locomotion?

Although V/V_{max} and myofilament overlap appear to be very important design constraints, they may not represent a complete description of muscle behavior during locomotion. The basic problem is that these constraints are based on the force–velocity and SL–tension curves, which are measures of steady-state mechanics of *maximally* activated cross-bridges. In measuring these parameters, the muscle is stimulated at a fixed length (isometric) and is allowed to shorten only after generating maximum tension (indicating complete activation). The muscle is then allowed several minutes to relax (complete relaxation) prior to being relengthened and stimulated again.

Although these steady-state measurements are useful for studying the mechanics of contraction, this is not how the muscle is used during locomotion, and thus they may ignore important constraints on muscle performance (Josephson, 1985a,b; Altringham and Johnston, 1990; Marsh, 1990; Rome, 1990; Rome and Swank, 1992; Rome *et al.*, 1992b). Unlike steady-state experiments, during cyclical locomotion muscle fibers alternately activate and relax, shorten and lengthen with no time interval between (as we have shown in swimming fish; see Figs. 3, 19, 20).

Muscles undergoing oscillatory contractions generate much less power than during steady-state contractions. For instance, in scup red muscle we found that during oscillatory contractions the maximum power outputs were 27.9 W kg^{-1} at 20°C and 12.8 W kg^{-1} at 10°C (Rome and Swank, 1992). These are much lower than values of 134 W kg^{-1} at 20°C and 71 W kg^{-1} at 10°C measured in force-clamp experiments (steady-state contractions) on the same preparation (Rome *et al.*, 1992b). Of course, muscle undergoing oscillatory contractions can shorten only during one-half the cycle, but this does not fully explain the much greater reduction observed. Hence, factors associated with activation and relaxation are no doubt involved.

Intuitively, it would appear to be beneficial for fibers to be fully "activated" during shortening and fully "relaxed" prior to relengthening. If the muscle were able to instantaneously activate and instantaneously relax, then the muscle behavior during shortening could be fairly well described by steady-state mechanical properties. During very low oscillation frequencies, muscle activation and relaxation requires a small fraction of the cycle, and can be viewed as instantaneous (Fig. 17A). However, our extensive analysis on swimming fish suggests that during oscillation frequencies that animals *actually use* during locomotion (7.5 Hz), the muscle may run into some important limitations (i.e., does not have enough time to relax, as illustrated in Fig. 17B).

Because the muscle cannot instantaneously relax (by relax, here we require all the bridges to detach or go into a state where subsequent lengthening of the muscle does not produce force), the muscle may have to operate in such a way that *deactivation proceeds during shortening*. If this occurs, the muscle can not remain completely activated during the shortening period. Hence, force production will be below that described by the steady-state force-velocity curve.

Deactivation during shortening is achieved in two ways. First, the

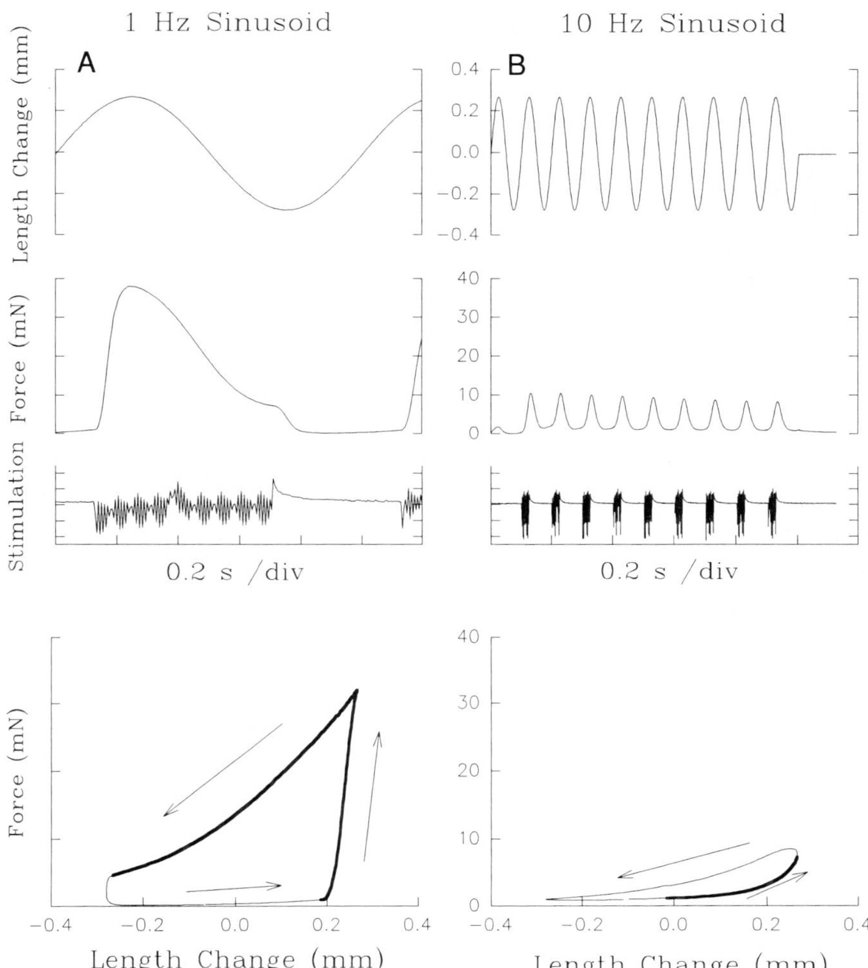

FIG. 17. Work performed by oscillatory contractions at 1 and 10 Hz. The length change, force, and stimulation are shown for the two oscillation frequencies. The force–length, or work loops are shown on the bottom. The heavy lines correspond to the stimulus being on. Note that at 1 Hz (A), the stimulus is turned on just prior to shortening and is turned off just prior to the end of shortening. Hence, activation and relaxation can be viewed as instantaneous, and they have little effect on muscle performance. At 10 Hz (B), however, the stimulus is started well before shortening begins, and it ends before shortening even begins. In this case, relaxation is limiting, as the muscle force never goes down all the way to zero as it does at 1 Hz. At 10 Hz, the kinetics of activation and relaxation impinge significantly on muscle performance. During locomotion (4–7.5 Hz), the muscle would be working somewhere between these conditions.

nervous system (or physiologist) can cut off the stimulus before the end of shortening (as in Fig. 17B; *N.B.*: stimulus ends before shortening begins). Second, an intrinsic property of the muscle called shortening deactivation (Fig. 18) causes the force-generating ability of the fiber to be reduced following shortening (Edman, 1980). Experiments using aequorin (Ridgeway and Gordon, 1984; Cannell, 1986) have demonstrated a release of calcium (Ca) from troponin (Tn) during shortening. Gordon and Ridgeway (1991) have postulated that the decrease in force associated with shortening increases the K_{off} of CaTn. To the extent that K_{off} limits the rate of relaxation, this might be an important adaptation necessary for muscles that undergo cyclical length changes.

Figure 18A shows isometric scup red muscle receiving short tetanic stimuli at 7.5 Hz. Clearly, the muscle does not relax between tetani, and from this, one would conclude that the fish could not possibly swim with a tail beat frequency of 7.5 Hz, although it does. In Fig. 18B, the muscle receives the same stimulus, but undergoes cyclical length changes, and in this case it relaxes almost completely between cycles (and generates positive work). Thus, it appears that the increased rate of relaxation caused by shortening deactivation allows the muscle to be used at higher frequencies.

Shortening deactivation has been hypothesized to be the cause of curved length records during force clamps. We have found that this effect is larger in fish muscle than in frog muscle, correlating with a fish's need for rapid relaxation because it employs cyclical locomotion

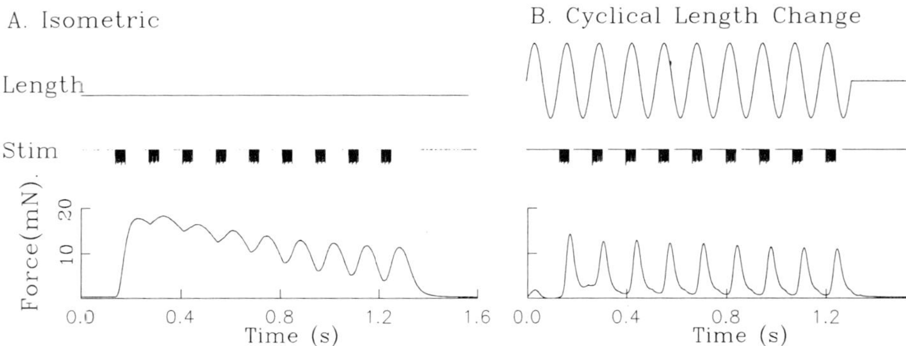

FIG. 18. Shortening deactivation during cyclical length changes allows the muscle to relax more rapidly. Whereas during isometric contractions (A) the scup red muscle does not relax between tetani, when given the same stimulus (40 ms, 50 pps, 7.5 Hz) while undergoing ±5% length changes (B), the muscle relaxes almost completely between tetani. Note that scup can swim with a tail beat frequency of 7.5 Hz.

(Julian *et al.*, 1986; Rome and Sosnicki, 1990; Rome *et al.*, 1992b). Because of this effect, velocity measurements in the force–velocity curve are only taken at the initial stage of shortening. Thus, even in a continually stimulated muscle, the cross-bridges may not be fully activated, and the force generated during shortening at a given velocity will fall below the force–velocity curve.

Another factor that influences muscle performance is the phase of stimulation. If the muscle is stimulated isometrically for 300 ms prior to release, it becomes completely activated, and attachment of the cross-bridges, force generation, and stretching of the series elastic component occur. If the muscle is allowed to shorten after only a short duration of stimulation, some of these processes may not be complete, hence the force production may be less than that predicted by the force–velocity curve. To obtain sufficiently high force during cyclical contractions, the stimulation may start prior to shortening (i.e., during the lengthening phase of the cycle; see Figs. 17B and 19). This presumably allows time for cross-bridge attachment (or transition from low-force-generating bridges to high-force-generating ones; Bagni *et al.*, 1988) and stretching of the series elastic component to occur.

A useful method for quantifying competing effects in a shortening–lengthening cycle is to measure the amount of *net* work in the cycle (positive minus negative work), which is graphically equivalent to the area contained within the force–length loop (Figs. 17A, and 17B), a technique developed by Josephson (1985a). Net power in cyclical contractions is described by the formula

$$\text{Net power output} = (\text{work done by muscle}_{\text{shortening}} - \text{work done on muscle}_{\text{lengthening}}) \text{ per cycle} \times \text{cycle frequency.}$$

Incomplete relaxation results in increased negative work (Fig. 17B). Incomplete activation or deactivating the fiber too early in the cycle results in lower positive work. From these observations we can anticipate that power output during swimming may involve a complex interplay between different muscle properties. The effect of activation and relaxation kinetics on the mechanical performance of the muscle *depends greatly on the work loop conditions* (compare Fig. 17A at 1 Hz to Fig. 17B at 10 Hz).

B. Influence of Kinetics of Activation and Relaxation on Muscle Performance during Locomotion

Relatively little is known about whether (a) muscle shortens with maximally activated cross-bridges and completely relaxes prior to

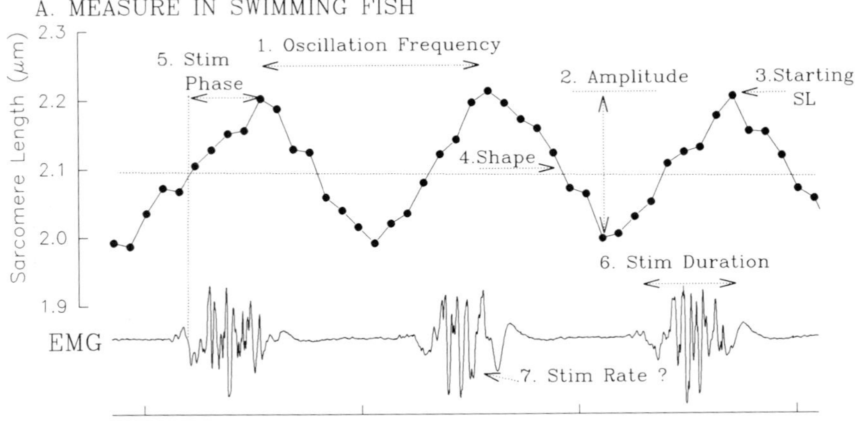

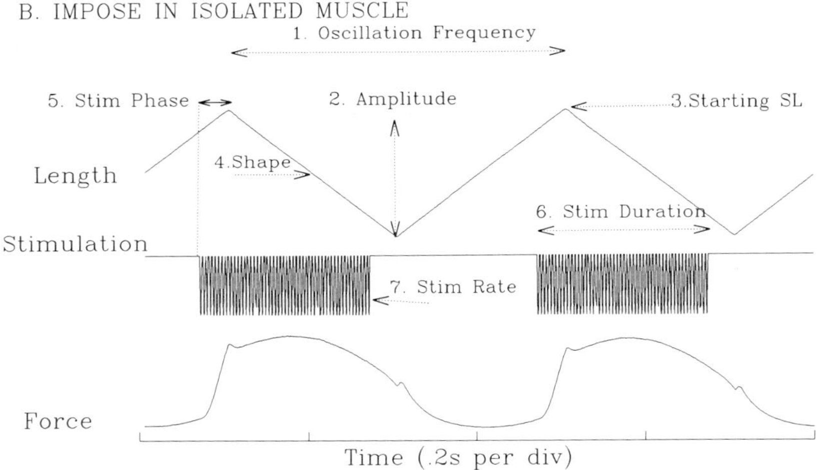

FIG. 19. (A) SL changes and EMG of the red muscle during steady swimming at 50 cm s^{-1}. (*N.B.*: Synchronization of EMG with SL is only approximate.) Six of the seven parameters that define the length change and stimulation pattern can be readily measured. Note that "shape" refers to the shape of the SL change, which can be a ramp, sinusoid, or an arbitrary waveform in between. (B) This shows the next step of imposing the seven defining parameters on a muscle and measuring the resultant force production. Note that in this case a ramp is used for shape and the stimulation rate was chosen to give a fused tetanus. Work production is calculated from resultant force–length loops as illustrated in Fig. 17. *Note that B is not meant to try to reproduce the results in A.*

lengthening or (b) the kinetics of activation and relaxation impinge on the mechanical performance of the muscle. The basic problem is that these can only be discerned by imposing the exact length changes and stimulation pattern that the muscle undergoes *in vivo* on isolated muscle and by measuring force production (Figs. 19A and 19B), but this had never been done until recently.

Several investigators (Josephson, 1985a,b; Stevens, 1988; Altringham and Johnston, 1990) have applied sinusoidal length changes to muscle and have varied the frequency and amplitude of the sinusoid, as well as the phase and duration of the stimulation. Typically, investigators have simply optimized as many parameters as possible to obtain maximum power output of the muscle, *without particular regard as to how the animal is using the muscle.* Although this work points to some *potential* problems that muscle *may* face (e.g., insufficient time to relax at high oscillation frequencies, as in Fig. 17B), it is not possible to know whether performance is limited in any way during locomotion, because the details of the length changes, oscillation frequencies, and phase and duration of stimulus during locomotion were not measured.

Marsh and colleagues have argued that the time for activation and relaxation (t_c, time from onset of *isometric* twitch to one-half relaxation) should limit the minimum time between alternating movements (i.e., the time for one-half cycle; Marsh and Bennett, 1986; Marsh, 1988, 1990). They reasoned that if an animal tries to locomote at a higher frequency, then the muscles may not be relaxed when the opposing muscle starts to stretch it, thereby offering too much resistance to overcome. They have found considerable evidence for this limitation at low temperatures. Marsh (1988) further found that because small lizards use faster stride frequencies, their muscles have faster twitch kinetics. Although the measure t_c is supposed to account for increased relaxation rate associated with shortening deactivation, it appears that in some cases the increased relaxation is greater than predicted (Fig. 18). Thus, animals can often locomote at much higher frequencies than those predicted from *isometric* experiments (Rome and Swank, 1992).

C. Molecular Basis of Activation and Relaxation: Implications for Locomotion

The rate of muscle relaxation shows tremendous variation. It varies greatly between slow muscle and fast muscle within the same animal, changes during development, is different in different types of muscle (reviewed in Appelt *et al.*, 1991; Marsh, 1990), and it varies with temperature and thermal acclimation (Johnson and Johnston, 1991).

Unlike the cases of V_{max} and the SL–tension curve, for which the underlying mechanisms of variation are straightforward (myosin heavy and light chains and myofilament lengths, respectively), a change in activation or relaxation kinetics cannot be uniquely assigned to one molecular component. The kinetics of activation and relaxation are very complex (Marsh, 1990). Activation is usually thought to consist of Ca^{2+} release from the SR and Ca^{2+} binding to Tn. Measurements and models show that these processes are quite rapid (10 ms; Huxley, 1985; Baylor and Hollingworth, 1988). However, for purposes of understanding muscle function during locomotion, maximal activation is not achieved until cross-bridges attach, generate force, take up the series elastic component, and follow the force–velocity curve. This series of events takes considerably longer and lags well behind formation of CaTn (Baylor and Hollingworth, 1988; Bagni *et al.*, 1988).

Relaxation is even more complex. During and following Ca^{2+} release, Ca^{2+} is sequestered by the SR as well as bound to parvalbumin, which is in quite high concentration in frogs and in white muscle of fish (Gillis *et al.*, 1982; Cannell, 1986; Hou *et al.*, 1990). Myoplasmic Ca^{2+} drops, causing Ca^{2+} to unbind from Tn and some cross-bridges to detach. Sarcomere shortening during isometric relaxation causes a decrease in the affinity of Tn for Ca^{2+} and causes increased Ca^{2+} release (Gillis *et al.*, 1982). Eventually, the Ca^{2+} is sequestered and the force drops to zero, and the muscle is considered "relaxed." However, from a functional viewpoint, cross-bridges must all be detached, and the drop in muscle stiffness, a measure of attached cross-bridges, is known to lag behind the drop in force during relaxation (Bagni *et al.*, 1988). An additional complexity is that cyclical length changes greatly increase the relaxation rate (Figs. 18 and 21) compared to in the isometric case.

This discussion of the mechanisms of "activation" and "relaxation" illustrates several important points. First, classical definitions of "activation" and "relaxation" do not include all the events that are important for understanding muscle function during locomotion (Marsh, 1990). Second, these mechanisms are sufficiently complex and encompass too many unknowns to be able to model the *whole* process accurately, especially when the muscle is shortening and lengthening.

D. What Are the Design Constraints?

Finally, given what we know about muscle, we can not convincingly argue how the kinetics of activation and relaxation should be set for the muscle to function optimally. Unlike the SL–tension and force–, power–, and efficiency–V/V_{max} curves, there is no obvious relationship from which we can determine an optimal value. We do not know whether the muscles' work loop and stimulation pattern should look like that in Figs. 17A or 17B when swimming with a tail beat frequency of 7.5 Hz. Although it might seem beneficial to design muscles so they can be stimulated throughout shortening and are completely relaxed during lengthening [i.e., instantaneous activation and relaxation (small duty cycle) as in Fig. 17A], this design may come with high energetic costs. Rapid relaxation requires a high SR pumping rate, and thus additional Ca^{2+} release from the SR. Rapid relaxation might also require a relatively high cross-bridge detachment rate and V_{max}, thus the V/V_{max} *might be too low for maximum efficiency.* Alternatively, the energetic cost could be reduced with a relatively slower relaxation rate and low V_{max}, but this would yield considerable negative work, similar to what occurs in Fig. 17B. Thus, to learn how muscle has been designed, empirical measurements are needed in which *in vivo* muscle performance is exactly reproduced *in vitro* (Fig. 19).

VI. What Needs to Be Done?

A. New Experiments, New Questions?

Three classes of experiments must be performed to improve our understanding of the design of the muscular system. The basic goal should be to identify new design constraints of the muscular system and then to test them.

1. Reproducing in Vivo Length Changes and Stimulation on Isolated Muscle

First, we must be able to measure the exact length changes and stimulation pattern that occur *in vivo* (Fig. 19A), then impose these on isolated muscle while measuring force (and work) production (Fig. 19B). Seven parameters define the length change and stimulation pattern:

1. Oscillation frequency;
2. Amplitude;
3. Starting SL;
4. Shape of length change (*i.e.*, triangle wave or sinusoid);
5. Phase of stimulation with respect to the length change;
6. Stimulation duration;
7. Stimulation rate.

Of these seven parameters, we can presently measure only six. Unfortunately, there is no easy way to ascertain the rate at which a given motor unit is being stimulated in a swimming fish or any locomoting animal, with the exception of some invertebrates. Quantitative EMG analysis gives the rate at which electrical spikes are being picked up by the electrodes, but formally, one does not know if:

1. there is one motor unit being stimulated at that rate;
2. ten motor units are being stimulated in an asynchronous manner at one-tenth the rate;
3. ten motor units are being stimulated in a synchronous manner at the high rate.

It is of interest that our previous data suggested that stimulation is synchronous in scup swimming (Rome *et al.*, 1992a).

A direct measurement of stimulation rate would involve recording from motor neurons during swimming, but this is technically difficult. Alternatively, we can gain some insight into how closely the muscle approaches a fused tetanus during swimming by stimulating the appropriate motor nerve at the fusion frequency and recording the EMG in an *in situ* preparation. Despite this lack of information, we can fix six of the seven parameters and vary stimulation rate to determine what effect it has on power output.

This approach will permit important issues to be addressed, such as whether fiber behavior during locomotion is adequately described by the steady-state SL–tension and force–velocity relationships, or how the kinetics of activation and relaxation affect the mechanical power of muscle during locomotion. Hopefully, it will also help us define the design constraints for activation and relaxation. Insight into this issue will be gained by reproducing *in vitro* on a number of species the *in vivo* fiber behavior, and trying to observe a pattern.

2. *Manipulate Different Components of the Muscular System*

To test hypotheses about muscle design, one must be able to manipulate different components of muscle and measure how these alterations affect power output. For instance, to assess the relative importance of V_{max} and kinetics of relaxation, it would be instructive to vary one while holding the other constant. A simple procedure, such as altering temperature, is not appropriate because it alters both components at the same time. However, by using pharmacological agents on isolated muscle, relaxation rate can be slowed while keeping V_{max} and cross-bridge kinetics constant.

Using this approach will enable us to address some basic issues of muscle design. For example, does slowing relaxation rate alter a muscle's functional ability to power locomotion? Fish white fibers have higher V_{max} and a faster relaxation rate than red ones, but the relaxation rate can be slowed pharmacologically (down to that of the red fibers). If a fiber has the V_{max} of white muscle, but a relaxation rate of red muscle, can it still power the fast swimming movements that white muscle does? We can also address which muscle property limits performance at low temperatures. The rate of relaxation of 20°C red muscle can be reduced pharmacologically to that of 10°C red muscle. Can red muscle with a 20°C V_{max}, but a 10°C relaxation rate, still power the faster swimming movements that occur at 20°C?

3. Integrated Models of Muscle Function

To better appreciate how the muscular system is designed, we must also develop a quantitative understanding of how finely it is tuned and what parameters are most important for this tuning. This requires the development of models that simulate muscle force production during arbitrary length changes and stimulation patterns. This modeling is especially important because our ability to manipulate components and measure the effects is limited. To improve our understanding of muscle design these models must be mechanistically based, as opposed to phenomenologically based. That is, the model must have all the mechanistic features that occur in the muscular system (e.g., K_{off} for CaTn) so that we can tell which of these are important design parameters. Development of such models will require far better understanding of the kinetics of activation and relaxation then we currently possess, and will be some years in developing.

The basic issue that needs to be addressed with the models is how would locomotor performance be altered by changes in muscle properties? One animal model that is particularly well suited for this analysis is frog jumping. Will a frog jump approximately the same distance with a threefold change of V_{max}, or will a 10% change in V_{max} cause a 70% change in jump distance? Will a frog jump further if V_{max} is kept constant, but the activation rate is doubled? With a good model, we can alter individual parameters and predict how performance would vary. It would also be useful to test the optimality of muscle, that is, if we altered some parameters, could we improve on evolution and get the frog to jump farther?

B. First Steps toward Answering These Questions

Our laboratory has taken a first step toward using these approaches. Other laboratories, for example, those of Altringham (fish, mammals),

Ellington (insects), Johnston (fish), Josephson (crabs), and Marsh (scallops) are beginning to use this approach on a variety of species as well.

1. Reproducing on Isolated Muscle in Vivo Length Change and Stimulation Pattern from Swimming Fish

Rome *et al.* (1993) have recently applied the technique described in Fig. 19 to swimming scup, providing a number of new insights into how the muscular system is designed. Figure 20 shows the results of the overall experiment. They first swam fish at 80 cm s^{-1} and measured EMGs and muscle length changes of the red muscle at 4 places along the length of the fish (ANT1, ANT2, MID, POST). It was found that moving caudally along the length of the fish, the length change of the muscle becomes larger. Strains were ± 1.5% of resting muscle length at ANT-1 increasing to ± 5.7% at the POST (Fig. 20B). In addition, the EMGs had a longer duty cycle in the anterior of the fish than in the posterior (Fig. 20A). This is due primarily to a large difference in the onset time of the EMG in the anterior compared to the posterior, but nearly simultaneous off times. Finally, EMGs precede the length changes by increasing amounts along the length of the fish towards the tail (Fig. 20A,B; Fig. 21A,B).

To determine the mechanical performance of the muscle during this swimming behavior, Rome *et al.* drove red muscle bundles isolated from the four positions on the fish through the length changes and stimulation conditions that they undergo during swimming and measured the resulting force and power the muscle generated. They found that despite the fact that the POST muscle is stimulated primarily during lengthening (Fig. 21I), it generated large amounts of work. This finding demonstrates that the time course of muscle activation differs dramatically from the time course of EMG activity; assuming that they are the same (which is typically done) can lead to qualitatively incorrect conclusions. The anterior muscle, on the other hand, generated significantly less work under its *in vivo* conditions (Figs. 20 and 21II).

The cause of the low power generated in the anterior region of the fish is not due to an intrinsic deficiency of the anterior musculature, rather it is due to the specific length change and stimulation pattern that the anterior muscle undergoes. For instance, when the ANT-1 muscle is driven through the length change and stimulation pattern that the POST muscle normally undergoes *in vivo* (Fig. 21), it generates just as much power as the POST muscle. The main reason for the low power output in the anterior region is the small strain (dl). A small dl reduces work output (Fxdl) per tailbeat. In addition, muscles under-

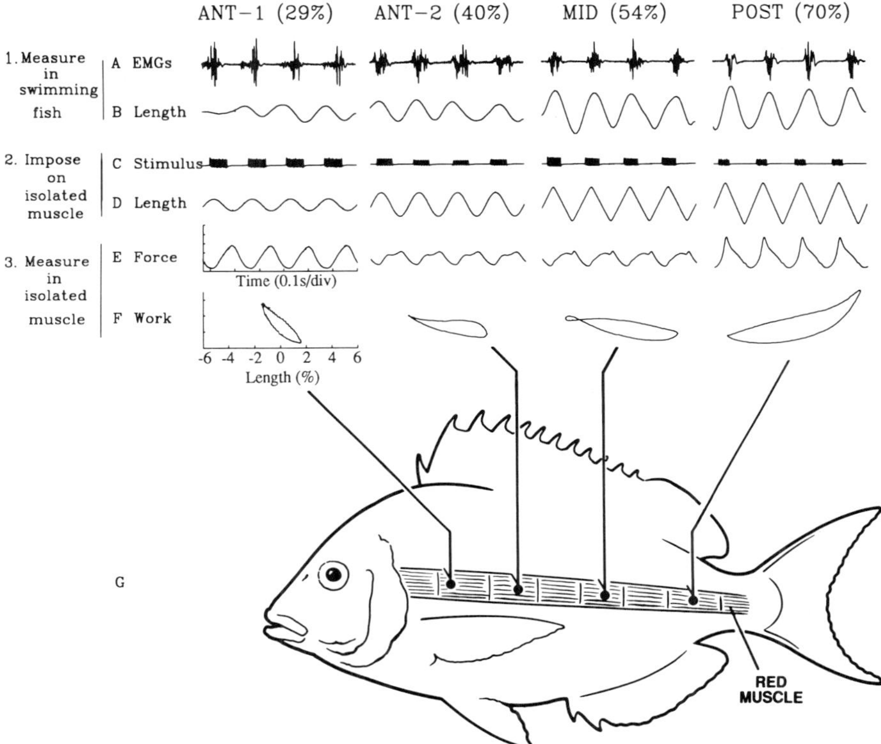

FIG. 20. Length changes, stimulation pattern, force production, and work output of red muscle of scup during swimming. Step 1 was to measure in a swimming fish (80 cm/s) the EMGs (A) and length changes (B) for the red muscle at four places along the length of the fish pictured in (G). Step 2 was to impose on muscle bundles isolated from these four positions the length changes (D) and stimulation pattern (C) that were observed during swimming. Step 3 was to measure in the isolated muscle the resulting force production (E) and work production (F). Note that the reason for the apparent discrepancy between traces A and B and C and D is that A and B represent records from one of the fish (tailbeat frequency = 6 Hz) while C and D represent the record for a muscle driven through the *average* swimming values (i.e., tailbeat frequency = 6.4 Hz). Traces A–E are all functions of time. Trace F is a plot of force produced against length changes where the area of the enclosed loop is the work produced during a tailbeat cycle. This value is much larger in the POST than in the ANT-1 position. Adapted from Rome *et al.*, 1993.

going small strains do not relax sufficiently fast. Relaxation is sped up greatly by muscle shortening (shortening deactivation), but the shorter the strain, the smaller the enhancement of relaxation (Josephson and Stokes, 1989; Altringham and Johnston, 1990; Rome and Swank, 1992).

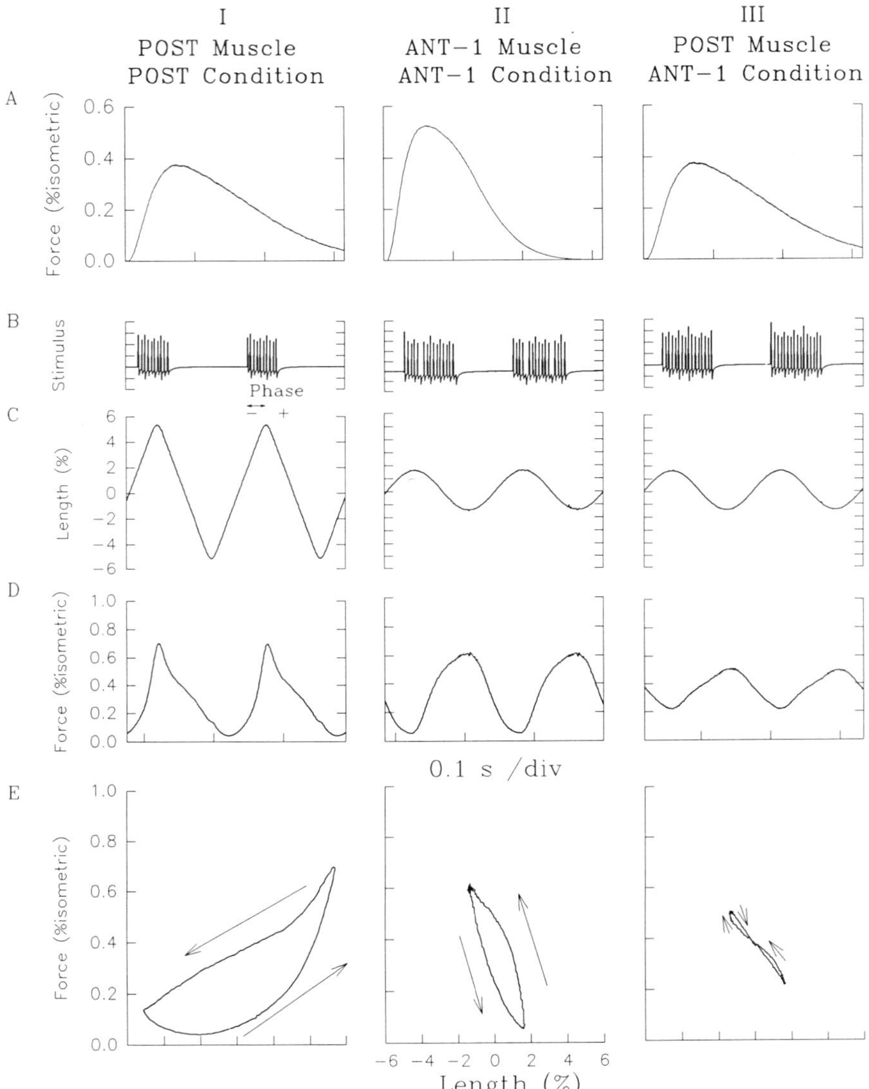

FIG. 21. Mechanical properties of ANT-1 and POST muscles. Columns I and II show a POST and an ANT-1 muscle bundle undergoing their respective length changes and stimulation pattern that the muscle undergoes during swimming. By contrast, column III shows a POST muscle undergoing the stimulation pattern and length changes that is encountered by the ANT-1 muscle during swimming. Trace A shows the isometric twitch of the muscles in question. Traces B and C show the imposed stimulation pattern and length changes determined during swimming experiments. (Note the phase of the stimulus is defined with respect to maximum muscle length.) The resulting force is shown in

The fact that the ANT-1 muscle can produce any work at all during swimming represents an adaptation. For instance, when the POST muscle is driven through the length change and stimulation pattern the ANT-1 muscle normally undergoes *in vivo*, it performs even worse than the ANT-1 muscle and generates no net work (compare Figs. 21II,E and 21III,E). The primary reason that the ANT-1 muscle can generate significant power despite undergoing only a small strain (Fig. 21II,E) is that it has the intrinsic ability to relax much more rapidly than the muscle in the posterior (Fig. 21A). Because of its slow intrinsic relaxation rate, POST muscle driven through the ANT-1 conditions does not come close to relaxing between stimulus trains, whereas the ANT-1 muscle relaxes almost completely (Figs. 21II,D and 21III,D). Although the mechanism for faster relaxation in the ANT-1 muscle has yet to be explored, these results show that relaxation is malleable and that it can greatly effect power production by the muscle.

How Is the Muscle Designed? As mentioned above, Fig. 17 provides two extreme cases by which evolution could have set the rates of activation and relaxation. The rates of activation and relaxation could have been set to be essentially instantaneous or slow. Thus the crucial question is *under the conditions scup actually use their muscle*, is the activation–relaxation rate instantaneous or slow? In this study Rome *et al.* (1993) found that scup swimming at 80 cm s^{-1} use a tailbeat frequency of 6.4 Hz. The work loops under these conditions (Figs. 20 and 21) show that the muscle is designed with a relatively slow activation–relaxation rate. The POST muscle (Fig. 21I) is stimulated only 27% of the cycle and is relaxing during most of the shortening phase. Therefore, during locomotion, activation–relaxation processes limit work production well below what the muscle is capable of in a steady state. To remove this limitation, and enable the muscle to relax "instantaneously" at 6.4 Hz, would require the Ca^{2+} pumping rate to be greatly increased, which would increase the energetic cost of the contraction. Therefore it is likely that evolution sets the activation–relaxation rate to be slow so that the mechanical power can be generated most efficiently.

To conclude, for prolonged cyclical locomotion, evolution appears to

---

trace D and the resulting work (area enclosed by loop) in E. A–D are functions of time, whereas E is force as a function of length. Note the large strain in the POST compared to the ANT-1 and the much larger work produced in that muscle. Note also that relaxation is much faster in muscle undergoing shortening (D, caused by shortening deactivation) than that being held isometrically (A). Finally note that relaxation is much faster in ANT-1 muscle that POST (A). This permits the ANT-1 muscle (II,E) to perform work under conditions where the POST muscle cannot (III,E). Adapted from Rome *et al.*, 1993.

have set a relatively slow activation–relaxation rate for the muscle and hence, during locomotion, these processes limit power production. We hypothesize that the design goal is maximizing efficiency of power production. To avoid the high cost of fast Ca^{2+} pumping, there is a compromise made that reduces power production. This hypothesis needs to be verified empirically by measuring efficiency of power production while the muscle is undergoing its *in vivo* length change and stimulation pattern.

2. *Reproducing on Isolated Muscle in Vivo Length Change and Stimulation Pattern from Jumping Frogs*

Lutz and Rome (1994) measured the SL change and stimulation pattern of the SM muscle during jumping (Fig. 7A,B), and then drove isolated SM muscle through this pattern (Fig. 7C,D) and measured the resulting force (Fig. 7E). Although the SM shortens at the correct SL's and V/V_{max} for maximum power generation, the question arises whether the muscle is stimulated for a sufficient time before shortening to be maximally activated during jumping. In terms of frog jumping, "activation" was considered to be maximal if the muscle generated the same force (and power) under *in vivo* conditions as it does at the corresponding V during force–velocity experiments (where maximal activation is achieved by stimulating for 60 ms before shortening).

Despite the short lags between the EMG and muscle shortening (18 ms), the force the muscle generates under the *in vivo* conditions (Fig. 7E), is greater than or equal to the force generated at the V on the force–velocity curve (dotted line), suggesting that the muscle is maximally activated. Further mechanical studies showed that the fibers were maximally activated by the *beginning* of shortening and hence *throughout* the shortening phase. Thus, although the rate of activation processes may vary slightly in amphibian type 1 and type 2 fibers (Lannergren *et al.*, 1982), activation processes are sufficiently fast in both fiber types to ensure maximal activation during jumping.

Comparison of isolated muscle and whole animal power production further support the conclusion that fibers were maximally activated and show that most of the extensor muscles in the frog hindlimb probably behaved similarly to the SM during jumping. The peak power required to accelerate the frog during jumping (determined from films of jumping frogs) was 67.2 (± 5.4) W per kilogram of body mass. Because the maximum power generated by the SM was 371 W/kg of muscle and only about 17% of the body mass can be involved in powering jumping, the maximum possible whole animal power = 371 (W/kg muscle) × 0.17 (kg muscle/kg of body mass) = 63 (W/kg of body mass). The close match between the power required to accelerate the frog (from films) and the maximum available from the muscle suggest that all the muscle fibers from each extensor muscle are probably recruited, and are stimulated at a sufficiently high frequency, and shorten over the appropriate SL's and V/V_{max} for maximum power generation during jumping. In addition, preliminary results for the cruralis muscle, a large knee extensor, show that it appears to be activated and shortened much like the SM. It should be noted that although some small muscles in the hindlimb apparently lengthen during jumping to help stabilize joints (Mai and Lieber, 1990), they represent a small fraction of the mass of the hindlimb.

Thus, the rate of activation (which includes calcium release, binding to troponin, crossbridge transitions, and stretching of the series elastic elements) are all sufficiently rapid so that the muscle is maximally activated prior to the beginning of shortening. This is a crucial point, because if the muscle starts to shorten before the muscle has been stimulated for sufficient time, then the muscle never becomes maximally activated during the jump. The inability to achieve maximal activation if shortening starts prematurely has also been demonstrated dramatically in experiments on single frog fibers (Cecchi *et al.*, 1978).

The time taken between the initiation of stimulation and the start of shortening involves the biomechanical arrangement of the muscles and the limbs, the inertia of the frog, and the kinetics of force generation. These parameters are set so that the muscle does not start to shorten appreciably prior to achieving maximal activation. The solution which gives this result would be well addressed by the type of modelling outlined above.

Another potentially important adaptation in the frog is that the force stays constant *during* shortening and so maximum power is generated (Fig. 7). This is opposed to the situation in fish muscle, in which force drops during shortening due to inherent shortening deactivation (see Fig. 18).

In short, the frog muscular system appears to have been designed to generate maximum power. Not only do the muscles shorten at the appropriate SL and at the appropriate V/V_{max}, but (1) the rate of activation is coordinated with the frog biomechanics, so that the muscle can maximally activate prior to shortening and (2) the muscle shows relatively little shortening deactivation and hence generates maximum power throughout the jump.

The success of the recent fish and frog experiments suggest that employing this technique (reproducing on isolated muscle the *in vivo* length change and stimulation pattern) in additional animals, will further our understanding of the design of the muscular system. It is also clear that new and different insights will be gained by perturbing single components of the muscular system and by modelling the system as described above.

Acknowledgments

I thank Cathy Kelly for help in preparing the manuscript and Gordon Lutz and Doug Swank for providing useful suggestions. This work was supported by NIH AR38404 and NSF IBN 9205397.

References

Alexander R. McN. (1969). The orientation of muscle fibres in the myomeres of fishes. *J. Mar. Biol. Assoc.* **49,** 263–290.

Altringham, J. D., and Johnston, I. A. (1990). Modeling muscle power output in a swimming fish. *J. Exp. Biol.* **148,** 395–402.

Altringham, J. D., and Young, I. S. (1991). Power output and the frequency of oscillatory work in mammalian disphragm muscle: The effects of animal size. *J. Exp. Biol.* **157,** 381–389.

Appelt, D., Shen, V., and Franzini-Armstrong, C. (1991). Quantitation of Ca-ATPase, feet and mitochondria in super fast muscle fibers from the toadfish, *Opsanus tau*. *J. Muscle Res. Cell Motil.* **12,** 543–552.

Bagni, M. A., Cecchi, G., and Schoenberg, M. (1988). A model of force production that explains the lag between crossbridge attachment and force after electrical stimulation of striated muscle fibers. *Biophys. J.* **54,** 1105–1114.

Baylor, S. M., and Hollingworth, S. (1988). Fura-2 calcium transients in frog skeletal muscle fibres. *J. Physiol. (London)* **403,** 151–192.

Calow, L. R., and Alexander, R. McN. (1973). A mechanical analysis of a hind leg of a frog (*Rana temporaria*). *J. Zool. (London)* **171,** 293–321.

Cannell, M. B. (1986). Effect of tetanus duration on the free calcium during the relaxation of frog skeletal muscle fibres. *J. Physiol. (London)* **376,** 203–218.

Cecchi, G., Colomo, F., and Lombardi, V. (1978). Force–velocity relation in normal- and nitrate-treated frog single muscle fibres during rise of tension in an isometric tetanus. *J. Physiol. (London)* **285,** 257–273.

Cutts, A. (1986). Sarcomere length changes in the wing muscles during the wing beat cycle of two bird species. *J. Zool.* **209,** 183–185.

Dimery, N. J. (1985). Muscle and sarcomere lengths in the hindlimb of a rabbit (*Oryctolagus cuniculus*) during a galloping stride. *J. Zool.* **205,** 373–383.

Eddinger, T. J., Cassens, R. G., and Moss, R. L. (1986). Mechanical and histochemical characterization of skeletal muscle from senescent rats. *Am. J. Physiol.* **251,** C421–C430.

Edman, K. A. P. (1980). Depression of mechanical performance by active shortening during twitch and tetanus of vertebrate muscle fibres. *Acta Physiol. Scand.* **109,** 15–26.

Edman, K. A. P., Reggiani, C., Schiaffino, S., and Tekronnie, G. (1988). Maximum velocity of shortening related to myosin isoform composition in frog skeletal muscle fibres. *J. Physiol. (London)* **395,** 679–694.

Gillis, J. M., Thomason, D., LeFevre, J., and Kretsinger, R. H. (1982). Parvalbumins and muscle relaxation: A computer simulation study. *J. Muscle Res. Cell Motil.* **3,** 377–398.

Gordon, A. M., and Ridgway, E. B. (1991). Evidence that increased cross-bridges and force decrease the rate of Ca dissociation from CaTn in barnacle muscle fibers. *Biophys. J.* **59,** 584a.

Gordon, A. M., Huxley, A. F., and Julian, F. J. (1966). The variation in isometric tension with sarcomere length in vertebrate muscle fibers. *J. Physiol. (London)* **184,** 170–192.

Goslow, G. E., Seeherman, H. J., Taylor, C. R., McCutchin, M. N., and Heglund, N. C. (1981). Electrical activity and relative length changes of dog limb muscles as a function of speed and gait. *J. Exp. Biol.* **94,** 15–42.

Heglund, N. C., and Taylor, C. R. (1988). Speed, stride frequency and energy cost per stride: How do they change with body size and gait? *J. Exp. Biol.* **138,** 301–318.

Henneman, E., Somjen, G., and Carpenter, D. O. (1965). Functional significance of cell size in spinal mononeurons. *J. Neurophysiol.* **28,** 560–580.

Hill, A. V. (1938). The heat of shortening and the dynamic constants of muscle. *Proc. R. Soc. London, Ser. B* **126,** 136–195.

Hill, A. V. (1964). The efficiency of mechanical power development during muscular shortening and its relation to load. *Proc. R. Soc. London, Ser. B* **159,** 319–324.

Hirano, M., and Rome, L. C. (1984). Jumping performance of frogs (*Rana pipiens*) as a function of muscle temperature. *J. Exp. Biol.* **108,** 429–439.

Hou, T. T., Johnson, J. D., and Rall, J. A. (1990). Effects of temperature on Ca and Mg dissociation rate from frog parvalbumin. *Biophys. J.* **57,** 550a.

Huxley, A. F. (1957). Muscle structure and theories of contraction. *Prog. Biophys. Biophys. Chem.* **7,** 255–318.

Huxley, H. E. (1985). The crossbridge mechanism of muscular contraction and its implications. *J. Exp. Biol.* **115,** 17–30.

Johnson, T. P., and Johnston, I. A. (1991). Power output of fish muscle fibres performing oscillatory work: Effects of acute and seasonal temperature change. *J. Exp. Biol.* **157,** 409–423.

Josephson, R. K. (1985a). Mechanical power output from striated muscle during cycle contraction. *J. Exp. Biol.* **114,** 493–512.

Josephson, R. K. (1985b). The mechanical power output of a tettigoniid wing muscle during singing and flight. *J. Exp. Biol.* **117,** 357–368.

Josephson, R. K., and Stokes, D. R. (1989). Strain, muscle length and work output in crab muscle. *J. Exp. Biol.* **145,** 45–61.

Julian, F. J., Rome, L. C., Stephenson, D. G., and Striz, S. (1986). The maximum speed of shortening in living and skinned frog sartorius muscle fibres. *J. Physiol. (London)* **370,** 181–199.

Lannergren, J. (1987). contractile properties and myosin isoenzymes of various kinds of *Xenopus* twitch muscle fibres. *J. Muscle Res. Cell Motil.* **8,** 260–273.

Lannergren, J., Lindblom, P., and Johansson, B. (1982). Contractile properties of two varieties of twitch muscle fibres in *Xenopus laevis*. *Acta Physiol. Scand.* **114,** 523–535.

Lutz, G., and Rome, L. C. (1993). Built for jumping: The design of the frog muscular system. *Science* **263,** 370–372.

Mai, M. T., and Lieber, R. L. (1990). A model of semitendinosus muscle sarcomere length, knee and hip joint interaction in the frog hindlimb. *J. Biomech.* **23,** 271–279.

Marsh, R. L. (1988). Ontogenesis of contractile properties of skeletal muscle and sprint performance in the lizard *Dipsosaurus dorsalis*. *J. Exp. Biol.* **137,** 119–139.

Marsh, R. L. (1990). Deactivation rate and shortening velocity as determinants of contractile frequency. *Am. J. Physiol.* **259,** R223–R230.

Marsh, R. L., and Bennett, A. F. (1986). Thermal dependence of sprint performance of the lizard *Sceloporus occidentalis*. *J. Exp. Biol.* **126,** 79–87.

Metzger, J. M., and Moss, R. L. (1987). Greater hydrogen ion-induced depression of tension and velocity in skinned single fibres of rat fast than slow muscle. *J. Physiol.* **393,** 727–742.

Moss, R. L. (1986). Effects of shortening velocity of rabbit skeletal muscle to variations in the level of thin-filament activation. *J. Physiol. (London)* **377,** 487–505.

Reiser, P. J., Moss, R. L., Giulian, G., and Greaser, M. L. (1985). Shortening velocity in single fibres from adult rabbit soleus muscle is correlated with myosin heavy chain composition. *J. Biol. Chem.* **260,** 9077–9080.

Ridgway, E. B., and Gordon, A. M. (1984). Muscle calcium transient: Effect of post-stimulus length changes in single fibers. *J. Gen. Physiol.* **13,** 75–103.

Rome, L. C. (1990). The influence of temperature on muscle recruitment and function *in vivo*. *Am. J. Physiol.* **259,** R210–R222.

Rome, L. C. (1992). Scaling of muscle fibers and locomotion in mammals. *J. Exp. Biol.* **168,** 243–252.

Rome, L. C., and Kushmerick, M. J. (1983). The energetic cost of generating isometric force as a function of temperature in isolated frog muscle. *Am. J. Physiol.* **244,** C100–C109.

Rome, L. C., and Sosnicki, A. A. (1990). The influence of temperature on mechanics of red muscle in carp. *J. Physiol. (London)* **427,** 151–169.

Rome, L. C., and Sosnicki, A. A. (1991). Myofilament overlap in swimming carp. II. Sarcomere length changes during swimming. *Am. J. Physiol.* **260,** C289–C296.

Rome, L. C., and Swank, D. (1992). The influence of temperature on power output of scup red muscle during cyclical length changes. *J. Exp. Biol.* **171,** 261–281.

Rome, L. C., Loughna, P. T., and Goldspink, G. (1984). Muscle fiber recruitment as a function of swim speed and muscle temperature in carp. *Am. J. Physiol.* **247,** R272–R279.

Rome, L. C., Funke, R. P., Alexander, R. McN., Lutz, G., Aldridge, H. D. J. N., Scott, F., and Freadman, M. (1988). Why animals have different muscle fibre types. *Nature (London)* **355,** 824–827.

Rome, L. C., Funke, R. P., and Alexander, R. McN. (1990a). The influence of temperature on muscle velocity and sustained performance in swimming carp. *J. Exp. Biol.* **154,** 163–178.

Rome, L. C., Sosnicki, A. A., and Goble, D. O. (1990b). Maximum velocity of shortening of three fibre types from the horse soleus: Implications for scaling with body size. *J. Physiol. (London)* **431,** 173–185.

Rome, L. C., Choi, I., Lutz, G., and Sosnicki, A. A. (1992a). The influence of temperature on muscle function in fast swimming scup. I. Shortening velocity and muscle recruitment during swimming. *J. Exp. Biol.* **163,** 259–279.

Rome, L. C., Sosnicki, A. A., and Choi, I. (1992b). The influence of temperature on muscle function in the fast swimming scup. II. The mechanics of red muscle. *J. Exp. Biol.* **163,** 281–295.

Rome, L. C., Swank, D., and Corda, D. (1993). How fish power swimming. *Science* **261,** 340–343.

Smith, R. S., and Ovalle, W. K., Jr. (1973). Varieties of fast and slow extrafusal muscle fibres in amphibian hind limb muscles. *J. Anat.* **116,** 1–24.

Sosnicki, A. A., Lutz, G. J., Rome, L. C., and Goble, D. O. (1989). Histochemical and molecular identification of fiber types in single, chemically skinned equine muscle cells. *J. Histochem. Cycochem.* **37,** 1731–1738.

Sosnicki, A. A., Loesser, K., and Rome, L. C. (1991). Myofilament overlap in swimming carp. I. Myofilament lengths of red and white muscle. *Am. J. Physiol.* **260,** C283–C288.

Stevens, E. D. (1988). Effect of pH and stimulus phase on work done by isolated frog sartorius muscle during cyclical contraction. *J. Muscle Res. Cell Motil.* **9,** 329–333.

Sweeney, H. L., Kushmerick, M. J., Mabuchi, K., Sreter, F. A., and Gergely, J. (1988). Myosin alkali light chain and heavy chain variations correlate with altered shortening velocity of isolated skeletal muscle fibers. *J. Biol. Chem.* **263,** 9034–9039.

Taylor, C. R., Heglund, N. C., McMahon, T. A., and Looney, T. R. (1980). The energetic cost of generating muscular force during running. *J. Exp. Biol.* **86,** 9–18.

van Leeuwen, J. L., Lankheet, M. J. M., Akster, H. A., and Osse, J. W. M. (1990). Function of red axial muscle of carp (*Cyprinus carpio* L.): Recruitment and normalised power output during swimming in different modes. *J. Zool.* **220,** 123–145.

Williams, T. L., Grillner, S., Smoljaninov, V. V., Wallen, P., Kashin, S., and Rossignol, S. (1989). Locomotion in lamprey and trout: The relative timing of activation and movement. *J. Exp. Biol.* **143,** 559–566.

ADVANCES IN VETERINARY SCIENCE AND COMPARATIVE MEDICINE, VOL. 38A

Relating Mechanics and Energetics during Exercise

C. RICHARD TAYLOR

Harvard University, Bedford, Massachusetts 01730

I. Overview

When we exercise, we breathe harder, our metabolism increases, and we feel like we are keeping fit. Perhaps we are even doing some useful work in the process. Our muscles consume this extra energy produced by metabolism to power the exercise. This paper looks at how the exercising muscles use energy: how they work and how, in some cases, they do not work at all (in the physical sense). Contrary to our intuition, muscles often consume lots of energy and fatigue without doing any work. Some of their most important jobs are as brakes, shock absorbers, and springs. We will focus on some of the most common exercises that many animals, including humans, perform: walking, running, hopping, and climbing. Our goal will be to develop some general rules for relating whole-animal performance to the performance of muscle cells. Finally, we will see how these rules might, or might not, apply to other exercises that humans and other terrestrial animals do not do so well, such as flying and swimming.

II. The Basic Machine

Let's start by considering the muscle cells that consume energy during exercise. They can be thought of as "biological machines" for converting metabolic energy into force, work, and power. These machines are remarkably similar among vertebrates, as has been discussed in previous paper in this volume. Our muscles are not much different from those that power a swimming fish, a flying bird, or a hopping frog. In fact, much of what we know about how muscles work comes from studies on frog muscles, which turn out to be much easier to study in the laboratory than those of birds and mammals.

What exactly do we mean by work and power? These terms have many meanings in common usage, but we will use the physical definitions. Work is the product of force and distance, and power is energy per unit time. The metabolic power of the muscles is the rate at which they consume chemical energy, and mechanical power is the rate at which they perform work. Therefore, work is force × distance and power is work/time. Using these definitions, it immediately becomes obvious how our muscles can consume energy and fatigue without doing any work. If I lift a 10-kg weight (98 N) from the ground to above my head (a height of about 2 m), then my muscles do about 200 J of work and consume 800–1000 J of chemical energy. About 20–25% of the chemical energy generated by metabolism is converted into gravitational potential energy, and the rest is lost as heat. Now if I continue to support the weight above my head, my muscles would continue to consume energy and eventually fatigue, but they would not be doing any work.

Among vertebrate muscles, the maximal stress (force per cross-sectional area of fibers) and strain (distance over which the muscle can exert maximal force divided by resting length) are generally considered invariant properties (see papers by Billeter *et al.* and Rome, this Volume). These invariant properties are important because they provide us with a starting point for developing some general rules relating the energetics and mechanics of muscle. It is helpful to briefly review some of the similarities in muscle structure that underlie its invariant properties before considering the role these properties play in linking energetics and mechanics.

Why are these properties of muscles invariant? The fundamental reason lies at the molecular level, as discussed in previous papers. Vertebrates use the same two proteins, actin and myosin, to generate force with muscle. When a muscle is activated, a chemical interaction occurs between these two proteins. Millions of tiny molecular heads projecting from the myosin molecules attach and form cross-bridges with the actin molecules, then ratchet and detach from the actin molecules, generating force. If resisting forces are not too great, the two proteins slide past each other and perform work.

An ATP (adenosine triphosphate) is consumed each time a cross-bridge detaches. During the cycle the ATP is broken down to ADP (adenosine diphosphate) and P_i (inorganic phosphorus), providing the energy in the high-energy phosphate bond to power the movement. The ATP molecules are regenerated by oxidative phosphorylation in the mitochondria of the muscle cells. About 20.1 J of chemical energy are released for each milliliter of oxygen consumed by the mitochondria.

Three processes consume the chemical energy that powers muscle. First is the cost of converting the chemical energy contained in carbohydrates, fats, and proteins into small enough packages (ATP) so that it can be used by the cell. This is costly, and about 60% of the chemical energy contained in the substrates is lost as heat in this process. Second, the cross-bridges must be activated. The activation–deactivation process uses the energy of ATP to pump Ca into the sarcoplasmic reticulum and consumes about 30% of the ATP (12% of the total chemical energy). Third, the cycling of cross-bridges between actin and myosin consumes about 70% of the ATP (28% of the total chemical energy) (Rall, 1985). If all the ATP energy consumed by the cross-bridges were converted into work, that is, if efficiency were 100% in converting chemical energy into work, then the efficiency of the conversion of substrate energy into work would be 28%. It is remarkable that both *in vivo* and *in vitro*, muscles reach maximal efficiencies of 20–25%, very close to this limit. The sarcomeres are clearly very efficient machines.

The actin and myosin molecules are arranged in a regular geometric pattern to form the basic element of force generation, the sarcomere (see paper by Billeter *et al.*). The regularity of the pattern in skeletal muscle gives rise to the name "striated" and has enabled physiologists to use microscopy and X-ray diffraction as tools to understand how muscles work and to follow the cross-bridge cycle. The dimensions of the thick and thin filaments and the number of cross-bridges/sarcomere are roughly the same in all vertebrate striated muscle, resulting in the same maximum force per cross-sectional area (stress), about 30–40 N cm^{-2}. The distance over which this force can be exerted is also set by the geometry of the sarcomere, about 0.3 of its length (strain of 0.3 cm cm^{-1} muscle), as discussed in the preceding paper by Rome.

Because maximum stress and maximum strain are so similar among the vertebrates, it would seem to follow that the maximum work per volume of muscle in a single contraction should also be the same (*i.e.*, stress of 30–40 N cm^{-2} × strain of 0.3 cm cm^{-1} = about 0.09–0.12 J cm^{-3} of work/volume). However, muscles generate their maximal force only when they do not shorten (isometric contraction). During such a contraction, strain is zero and no work is done.

The amount of force muscles generate depends on how rapidly they shorten. From everyday experience, we know we can lift light loads faster than heavy ones. In fact, as we add weights there comes a point where the load is too heavy to lift at all. The relationship between muscle force and velocity was first described for isolated muscles by Wallace Fenn while he was working in A. V. Hill's laboratory (Hill, 1970). He found that force falls in a regular hyperbolic manner as shortening velocity increases until, at the muscle's maximum shortening velocity, no force is exerted and no work is done by the muscle.

The relationship between force and velocity is similar for most vertebrate striated muscles. They generate their maximum force when the load just equals their ability to generate force and no shortening occurs. If the load is increased still further, the attached cross-bridges of the active muscle are stretched, and they absorb energy. Under these conditions, the forces developed can reach one and a half times the maximum force generated by the muscle alone. Some of the strain energy stored in the stretched cross-bridges can be recovered. We will see that in some exercises, like running, energy stored in cross-bridges as the foot lands is recovered on takeoff. The muscle, in effect, has become a spring. The cross-bridge can also store chemical energy in the cross-bridge high-energy phosphate bond as the muscle is stretched, and this can also be recovered (Cavagna and Citterio, 1974).

Even though force decreases as muscles shorten, we would still expect the volume-specific maximum work per contraction to be an invariant property of muscles, occurring when they shorten at their slowest velocities, close to isometric. The maximum work should be close to the 0.09–0.12 J cm^{-3} calculated here. But is this necessarily an important parameter for muscles involved in locomotion? Because of differences in the amount of time during the stride cycle during which muscles develop force, we might, a priori, think that maximum work *rate* would be more important than the total amount of work. Is this variable also invariant among vertebrates?

The answer is no; maximum work rate varies dramatically with body size. The mechanical power generated by the muscle ($\dot{E}_{mech}$) is simply force × shortening velocity. Mechanical power is zero when force is maximal (because the muscle does not shorten). It increases to a maximum when the muscle shortens at about 0.3 times its maximal velocity, and then falls to zero at the maximum shortening velocity (because force is zero). Maximal velocity (V_{max}) and maximal power ($\dot{E}_{mech,max}$) vary more than 1000-fold among vertebrates. V_{max} can even vary by more than 10-fold among different fibers of the same muscle, allowing faster muscle fibers to be recruited and more power to be generated as an animal runs faster (Rome *et al.*, 1990).

These differences in performance disappear when we standardize shortening velocity to V_{max} (Hill, 1950). Figure 1 shows that relative muscle force (F/F_{max}) and relative mechanical power ($\dot{E}_{mech}/\dot{E}_{mech,max}$) become invariant functions of velocity when we standardize them to relative shortening velocity (V/V_{max}). In general, muscles produce their maximum power over a narrow range of V/V_{max}, 0.15 to 0.40. Although this relationship may appear complicated at first, the point to remember is that mechanical performance is the same when V/V_{max} is the same. This property of muscle is important in understanding how animals use their muscles. For example, fish adjust the recruitment of muscle fibers so that they deliver close to maximal power over a wide range of swimming speeds and water temperatures, as Rome and collaborators have shown in their elegant experiments on swimming fish (see paper by Rome). It seems likely that other animals will also use their muscles to maximize power output when doing work on their environments.

How do we link mechanical power output to metabolic power input?

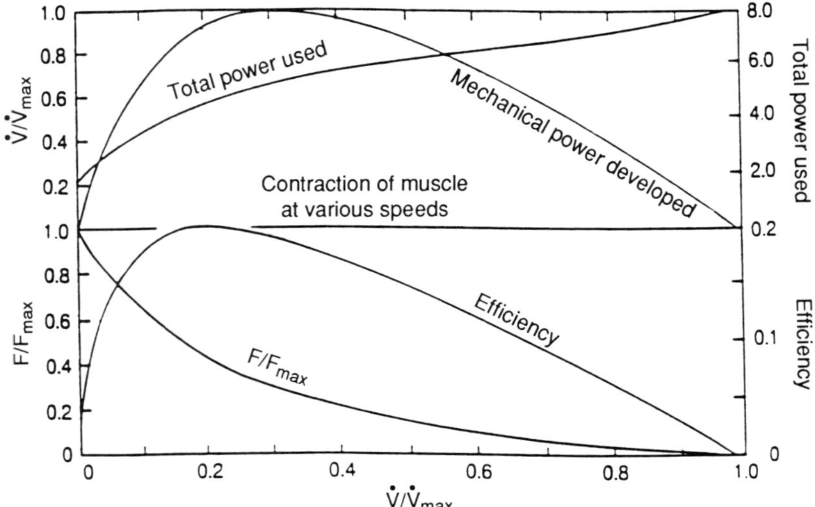

FIG. 1. Hill's classic explanation of how muscles work efficiently during locomotion can also be used to explain how they could operate economically to generate force. Shortening velocity of the muscle, V, is standardized to its maximum shortening velocity without a load, V_{max}. Force, F, is plotted as a fraction of maximum force, F_{max}. Efficiency is mechanical work done/total metabolic energy used. Mechanical power developed by the muscle is calculated as relative force × relative velocity. Total power used is calculated from measurements of heat produced by the muscle. The highest force is developed and the least amount of metabolic power is used when the muscle does not shorten and V/V_{max} is zero. Force decreases and metabolic cost increases as shortening velocity increases.

Again, it was A. V. Hill (1938) who established the link with his classic muscle experiments. He measured the heat produced by muscles over their range of shortening velocities. From these measurements he calculated the metabolic power consumed to power the muscle, and its efficiency. These calculations revealed another invariant property of striated muscle. The efficiency of converting metabolic power into mechanical power is the same when muscles shorten at the same relative velocity. If the maximum shortening velocity of a muscle is ten times faster, it both works and consumes energy at ten times the rate as at any given V/V_{max}.

III. The Cost of Generating Force, Work, and Power

Now we can return to the pursuit of our goal: applying these invariant properties of muscles to develop simple rules that link mechanics and energetics of muscle during some common exercises. For muscularly equivalent activities, that is, those with the same ratio of V/V_{max}, the rates of energy consumption and work will be directly proportional to V_{max}, and efficiency will be the same. It seems reasonable to assume that over the course of evolution, muscles have been optimized to perform their jobs economically, whether they are used for swimming, running, walking, hopping, or flying. They should operate at a V/V_{max} for which the least amount of muscle is used at the lowest metabolic cost. The swimming fish, for example, uses a narrow range of V/V_{max} over which mechanical power and efficiency are maximized. Different fibers are recruited over the range of speeds and muscle temperatures so that the muscles perform most economically.

How can we apply these invariant properties of muscles to link the energetics and mechanics of running? If there were a single V/V_{max} at which performance is optimized, then animals would need to manipulate V_{max} in order to economize. To understand how animals might do this, let's start by considering muscles as simple machines for generating force. We can then write an equation expressing rate of energy consumption of a muscle ($\dot{E}_{metab}$, in W) in terms of the force it generates ($\bar{F}_{muscle}$, in N), its fiber length [l_{fiber}, in m (fiber length)$^{-1}$], and its V_{max} (in fiber length s^{-1}) for exercises in which V/V_{max} is the same:

$$\dot{E}_{metab} = \mathrm{k}_1 \cdot \bar{\mathrm{F}}_{muscle} \cdot \mathrm{l}_{fiber} \cdot \mathrm{V}_{max}. \tag{1}$$

Sarcomeres of long and short fibers have the same length and are invariant; long muscles, with long fibers, just have more of these basic units in series. Both V and V_{max} are usually expressed in units of fiber length s^{-1}, therefore, dividing shortening velocity by fiber length gives a measure of sarcomere shortening velocity, which is why it is expressed this way. However, because we are interested in muscle energetics, we must consider muscle length, as a muscle that is ten times as long will shorten ten times as rapidly and consume energy at ten times the rate for the same sarcomere shortening velocity. For this reason, fiber length must be included in the equation.

We can rearrange Eq. (1) to express the metabolic cost of generating a unit of force:

$$\frac{\dot{E}_{metab}}{\bar{F}_{muscle}} = k_1 \cdot l_{fiber} \cdot V_{max}. \tag{2}$$

The available evidence supports the idea that the cost of generating force is directly proportional to the number and V_{max} of the active sarcomeres when V/V_{max} is the same. The best data are from isometric muscles (V/V_{max} is 0), for muscles operating at maximal efficiencies (V/V_{max} is about 0.2), and for muscles shortening at V_{max} (Woledge *et al.*, 1985; Heglund and Cavagna, 1987). Although the data are quite variable, Eq. (2) seems to be a reasonable first approximation that can be accommodated in the Huxley (1957) model of cross-bridge dynamics (see Rome, 1992).

Although k_1 is constant for the same V/V_{max}, it changes dramatically with the relative speed of shortening. It is cheapest to generate force when the muscle is isometric, as seen in Fig. 1. The muscle generates its highest force and its metabolic rate is lowest when velocity is zero. The cost of generating force increases dramatically as the muscle shortens faster, because force falls and metabolic rate increases. Cost has increased by about threefold when V/V_{max} reaches 0.2 and efficiency approaches a maximum, and by about four- to fivefold when V/V_{max} approaches 0.3 and mechanical power reaches a maximum.

We have a simple equation relating energy cost and force to muscle performance, but how can we apply this to exercises like running? When we run, our muscles do not shorten at a constant velocity; they are stretched when the foot lands, reverse direction, and then shorten during takeoff. In fact, our muscles will only exercise at a constant velocity when we exercise on a Nautilus machine, or some other sort of "muscle lever." If our muscles do not shorten at a constant V/V_{max} during running, can we use Eq. (2)? Fortunately, it can be shown that the energy consumed for any pattern of shortening velocities should be a constant multiple of the energy cost of generating isometric force (Alexander, 1991).

IV. How Do Muscles Work during Running?

The simplest assumption possible would be that muscles of running animals, like those of swimming fish, work close to their maximal power outputs and efficiencies during locomotion. If this were the case, we would expect the efficiency of the muscles to be about 20–25% over the entire range of speeds for all running animals. Measurements of the efficiency of running show that this assumption is clearly wrong. In fact, we shall see that it may be more reasonable to assume the muscles operate at near zero efficiency, that is, they do not work during running. Let's review the evidence that leads us to such a surprising conclusion.

Speed, size, and limb design are three variables that provide at least 10-fold differences in energetics and/or mechanics of running. These large differences enable us to develop and test general rules about how muscles of running animals work. Let's start our review with energetics, proceed to mechanics, and then try to decide what the muscles of running animals are actually doing.

V. Energy Cost of Running

A. Speed

How does the energy cost of running change with speed? A hundred years ago this was an important practical question because most forms of land transportation were still powered by muscle. It was important to know how to minimize cost and increase economy. We face similar problems today in designing fuel-efficient motorized vehicles. A century ago a German scientist, Zuntz (1897), developed a convenient way to measure the energetics of running. He built a treadmill and measured the rate of oxygen consumption while humans, horses and dogs ran at constant speeds. Using energetic equivalents for the fuels being oxidized (about 20.1 J ml^{-1} O_2), he calculated their rates of energy consumption. His findings were surprising, but simple. The first finding he made was that the rate of energy consumption increased linearly with speed in all the animals.

A linear increase in oxygen consumption with speed is the first general rule for energetics of running. It leads to the remarkable conclusion that the amount of energy our muscles use to run a distance is independent of how fast we run. This is counterintuitive, because we are much more out of breath and think we have used much more energy if we run a mile as fast as we can than if we jog it at a leisurely pace. The difference, however, is not in the amount of energy used, but how fast we use it. The slope of the linear relationship (a constant) is simply the increment in energy consumed per increment in speed. Expressed on a mass-specific (per gram) basis, it represents the amount of energy consumed by the muscles in moving a unit body mass over a unit distance, and is usually referred to as the cost of transport. The linear increase has been found in hundreds of species, with only a few exceptions. The kangaroo is a remarkable exception to which we will return later.

B. Size

The second general rule of running energetics is that cost of running decreases with body size, again expressed on a mass-specific basis. This rule was also noticed a hundred years ago by Zuntz. Slowtzoff (1903), working in Zuntz's laboratory, quantified the relationship for dogs of different size. The generality of the relationship was shown 50 years later for mammals spanning an 850-fold range of body weight (Taylor *et al.*, 1970). It turns out that cost of running decreases in such a predictable way with size that the "size rule for running" can be expressed quantitatively by the simple equation

$$\frac{\dot{E}_{\mathrm{metab}}}{M_{\mathrm{b}}} = 10.7 M_{\mathrm{b}}^{-0.316} \cdot V, \tag{3}$$

where $\bar{E}_{\mathrm{metab}}$ is in W, M_{b} is body mass in kg, and V is running speed in m s^{-1}. This equation is based on data from 62 species of running birds and mammals spanning a 500,000-fold range of body weight (Taylor *et al.*, 1982). On a per gram basis, a mouse uses about 20 times as much energy to travel a mile as a pony. If we consider the size extremes of living land mammals, from shrews to elephants, the difference reaches 100-fold.

The size differences in cost of running are enormous! They provide a powerful tool for unraveling the link between energetics and mechanics of muscles during locomotion. How can we explain this 100-fold difference in cost in terms of what the muscles are doing? Are the mouse muscles working 100 times as hard as the elephant's? Any explanation of the energetics of running must account for these differences!

C. Two Legs versus Four Legs

Humans use about twice as much energy to run a mile as a four-legged mammal of the same size. This raises the interesting question of whether or not there is a fundamental difference between the energetics of running on two versus four legs. Does economy increase directly with the number of legs that an animal uses when running? A study of seven species of birds (Fedak *et al.*, 1974) initially seemed to confirm this difference. However, as more data became available on more species of birds, the difference disappeared (Fedak and Seeherman, 1979; Paladino and King, 1979). Although waddling is more costly than running (Pinshow *et al.*, 1977), there is no difference in cost of running between birds and mammals. This is an important result, because the bipedal birds have longer legs and take fewer strides over a given distance than quadrupedal mammals of the same size. As a consequence, we might have expected the birds to run more economically, and we will return to discuss why this is not the case later.

Not only birds and humans have become specialized for two-legged locomotion. Many mammals travel bipedally, but most hop rather than run. A priori, we might guess that the energetics of hopping would be very different from that of running, and it is, at least in kangaroos! Instead of increasing as it hops faster, the kangaroo's rate of energy consumption stays constant or decreases slightly (Dawson and Taylor, 1973). Fuel economy improves with speed, and at moderate speeds it surpasses the economy predicted by Eq. (3) for running animals. Kangaroos appear to travel nearly effortlessly at high speeds, bouncing along the ground as if they were powered by springs instead of muscles. In their natural habitat, they travel long distances at high speeds, staying ahead of fast-moving brush fires or moving between patches of green pasture that follow patchy rains. Although kangaroos are clearly an extreme case, they dramatically demonstrate that we must consider the possible role that springs might play in powering all locomotion.

Springs do not seem to be as important for small hoppers as they are for kangaroos. In fact, kangaroo rats, spring hares, and rat kangaroos use the same amount of energy to travel a mile as a runner of the same size, despite the remarkable difference in gaits (Thompson *et al.*, 1980). It is only the large hoppers, wallaby size (4–5 kg) and up, that increase their speed without using more energy. A quail, a chipmunk, and a kangaroo rat use the same amount of energy to travel a mile. Furthermore, the chipmunk and the kangaroo rat use the same number of strides, twice as many as the quail. We will see that stride length and frequency are important in determining the energetics of running, and the difference between these animals provides a good test of the general rules we are developing.

The question of whether or not it costs more to run on two or four legs can be asked directly in some primates. For example, organ grinder monkeys and chimpanzees routinely do both. Furthermore, they seem to like to perform. With the help of a few cookies, they can be easily trained to run on a treadmill and to switch between moving on two or four legs with a simple voice command. If there is a fundamental difference between energetics of bipedal and quadrupedal running, a comparison in the same animal would seem to be a good way to reveal it. We find that both species use the same amount of energy whether running on two or four legs (Taylor and Rowntree, 1973).

The conclusion seems to be that humans are not as economical runners as most other animals. However, their twofold higher cost is not simply a result of running on two legs instead of four. Other factors are involved. Despite this poor fuel economy, things could be worse for our species. We shall see that our economy improves dramatically when we walk. In fact, we consume about the same amount of energy to walk a mile as a pony or ostrich does to run it.

D. Limb Design

The third general rule of running energetics is that energy cost of running does not depend on limb design. Functional anatomists have argued convincingly that cursorial animals are "efficient runners," the limbs of which are designed for running economically (Howell, 1944; Hildebrand, 1960; Gray, 1968). Such a design involves lightening the distal parts of the limbs and/or concentrating the muscle mass of the limbs close to the hip/shoulder joint, thereby minimizing the distance it moves during a stride and the amount of energy needed to rotate the limb. Graviportal animals are at the opposite end of the continuum in terms of limb design. They require massive pillarlike limbs simply to support their weight, and these limbs should make them "inefficient runners."

The gazelle limb is often cited as one of the best examples of this type of economic design. It is so light and fragile that great care must be taken when handling these animals so as not to break their legs. Furthermore, the muscle mass of the legs is centered around the proximal joints, a design that would also help to minimize the work of swinging the limbs. The cheetah is an equally good runner (at least in terms of speed) of about the same size that has been built with different design constraints. It accelerates rapidly and uses its front legs to knock down prey. Moving cheaply certainly is not as important for the cheetah as having limb bones that are able to withstand the stresses involved in rapid acceleration and clubbing down prey.

These design arguments assume that much of the energy consumed during running is used to accelerate and decelerate the limbs during each stride. Gazelles, goats, and cheetahs provide a nice gradation of limb configurations in animals of the same body mass and limb length, and therefore provide the opportunity to quantify the effect of limb design on energetics of running.

Contrary to expectations, the three species use nearly the same amount of energy to run over a wide range of speeds, despite a 30-fold difference in the moments of inertia of their limbs (Taylor *et al.*, 1974). Furthermore, their cost of running is nearly identical to that predicted on the basis of body mass by Eq. (3). The amount of muscular energy used for swinging the limbs in these animals appears to be small relative to the total amount of energy consumed during running, at least over the speed range at which energy is supplied by oxidative pathways and can be measured. This finding is important in working out the relationship between energetics and mechanics of running.

Even when we consider the extreme of graviportal limb design, the elephant, we still do not find higher energetic costs of locomotion. Elephants are the largest living land animals; an adult male may reach a weight of 6 metric tons. These animals have massive pillarlike limbs that are needed to support their weight. Elephants can be trained to perform remarkable feats, as we know from watching a circus or a documentary on television, in which they may outperform bulldozers and other mechanical devices in harvesting large trees in the forests of Asia. They can even be trained to wear a gas mask and chase after a golf cart while their oxygen consumption is measured. These measurements show the elephant to be the most fuel-efficient living land animal, and that its energetic cost of locomotion matches that predicted by Eq. (3) for an animal of its weight remarkably well (Langman *et al.*, 1990). This finding adds further evidence that energetics appears to be independent of limb design.

VI. The Work of Running

A. What Work Do Muscles Do?

In contrast to swimming or flying animals, running animals do little work on their environment. The work required to overcome air resistance during running is small compared to when flying because the speeds are slow, and it is small compared to when swimming because the air imposes less drag than water. Humans provide a "worst case" example for the work needed to overcome air resistance. We rank among the fastest athletes in the animal kingdom, yet we are not streamlined to reduce our frontal area. Despite these design limitations, our muscles do a negligible amount of work in overcoming air resistance at comfortable running speeds (Pugh, 1971); they also do a negligible amount of work on the ground, unless we are running in sand or mud. As a result, we can safely ignore the work done on the environment in our considerations of the work that muscles do during running.

Once humans reach a constant speed, most of the muscular work of running is done on the body itself. The mechanical energy (kinetic energy + gravitational energy) of the body oscillates during each stride. The center of mass rises and falls, and speeds up and slows down in a regular manner. For example, at a speed of 3.3 m s^{-1} (an average speed run), the center of mass rises and falls 9 cm, and accelerates and decelerates by 0.25 m s^{-1} during each step. The energy of the limbs and other body parts, relative to the center of mass, also oscillates as a human swings its limbs and bends its trunk. These oscillations can be easily visualized by filming humans or animals while they run on a treadmill.

We can set a reasonable upper limit to the work the muscles do during running by adding all the increases in mechanical energy that occurr within the body during a stride. Then, if we assume all the decreases are lost as heat, the muscles have to supply the lost energy. This analysis assumes there is no storage and recovery of other forms of energy during the stride. If we analyzed a perfectly elastic bouncing ball in this way, we would estimate that its "muscles" were providing the energy to lift and accelerate the ball during each bounce. Muscles are attached to tendons, structures that store and release elastic energy. Indeed, even the cross-bridges of muscles can serve as elastic springs. In fact, there is good evidence that cross-bridges can store energy in chemical bonds when they are stretched, and this energy can be recovered on shortening (Cavagna and Citterio, 1974). The lower limit to the work that muscles do is zero. This would occur if all of the decreases in mechanical energy were stored and recovered during the stride, and the animal bounced like a ball.

At about the same time that Zuntz developed the treadmill technique for measuring the energetics of running, Marey and Demeny (1885) designed a number of ingenious devices to quantify the mechanics of running. They used a combination of accelerometers, air bellows shoes, articulated lever systems, and force platforms to measure the oscillations in kinetic and potential energy during running. Like Zuntz, they arrived at a simple and general rule: *mechanical energy is stored and recovered during each stride*. From their measurements, they concluded that the work done in lifting and accelerating the body and the limbs overestimates the work done by the muscles. The question is by how much?

B. Speed

Thirty years ago Cavagna and Margaria perfected a force platform/film technique to accurately measure the work of running humans (Cavagna *et al.*, 1964). Their question was exactly the same as ours: Do muscles operate at a V/V_{max} that provides maximal power and efficiency during running? They expected to find that the work to run a mile, like energy cost, would be independent of speed and account for about 25% of the energy consumed by the muscles. What they did find was not so simple and straightforward, but much more interesting. Like Marey, they were forced to conclude that muscles do not do all of the work during running.

Even at low running speeds, the work rate exceeded the capacity of muscles to convert chemical energy into work (25% efficiency). Then, in marked contrast to the linear increase in energy consumption, *work rate increased curvilinearly with speed* (Cavagna and Kaneko, 1977). This turns out to be a general rule for the mechanics of running. The work to run a mile increased by two- to threefold with speed, whereas the energy cost did not change. Muscles could account for a smaller and smaller fraction of the work as speed increased. Clearly, we have to conclude that not all the decreases in kinetic and gravitational energy were lost as heat. A progressively larger fraction of energy must have been stored as elastic/chemical energy in one part of the stride and recovered in another part as speed increased. At the highest speeds this recovery accounted for most of the work of running!

C. Size

Not only is there a mismatch between energetics and mechanics of locomotion with speed, there is also a mismatch with body size. A diverse assortment of birds and mammals, ranging in size from 30-g quails and chipmunks to 100-kg ostriches and rams, have been run across force platforms and filmed (Cavagna *et al.*, 1977; Heglund *et al.*, 1982). Calculations of their work rates as functions of speed and size show that *the general rule for mechanical work rate is that it is independent of body size* (in marked contrast to the size dependence of energetics). On a mass-specific basis, work rate changes in about the same way with speed in all animals, independent of size. We can express the "size rule for work" as a simple equation, just as we have done for energetics:

$$\frac{\dot{E}_{\mathrm{metab}}}{M_{\mathrm{b}}} = (0.478 \cdot V^{1.53}) + (1.21 \cdot V) + 0.72\,, \qquad (4)$$

in which $\dot{E}_{\mathrm{metab}}$ is in W, M_{b} is in kg and V is running speed in m s^{-1}. This equation describes the increase in work rate with speed equally well for birds running on two legs and mammals running on four.

As with the curvilinear increase with speed, the size independence of work is both surprising and informative. Taken together with the 100-fold difference in energetics that occurs in the same animals, it

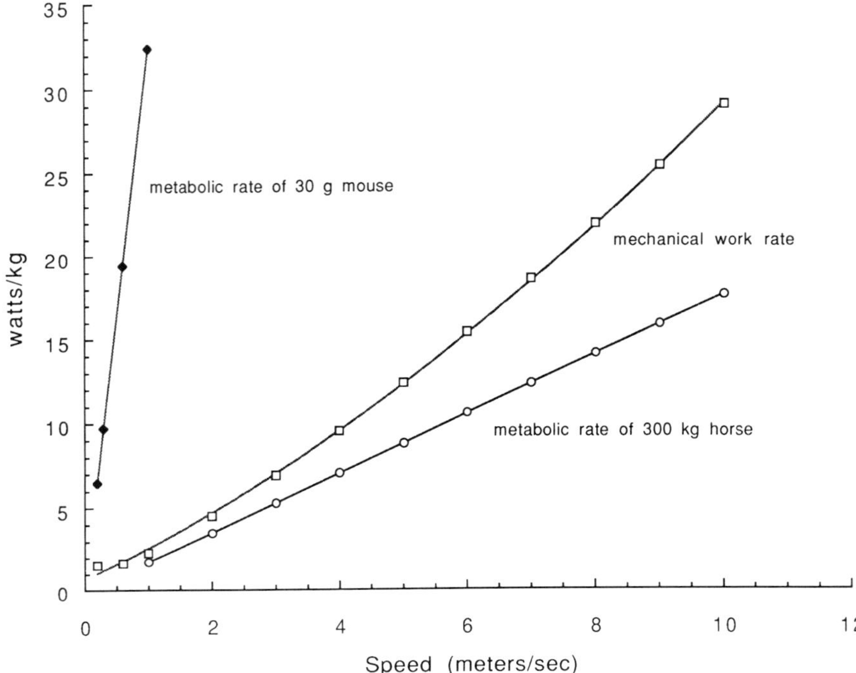

FIG. 2. Mechanical work rate (power) of running animals increases in the same way with speed over the entire size range of birds and mammals; it is identical in a 30-g mouse and 300-kg pony. In contrast, the metabolic rate of muscles increases much more rapidly with running speed in small animals, as illustrated for a mouse and pony. The mechanical work rate of the horse exceeds the rate at which its muscles consume energy over its entire range of speeds, indicating that springs must supply the difference between energetic input and output.

leads to the conclusion that muscles of running animals do not work at their maximal efficiencies over the range of speed and body size. We have to look for another explanation of the energetics of running! This can be appreciated by using our "size rules for work and mechanical energy" [Eqs. (3) and (4)] to calculate energy, work, and efficiency of running. Let's compare values for a 300-kg pony and a 30-g mouse running at their fastest sustained speeds (Fig. 2).

D. Running Springs

The pony consumes about 1.9 J kg^{-1} m^{-1} of metabolic energy over its range of speeds, while its mechanical work increases from 1.7 to 2.8 J kg^{-1} m^{-1}. At its slowest speed, the work accounts for almost 90% of the energy consumed. The muscles could only provide less than one-third of the work, even if they operated at their highest possible efficiency of 25%. At the fastest speed, work reaches 150% of the energy consumed. Now the muscles can only account for one-sixth of the total work done, indicating that the remainder must be done using energy that was stored during the stride. Recovery of stored energy is even more important in ponies than in humans. The conclusion is simple: "springs" do most of the work in large animals at high speeds. Are they equally important in small animals?

Let's consider similar relationships in the mouse. We calculate that the mouse muscles consume about 30 J to move a gram a kilometer as it increases its speed from its slowest trot to its fastest sustained gallop. The mouse does not run nearly as fast as a pony and has a much smaller speed range, thus the work done is smaller and changes less. Over its speed range, the mechanical work done by the mouse remains almost unchanged at about 1.8 J g^{-1} km^{-1}. The maximum amount of energy that could be recovered from elastic/chemical stores is only 6% of the energy consumed, in contrast to the minimum of 70 to 83% for the pony. Clearly, there is not a large amount of metabolic energy to be saved by the mouse, even with perfect springs!

Our calculations show that recovery of elastic/chemical energy increases with speed and size and accounts for most of the work over the entire speed range in larger animals. It is possible that mice, like ponies, have good springs that conserve most of the mechanical energy during a stride, but there is not much of a saving to be had in terms of metabolic energy. Finally, we can conclude that mouse muscles are very inefficient in terms of converting chemical energy into the work of running. Even if there were no springs and mouse muscles did all the work, their efficiency would be only 6%.

R. McNeill Alexander and his colleagues have used a more direct approach to evaluate the importance of springs in running animals (see Alexander, 1988). They have calculated the amount of energy stored during locomotion in some of the large tendons of kangaroo rats, kangaroos, humans, donkeys, and deer and compared them to the energy changes of the centers of mass. They arrive at similar conclusions to those we have considered, namely, storage and recovery of elastic energy are very important. In fact, they show that when a 70-kg human runs at 4.5 m s^{-1}, more than half of the energy can be stored in just two springs, the Achilles tendon and the arch of the foot.

Springs may well store all the mechanical energy at slow speeds and in small animals, and this could be very important for minimizing the cost of generating force, as we shall see. There is good indirect evidence supporting this idea. Mathematical models of running that treat animals as simple spring–mass systems bouncing along the ground (McMahon, 1984; Blickhan, 1989) do a remarkably good job of explaining the mechanics of running, galloping, and hopping over the entire range of speeds and animal size. Measurements of peak forces and displacements of the legs can be used to calculate stiffnesses of the "leg springs" for animals spanning the size range of hoppers from kangaroo rats to kangaroos, and runners from white rats to humans and horses. Over the entire range of animal size, stiffness is adjusted to match the resonant frequency of the spring and to maximize elastic storage (Farley *et al.*, 1993). This finding certainly suggests that elastic storage may be important even in the small animals for which the magnitude of the metabolic savings would be small.

VII. Economic Force Instead of Efficient Work?

We have used our comparative tools to formulate some general rules about the energetics and mechanics of running. These rules disprove the simplest hypothesis for linking energetics to mechanics of running. *Muscles do not maximize mechanical power and efficiency!* Where do we go from here? Let's stop and think about what the muscles of running animals do, formulate a new hypothesis, and try again.

When animals walk, run, hop, trot, or gallop, muscles do not have to work against the environment, but they do have to support the body's weight. Otherwise, the limbs would collapse and that would be the end of the run. Even in the best of all possible worlds, one in which the tendons are "perfect springs" that alternately store and release all the mechanical energy changes, the muscles do not have to shorten and do work, but they still have to be active and generate force each time the foot touches the ground. The magnitude of the force is fixed by the animal's weight, regardless of the speed, size, number of limbs, or limb design. It has been proposed that the cost of generating this force economically sets the cost of running. Let's apply our rules relating to the energetics and mechanics of locomotion to test this hypothesis.

When an animal runs at a constant speed, the average force exerted on the ground, $\bar{F}_{\mathrm{ground}}$, must equal its body weight, F_{bw}. If the force exerted were greater than body weight, the animal would accelerate (upward and/or forward); if it were less, the animal would sink into the ground and/or slow down. How can muscles generate this force economically? As we have seen (Fig. 1), the highest forces are developed by muscles when they are isometric and do not shorten, when the metabolic cost of generating force is lowest. Is it possible for the muscles to operate isometrically during locomotion and generate force most economically?

We know that the muscles and their attached tendons are alternately stretched when the foot lands, and then shorten on takeoff. However, it is difficult to determine how much of the changes in length occur in muscle and how much in tendon. For maximal storage and recovery of energy, the muscles should undergo very small length changes, just enough to let the attached cross-bridges store and recover elastic and chemical energy without detaching; the tendons should undergo the remainder of the length change. Isometric activity should maximize energy storage, as well as minimize cost of generating force. It seems reasonable to frame our new hypothesis by assuming isometric activity.

It is average ground force/body weight that remains the same over the range of speed and body size. How do we relate ground force to muscle force? Sir James Gray pointed out more than 50 years ago that small animals run in a more crouched posture than do large animals, and have a poorer mechanical advantage for supporting their weight (see Gray, 1968). A more upright posture improves the mechanical advantage of the muscle by decreasing the moment arm of the ground force acting about the joints. Thus, larger animals need to generate less muscle force to support their weight. Biewener (1989a,b) has measured the ratio of peak ground force to peak muscle force (the effective mechanical advantage of the limb muscles) over a range of speeds and in a large size range of quadrupeds and small bipedal hoppers. He found that it changes very little with speed and varies in a regular way with body size. The average effective mechanical advantage of both legs over the stride, EMA, is

$$EMA = k_2 \cdot M_b^{0.26}. \tag{5}$$

We need to incorporate these changes in ground force/muscle force with body size into our considerations of the energetics of running. The average muscle force during a stride, $\bar{F}_{\text{muscle}}$, is

$$\bar{F}_{\text{muscle}} = \frac{F_{\text{bw}}}{EMA} = \frac{F_{\text{bw}}}{k_2 \cdot M_b^{0.26}}. \tag{6}$$

The muscles of a 30-g mouse, as it runs on its bent legs, have to generate a force nearly 12× body weight during each stride. With its straighter legs, the mechanical advantage of a 300-kg pony approaches 1, and its muscles have only to generate a force equal to body weight.

Could this 12-fold difference in muscle force account for the size rule for energetics of running [Eq. (3)]? This possibility is certainly suggestive!

Now let's return to the question of how much energy is required to support body weight. We can calculate the muscular cost of supporting weight during running by combining Eq. (2) with Eq. (6):

$$\frac{\dot{M}_{\text{metab}}}{F_{\text{bw}}} = \frac{k_1}{k_2 \cdot M_{\text{b}}^{0.26}} \cdot l_{\text{fiber}} \cdot V_{\text{max}} . \tag{7}$$

This simple relationship applies across the range of speeds and body sizes for the quadrupeds and hoppers for which mechanical advantage has been measured. Now all we need to test our hypothesis of economic force generation are measures of fiber length and $V_{\max}$ across speeds and sizes.

It seems reasonable to assume that fiber length does not change with speed, because the same muscles are used for running at different speeds; however, larger animals with longer legs will have longer fibers. Alexander and co-workers (1981) have found that fiber length changes with body size in the same regular manner as *EMA*:

$$l_{\text{fiber}} = k_3 \cdot M_{\text{b}}^{0.26} . \tag{8}$$

If we now combine Eqs. (7) and (8), we find

$$\frac{\dot{E}_{\text{metab}}}{F_{\text{bw}}} = \frac{k_1}{k_2 \cdot M_{\text{b}}^{0.26}} \cdot k_3 \cdot M_{\text{b}}^{0.26} \cdot V_{\text{max}} , \tag{9}$$

which becomes

$$\frac{\dot{E}_{\text{metab}}}{F_{\text{bw}}} = k_4 \cdot V_{\text{max}} \tag{10}$$

in which k_4 combines the proportionality constants for length and mechanical advantage with k_1 from Eq. (2) (which is only constant at any particular ratio of $V/V_{\max}$). The size effects of fiber length and *EMA* cancel each other; the better mechanical advantage of larger animals is almost exactly offset by their having longer fibers, and the volume of muscle needed to support an animal's body weight is constant across size. This outcome makes sense in terms of design and is consistent with the concept that animals use the same fraction of muscle to support weight. Skeletal muscle accounts for nearly 40% of a mammal's body weight, regardless of size (Munro, 1969). It is difficult to imagine a mouse using 12× as much muscle to support its weight as a pony, as would be necessary if both animals had fibers of the same length. We can conclude that differences in *EMA* cannot account for the change in energetics with body size.

Equation (10) connects our "basic muscle" into the lever system of the limb. It predicts that the cost of generating force to support a running animal's weight will increase directly with the V_{max} of active fibers. To minimize this cost, the muscles must shorten very slowly or remain isometric; furthermore, this equation neglects the costs of swinging the limbs. If these assumptions are reasonable, then all we need is a measure of V_{max} to begin to use the equation to test our force-generation hypothesis.

A. Can Muscles Shorten Slowly?

We have already seen that there is good evidence to support the idea that muscle activity in small animals is nearly isometric. We calculated that the efficiency of a mouse's muscles never exceeds 6%. If we refer to the force–velocity–efficiency curve (Fig. 1), we see that such low efficiencies are consistent with very slow shortening velocities and economical force generation. In larger animals, where springs do most of the work during locomotion, the muscle has to function very nearly isometrically for the cross-bridges to work as springs or chemical stores. The length change that can occur cannot exceed the distance that the actin and myosin filaments can be stretched and recoil while remaining attached. Although the conclusion is counterintuitive and difficult to test, it seems likely that muscles of running animals remain nearly isometric over the entire range of body size and running speeds.

B. Can We Ignore the Limbs?

The cost of generating force during the time that the foot is in contact with the ground can account for the costs of operating the springs to lift and accelerate the center of mass and moving the limb backward, however, it only accounts for part of the energy needed to reverse limb direction and return it to the next support position. The amount of work involved in this process is small enough to be neglected in small animals and at slow speeds, but it is proportional to $V^{1.53}$ and becomes important in larger animals at moderate speeds. It seems likely that this energy is also stored and recovered, because energy consumption does not increase with this increase in work rate. Furthermore, we have seen that limb design has little effect on energetics of locomotion. This also suggests that energy cost of the limbs is small and can be ignored.

It is possible, however, to do a simple experiment to test this assumption directly. The force needed to support the weight of an animal can be increased without changing the work of swinging the limbs by simply having the animal carry a load. Animals ranging in size from rats to ponies can carry weights up to about 30% of their body mass without altering the accelerations of the center of mass, stride frequency, or the time of foot contact. Because force is simply mass × acceleration, a 30% increase in mass increases the muscle force for supporting the mass and the work on the center of mass by exactly 30% over the time of foot contact. It has no effect, however, on the work of swinging the limbs (assuming the additional mass is affixed to the body, not the limbs). If the cost of swinging the limbs were important, for example, half of the total cost, then a 30% increase in force would increase cost by only 15%. If it were insignificant, then we would expect to find close to a 30% increase in cost. In all the animals evaluated, the increase in energetic cost was directly proportional to the increase in force across the animals' entire ranges of speeds (Taylor *et al.*, 1980). This finding indicates that the cost of swinging the limbs can safely be ignored.

C. Cost of Generating Force

We can learn a great deal more about the energetics of locomotion from the results of this simple load-carrying experiment. The direct proportionality between load and cost shows that the cost of supporting each gram of load changes with speed and body size in exactly the same manner as the cost of running. It can cost a human or other animal 10x as much energy to generate 1 N of force at a high speed as at a low speed, and it can cost a small animal 10× as much as a large animal to generate 1 N of force at the same speed. These results suggest that the cost of generating force against the ground may play a major role in determining the energetics of locomotion. This simple experiment provides strong support for the force hypothesis.

It is also possible to reduce the force necessary to support body weight during locomotion by suspending a treadmill runner by long compliant springs. This type of experiment is of practical interest, particularly to astronauts, because it provides a good first approximation of running in reduced gravity. However, the springs do not change the amount of work required to swing the limbs, as would occur on the moon, and they can be used to test the force hypothesis. The results from this experiment (Farley and McMahon, 1992) lead to the same conclusion as with the load-carrying experiments. When force is reduced by 25%, energy cost of running is reduced by 25%; when force is reduced by 50%, cost is reduced by 50%; and even when force is reduced by 75%, cost is reduced by about 75%. Thus, direct proportionality between the cost of running and the force required to support body weight is found over a range from 0.25 to 1.3 times body weight. This certainly seems to be the major factor involved in linking the energetics and mechanics of running.

D. Can We Measure V_{MAX}?

It seems reasonable to assume that the cost of generating force to support body weight can account for most of the energy consumed by the muscles of running animals, and that muscles operate at slow velocities where this can be done economically. All that remains to test our model is to measure V_{max}. Force can only be applied during the time a foot is in contact with the ground, and this time decreases with increasing speed. This time course of force generation is easily measured using high-speed films. The reciprocal of this time can be thought of as an "intrinsic rate" of force development for the muscle fibers that are active. Faster muscles, with faster rates of force development, are needed to generate force more rapidly as the animal runs at a higher speed.

There is evidence from isolated muscles to suggest that the rate of force development is directly proportional to V_{max} (Stevens and Renaud, 1985), and it seems reasonable to assume this is also the case for the muscles supporting the weight of a running animal. If we make this assumption, then we can use the simple measure of foot contact time, t_c, as a measure of the intrinsic rate of force development; $1/t_c$ should be proportional to V_{max}:

$$V_{max} = k_5 \cdot \frac{1}{t_c} , \tag{11}$$

where t_c is the time of force generation in s, and k_5 has the units of fiber length.

Combining Eqs. (10) and (11), we produce the testable relationship

$$\frac{\dot{E}_{metab}}{F_{bw}} = c \cdot \frac{1}{t_c} \tag{12}$$

in which c becomes a cost coefficient for generating force; it combines the proportionality constants for *EMA*, fiber length, and the particular V/V_{max} that is used, and has the dimensions of length.

This simple relationship provides us with a very rigorous test of our hypothesis of economic force generation. $\dot{E}_{metab}$, F_{bw}, and $1/t_c$ are easily measured, allowing us to calculate the cost coefficient, c. The cost coefficient should be a constant only if V/V_{max} and the (volume of active muscle) F_{bw}^{-1} are the same over the entire range of speeds and sizes. Using the variables of speed and size, we can test for the constancy of c over a 100-fold difference in $\dot{E}_{metab}\,F_{bw}^{-1}$. What do we find when we do this?

E. Speed

Films of running animals reveal a linear increase in $1/t_c$ with speed, paralleling the linear increase in $\dot{E}_{metab}\,F_{bw}^{-1}$ (Fig. 3; Kram and Taylor, 1990). This is observed in all the mammals and birds for which it has been measured, including humans, ponies, llamas, dogs, goats, ground squirrels, ostriches, turkeys, and quail. It is even found in small bipedal hoppers like kangaroo rats and spring hares. Unfortunately, we do not yet have good measurements on kangaroos, and they may be the exception that proves the rule. This simple relationship provides us with an important general rule for the mechanics of running that provides the key to linking energetics and mechanics: *$1/t_c$ increases linearly with speed, paralleling almost exactly the increase in the rate of energy consumption.*

At first, the linear increase in $1/t_c$ with speed seems surprising, particularly in hopping and galloping animals in which stride frequency is nearly constant over the entire speed range. Stride frequency changes with running speed in a totally different way than does $1/t_c$, and this has proven problematic in attempts to link energetics with mechanics of running. It has sometimes been assumed that V_{max} parallels stride frequency, but this is not the case across speed, nor at the same speed across size. It is the case only at "equivalent speeds," where the feet spend the same fraction of the stride in the air and on the ground (Taylor, 1985; Rome, 1992). These equivalent speeds have been useful for comparing animals of different size, but the $1/t_c$ relationship enables us to look across both speed and size.

To understand the relationship between $1/t_c$ and speed we need to understand the relationship between steps and strides. So far, we have considered the cost of generating ground force averaged over the entire stride; this force does not change with speed. However, we know that our muscles generate greater force as we run faster. This is simply because our feet spend less time on the ground and more time in the air. Our muscles have to generate enough force during contact time to

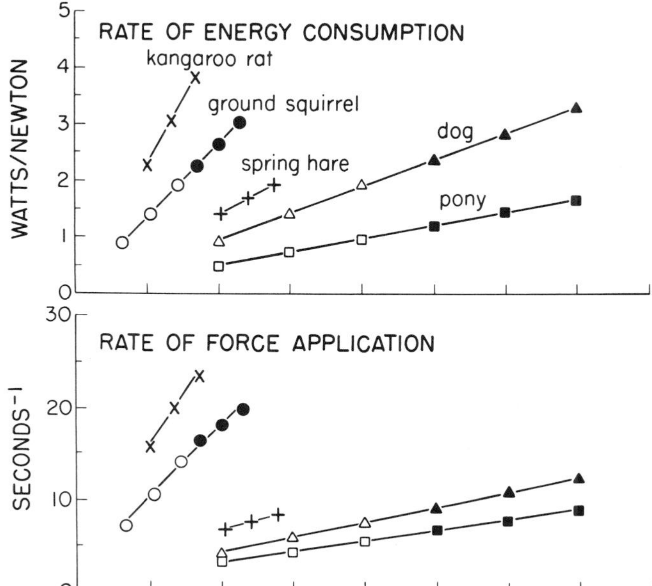

FIG. 3. Rate of force application of the foot on the ground ($1/t_c$, lower graph) increases linearly with running speed (abscissa), paralleling the increase in metabolic rate of the muscles (upper graph) in a diverse assortment of two- and four-legged runners. Data are from Kram and Taylor (1990).

make up for the time in the air. The fraction of the stride that a foot spends on the ground is called the duty factor ($t_c\ t_s^{-1}$). At high speeds, muscles might have to generate more than twice as much force while they are active on the ground as they do at slow speeds. How do these large differences in the ratio of ground force to body weight ($\dot{E}_{\text{metab}}\ F_{\text{bw}}^{-1}$) affect the rate of energy consumption?

The effects of duty factor on muscle force and metabolism cancel each other, so that Eq. (12) applies regardless of whether animals increase speed by increasing stride frequency, stride length, or a combination of both mechanisms. For simplicity, let's imagine a one-legged animal. The average ground force during contact time × duty factor is body weight:

$$F_{\text{bw}} = \bar{F}_{\text{ground,contact}} \cdot \frac{t_c}{t_s}\,. \tag{13}$$

However, the muscle only consumes energy to support the weight when the foot is on the ground, thus the average metabolism is

$$\dot{E}_{\text{metab}} = \dot{E}_{\text{metab,contact}} \cdot \frac{t_c}{t_s} . \tag{14}$$

Combining these two equations with Eq. (12):

$$\frac{\dot{E}_{\text{metab}}}{F_{\text{bw}}} = \frac{\dot{E}_{\text{metab,contact}} \cdot \dfrac{t_c}{t_s}}{\bar{F}_{\text{ground,contact}} \cdot \dfrac{t_c}{t_s}} = c \cdot \frac{1}{t_c} . \tag{15}$$

We can now return to the original question we asked about the energetics of running: Why is cost independent of speed? The answer is that it is a simple consequence of the linear increase in $1/t_c$ with speed. There is a good mechanical reason for a linear relationship: it allows mechanical advantage to be optimized at all speeds. The linear increase is a consequence of the animal using the same step length, L_c, over its entire range of speeds. Step length is the distance that the animal moves while its foot is in contact with the ground and is generating force. It can be expressed as

$$L_c = V \cdot t_c . \tag{16}$$

For L_c to be constant, $1/t_c$ has to increase linearly with speed. A constant step length enables the limb to swing through the same angle and to use the same "optimal mechanical advantage" at all speeds. This helps explain why effective mechanical advantage does not change with speed.

A constant step length results in a constant energetic cost for running a given distance! Remember that the energy cost to run a given distance is just metabolic rate/speed. Using Eqs. (12) and (16) this becomes

$$\frac{\dot{E}_{\text{metab}}}{F_{\text{bw}} \cdot V} = c \cdot \frac{1}{L_c} . \tag{17}$$

Longer legs can explain why larger animals can run more cheaply. Longer legs enable animals to take longer steps and use slower fibers at the same speed. This simple relationship is expressed quantitatively in Eq. (17).

VIII. Testing the Hypothesis

A. Speed and Size

When metabolism and intrinsic velocity of shortening are measured, we find that the simple relationship between metabolic cost and $1/t_c$ of

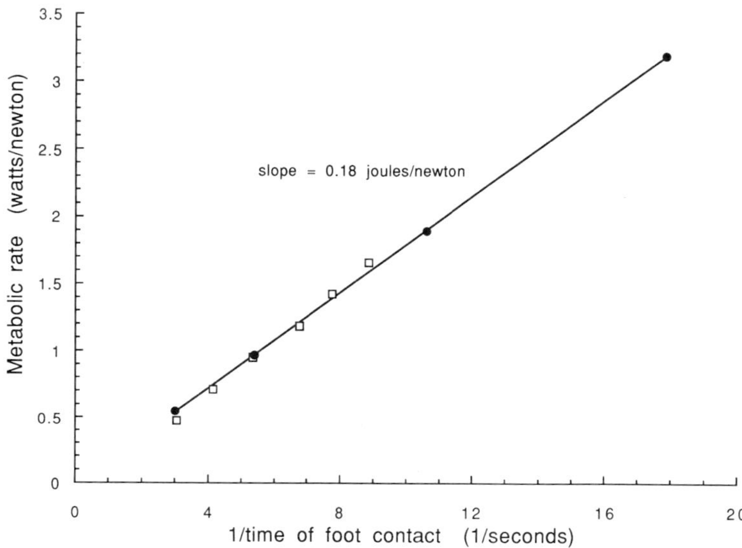

FIG. 4. Metabolic rates of the muscles of both the pony (open squares) and the mouse (solid circles) increase linearly with rate of force application (1/time of foot contact, t_c). The slope of this relationship is the constant c of Eq. (12).

Eq. (12) applies over the entire range of trotting and galloping speeds in quadrupeds, and hopping speeds in small hoppers, spanning the size range from chipmunks and kangaroo rats to ponies (Kram and Taylor, 1990). Energy cost of generating force increases linearly with $1/t_c$, and the slope of this relationship is 0.18 J N^{-1} (= 0.18 m), irrespective of whether the rate varies as the result of differences in speed or size (Fig. 4). This slope is the constant, c, for quadrupeds and small hoppers in Eqs. (12) and (17). At its fastest sustained galloping speed, a ground squirrel generates 1 N of force in 1/15th the time of the pony trotting at its slowest speed, and the squirrel consumes energy at 15 × the rate to generate 1 N of force.

Step length also appears to be nearly constant over the entire range of trotting, galloping, and hopping speeds (in small hoppers) (Kram and Taylor, 1990). Furthermore, it decreases in a regular manner with body size:

$$L_c = 0.18 \cdot M_b^{0.30}, \tag{18}$$

where L_c is in m and M_b is in kg.

Combining Eqs. (17) and (18) and dividing average body weight by 9.8 m s^-2 so as to express cost in W kg^{-1} of M_b instead of in terms of body weight, we arrive at the simple size rule for energetic cost of running that we started with in Eq. (3) when c equals 0.2 m:

$$\frac{\dot{E}_{\text{metab}}}{M_{\text{b}}} = 10 \cdot M_{\text{b}}^{-0.30} \cdot V \tag{19}$$

B. Two Legs versus Four Legs

So far everything seems to fit pretty well into our simple model of economic force generation, at least for quadrupedal runners and bipedal hoppers. But the bipedal birds and humans are a puzzle. An ostrich and a pony of the same weight use the same amount of energy to run a given distance. However, the ostrich has longer legs, takes steps that are twice as long, and has twice as long to generate the force. Other things being equal, it should run at half the cost. This is also true for other birds spanning the entire size range from the ostrich down to small quail.

Why is it so expensive for birds to run?

The answer is that they need more muscle to generate each N of force on the ground than a quadruped of the same weight (Chen *et al.*, 1992). As might be expected, the muscles in their longer legs have longer fibers and a different effective mechanical advantage than do those of quadrupeds of the same size. When these two factors are taken into account, the cost coefficient for generating force in birds is about twice as great as in mammals. This helps explain why they do not run more cheaply and indicates that the "economic force hypothesis" applies to birds as well as to mammals. Although the analysis has not been completed for humans yet, it seems likely that their higher cost is also the result of longer fibers and/or a difference in mechanical advantage.

C. Running Uphill

Running uphill provides one of the best tests of our simple model. We can quantitatively "remove the springs" by gradually increasing the incline until the muscles do all the work and reach their maximal efficiency of 25%. Using this simple approach, we can predict how efficiency will change with speed, size, and incline. Then we can test our predictions by taking advantage of the large differences in V_{max} provided by a comparative approach.

When an animal runs uphill, the average force on the ground is still body weight, just as it is on level ground. The difference is that the muscles have to shorten and work against gravity to lift the body up the incline. The rate at which they work can be increased by increasing either speed and/or incline, and it is easily measured as body weight times the vertical velocity (on a treadmill, vertical velocity is tread speed $\times$ sin ϕ, where ϕ is the angle of the incline). Thus, we can replace the work of the springs with work by the muscles and predict how energy cost and efficiency will change. We can calculate efficiency (in the same way we have done for level running) by assuming that muscles use the additional energy consumed during running and subtracting the metabolic rate at zero speed.

Our hypothesis is that force is generated as economically as possible. This applies over the entire range of speeds on any incline. How could this be accomplished? We would predict that for each incline there should be one $V/V_{\max}$ for which the cost of generating force is lowest, and one L_c for which mechanical advantage is optimized, just as we find on the level. If these were true, then our first prediction is that rate of energy consumption will increase linearly with speed. This is exactly what Zuntz found when he made the first measurements on humans, horses, and dogs at the turn of the century. It has also been observed in almost all the animals for which measurements have been made, and is a general rule. Our model makes the correct prediction. Furthermore, it enables us to explain an empirical observation that is otherwise very difficult to understand. A priori, we would expect that $V/V_{\max}$ would increase as the muscles shorten more rapidly at higher speeds, just as we find in isolated muscles (where recruitment is not possible).

Measurements of L_c have been made over a range of speeds, body sizes and inclines in a variety of birds and mammals (Weyand *et al.*, 1993). Again, as predicted, L_c does not change with either speed or incline (at least over the range of speeds where animals can run aerobically, and at inclines from 0 to 15°). The simple allometric relationship of Eq. (18) predicts L_c for quadrupeds and small hoppers, regardless of speed or incline. So far our model works pretty well!

Next, we predict that efficiency does not change with speed. Constant efficiency is a simple consequence of muscles working at a constant $V/V_{\max}$, at which economy of force generation is maximized. Again, this is exactly what we find. The linear increase in metabolic rate that we observe results in a constant efficiency, because work rate also increases linearly with speed for any incline. Animals work harder as they run faster, and accomplish this by recruiting faster fibers. We can use Eq. (12) to predict metabolic rate if we determine the appropriate c for the $V/V_{\max}$.

Although efficiency is independent of speed on any incline, it will increase with incline. Muscles must provide all the work against gravity; therefore, their V must increase in proportion to vertical velocity (work rate is $F_{bw} \times$ vertical velocity). Because L_c is unchanged as we increase incline, $1/t_c$ (V_{max}) is also unchanged at the same speed. From Eqs. (11), (16) and (18) we see that for the same speed,

$$V_{max} = k_6 \cdot \frac{1}{L_c} = k_7 \cdot M_b^{-0.30} . \qquad (20)$$

Because V is the same for all the animals running up an incline at the same speed, smaller animals, with their faster fibers, will have to run up steeper inclines to reach the same V/V_{max} and efficiency as larger animals (efficiency does not change with speed). The steepness of the incline should vary with $\sin \phi \times M_b^{0.30}$ in order for V/V_{max} and efficiency to be the same. Again, this is exactly what we find when we compare efficiencies of different-sized quadrupeds (Weyand *et al.*, 1993). A 300-kg pony reaches its maximal efficiency of 25% running up an 8° incline. However, the efficiency of a 30-g mouse running up the same incline is less than 2%. Efficiency on this incline for the two animals is proportional to $M_b^{0.30}$, just as our model predicts.

Why are two legs more efficient than four? We can also use our force hypothesis to predict that birds, with their longer legs, will reach higher efficiencies on the same incline than will quadrupeds. We have seen that at the same speed they use fibers that are only half as fast. According to our model, they should reach the same efficiency running up an incline at half the work rate. Again, this is exactly what we find if we compare 4-kg turkeys and 4-kg dogs (Wu and Taylor, 1992). The turkeys reach an efficiency of 21% running up an incline of 12°, whereas the efficiency of dogs of the same weight is only about 11%. The turkeys are 11% efficient on an incline of only about 6°. Again, our model does a remarkably good job of predicting efficiency.

IX. Extending the "Force Hypothesis" to Other Exercises

A. Walking

It is cheaper to walk than to run. In fact, our muscles would use only about half as much energy if we decided to walk a given distance instead of running (see Margaria, 1976). If our goal were weight loss, rather than energy conservation, we would be much better off to run. Both the energetics and the mechanics of walking are fundamentally different from those of running. However, in both cases, our muscles have to generate enough force to support our weight, and it would seem advantageous to do this as economically as possible. Our economic force hypothesis should apply equally well to walking as to running.

The cost of walking a mile varies with speed, whereas cost of running does not. This can be easily understood in terms of the mechanics involved. Energy cost of walking is minimized by using pendulums, whereas cost of running is minimized by using springs (Cavagna and Kaneko, 1977). We vault over our leg in a walk, converting the decrease in kinetic energy as we slow down into gravitational potential energy as our body rises. Then we recover this stored energy as we regain our speed. In fact, 60–80% of the oscillations in kinetic and potential energy of the center of mass are alternately stored and recovered during each stride. Storage is maximized at one speed where the magnitudes of the kinetic and potential energy changes are about the same, and occur 180° out of phase. At faster or slower walking speeds, the exchange is not as complete and the muscles and tendons will have to do the work. The limbs also conserve energy in the recovery stroke by a pendulum mechanism when we walk, and little energy is needed to swing them (see McMahon, 1984).

Humans are not the only animals that use pendulums to conserve energy when they walk. In fact, this seems to be a general mechanism used by most birds and mammals (Cavagna *et al.*, 1977). The mechanism seems to work equally well in humans, birds, and quadrupeds, accounting for 60–80% of the energy oscillations in all the animals. Although the mechanical data on walking are much more limited than for running, we find that the mass-specific work rates of a 160-g quail, a 7-kg turkey, and a 23-kg rhea are about the same at speeds where the pendulum works most efficiently, and it seems likely that it is independent of size.

The mass-specific rate of energy consumption of the walking quail is 4× higher than that of the rhea, the same difference we found for running. In fact, there are enough data on energetic cost of walking across a range of body size (from quail to elephant) to conclude that it varies with the same $M_b^{-0.30}$ as we find for running. There may be a twofold difference between cost of walking and running in all animals (Stoffel *et al.*, 1992), but cost of walking and efficiency seem to change with size as we would expect if economic force generation, rather than maximal power and efficiency, sets the cost.

B. Flying

In marked contrast to walking and running, birds have to work against the environment when they fly; this energy is lost and cannot be recovered. Economic force generation might explain the energy cost of gliding, but it cannot explain the energetics of powered flight. It is reasonable to assume that flight muscles operate so as to maximize power output and efficiency, and this is generally done. The power required for flight can be calculated from aerodynamic models and compared to the power available from muscles (Pennycuick, 1989). These calculations can explain a great deal about the performance of flying birds.

Recently it has been possible to measure muscle force and velocity during flight using strain gauges and films. These measurements elegantly demonstrate how flight muscles are able to maximize power and efficiency when birds fly on the level, when they ascend, and even when they carry loads (Dial and Biewener, 1993). Using these measurements, we should be able to predict the energetics of flight using our simple model of cost of force generation by using an appropriate constant for maximum power or maximum efficiency in Eq. (12). It would be possible to use this approach to predict the metabolic power required for flight, then test the predictions against metabolic measurements. It seems likely that this approach will provide new insights into the link between energetics and mechanics of exercising muscles in the near future.

C. Swimming

Swimming fish, like flying birds, have to work against the environment. In the previous paper, Rome has shown how their muscles are able to maximize power and efficiency over a range of speeds and muscle temperatures. As in birds, this is a case where economic force generation cannot explain the link between energetics and mechanics of locomotion. It is now becoming possible to predict energetics of swimming from direct measurements of muscle performance. This approach has already provided us with general rules about how vertebrate muscles achieve maximal power output. It also provides the opportunity to explore the link between mechanics and energetics when muscles perform other activities.

X. Conclusions

It is possible to explain how the energy cost of running changes by more than 100-fold with speed and size in terms of generating force economically, but not in terms of working efficiently. Springs have to do most of the work when large animals run at high speeds, and it seems likely that they also do most of the work at slow speeds and in small animals. Muscles generate force most economically when they shorten at very slow velocities, close to isometric, and it seems possible that this is how they operate in running animals. This would maximize spring performance by allowing energy to be stored and recovered in the cross-bridges of muscles as well as in tendons. The evidence for economic force generation comes from using a comparative approach to explain the link between energetics and mechanics of running. It seems likely that there are other exercises where it might apply (such as walking), but during exercises where animals work against the environment, muscles operate to maximize power and efficiency. The link between energetics and mechanics will depend on the link between the animal and its environment.

References

Alexander, R. McN. (1988). "Elastic Mechanisms in Animal Movement." Cambridge Univ. Press, Cambridge, UK.

Alexander, R. McN. (1991). Energy-saving mechanisms in walking and running. *J. Exp. Biol.* **160,** 55–69.

Alexander, R. McN., Jayes, A. S., Maloiy, G. M. O., and Wathuta, E. M. (1981). Allometry of the leg muscles of mammals. *J. Zool.* **194,** 539–552.

Biewener, A. A. (1989a). Scaling body support in mammals: Limb posture and muscle mechanics. *Science* **245,** 45–48.

Biewener, A. A. (1989b). Design of the mammalian terrestrial locomotor system in relation to body size. *BioScience* **39,** 776–783.

Blickhan, R. (1989). The spring-mass model for running and hopping. *J. Biomech.* **22,** 1217–1227.

Cavagna, G. A. and Citterio, G. (1974). Effect of stretching on the elastic characteristics and the contractile component of frog striated muscle. *J. Physiol. (London)* **239,** 1–14.

Cavagna, G. A. and Kaneko, M. (1977). Mechanical work and efficiency in level walking and running. *J. Physiol. (London)* **268,** 467–481.

Cavagna, G. A., Saibene, F. P. and Margaria, R. (1964). Mechanical work in running. *J. Appl. Physiol.* **19,** 249–256.

Cavagna, G. A., Heglund, N. C. and Taylor, C. R. (1977). Mechanical work in terrestrial locomotion: Two basic mechanisms for minimizing energy expenditure. *Am. J. Physiol.* **233,** R243–R261.

Chen, M. S., Roberts, T. J., Goslow, T., and Taylor, C. R. (1992). Two steps for the price of one: Energetics of bipedal and quadrupedal runners. *Physiologist* **35,** 186.

Dawson, T. J., and Taylor, C. R. (1973). Energetic cost of locomotion in kangaroos. *Nature (London)* **246,** 313–314.

Dial, K. P., and Biewener, A. A. (1993). Pectoralis muscle force and power output during different modes of flight in pigeons. *J. Exp. Biol.* **176,** 31–54.

Farley, C. T., and McMahon, T. A. (1992). Energetics of walking and running: insights from simulated reduced-gravity experiments. *J. Appl. Physiol.* **73,** 2709–2712.

Farley, C. T., Glasheen, J., and McMahon, T. A. (1993). Running springs: Speed and animal size. %iiJ. Exp. Biol. (in press).

Fedak, M. A., and Seeherman, H. J. (1979). Reappraisal of energetics of locomotion shows identical cost in bipeds and quadrupeds including ostrich and horse. *Nature (London)* **282,** 713–716.

Fedak, M. A., Pinshow, B., and Schmidt-Nielsen, K. (1974). Energy cost of bipedal running. *Am. J. Physiol.* **227,** 1038–1044.

Gray, J. (1968). "Animal Locomotion." Norton, New York.

Heglund, N. C., and Cavagna, G. A. (1987). Mechanical work, oxygen consumption, and efficiency in isolated frog and rat muscle. *Am. J. Physiol.* **253,** C22-C29.

Heglund, N. C., Fedak, M. A., Taylor, C. R., and Cavagna, G. A. (1982). Energetics and mechanics of terrestrial locomotion. IV. Total mechanical energy changes as a function of speed and body size in birds and mammals. *J. Exp. Biol.* **97,** 57–66.

Hildebrand, M. (1960). How animals run. *Sci. Am.* **5,** 148–157.

Hill, A. V. (1938). The heat of shortening and the dynamic constants of muscle. *Proc. R. Soc. London, Ser. B* **126,** 136–195.

Hill, A. V. (1950). The dimensions of animals and their muscular dynamics. *Sci. Prog. (Oxford)* **38,** 209–230.

Hill, A. V. (1970). "First and Last Experiments in Muscle Mechanics." Cambridge Univ. Press, Cambridge, UK.

Howell, A. B. (1944). "Speed in Animals." Univ. of Chicago Press, Chicago.

Huxley, A. F. (1957). Muscle structure and theories of contraction. *Prog. Biophys. Biophys. Chem.* **7,** 255–318.

Kram, R., and Taylor, C. R. (1990). Energetics of running: A new perspective. *Nature (London)* **346,** 265–267.

Langman, V. A., Roberts, T. J., Black, J., Maloiy, G. M. O., Heglund, N. C., Weber, J.-M., Kram, R., and Taylor, C. R. (1990). African elephants: Energetics of walking in a large mammal. *Physiologist* **33,** A36.

Marey, J., and Demeny, G. (1885). Locomotion humaine, mecanisme du saut. *C. R. Hebd. Seances Acad. Sci.* **101,** 489–494.

Margaria, R. (1976). "Biomechanics and Energetics of Muscular Exercise." Oxford Univ. Press, (Clarendon) Oxford.

McMahon, T. A. (1984). "Muscles, Reflexes and Locomotion." Princeton Univ. Press, Princeton, NJ.

Munro, H.N. (1969). Evolution of protein metabolism in mammals. *In* "Mammalian Protein Metabolism" (H. N. Munro and J. B. Allison, eds.), Vol. 3, pp. 133–182. Academic Press, New York.

Paladino, F. V., and King, J. R. (1979). Energetic cost of terrestrial locomotion: Biped and quadruped runners compared. *Rev. Can. Biol.* **38,** 321–323.

Pennycuick, C. J. (1989). "Bird Flight Performance." Oxford Univ. Press, New York.

Pinshow, B., Fedak, M. A., and Schmidt-Nielsen, K. (1977). Terrestrial locomotion in penguins: it costs more to waddle. *Science* **195,** 592–594.

Pugh, L. G. C. E. (1971). The influence of wind resistance in running and walking and the mechanical efficiency of work against horizontal or vertical forces. *J. Physiol. (London)* **213,** 255–276.

Rall, J. A. (1985). Energetic aspects of skeletal muscle contraction: Implications of fiber types. *Exercise Sport Sci. Rev.* **13,** 33–74.

Rome, L. C. (1992). Scaling of muscle fibres and locomotion. *J. Exp. Biol.* **168,** 243–252.

Rome, L. C., Sosnicki, A. A., and Goble, D. O. (1990). Maximum velocity of shortening of three fibre types from the horse soleus: Implications for scaling with body size. *J. Physiol. (London)* **431,** 173–185.

Slowtzoff, B. (1903). Ueber die Beziehungen zwischen Korpengrosse und stoffverbrauch der Hunde bei Ruhe und Arbeit. *Arch. Gesamte Physiol. Menschen Tiere* **95,** 158–191.

Stevens, E. D., and Renaud, J. M. (1985). Use of *d*P/*dt* and rise time to estimate speed of shortening in muscle. *Am. J. Physiol.* **249,** R510-R513.

Stoffel, C. M., Bihldorff, J. E., Kram, R., Roberts, T., and Taylor, C. R. (1992). Is it cheaper to walk than to run? *Physiologist* **35,** 186.

Taylor, C. R. (1985). Force development during sustained locomotion: A determinant of gait, speed and metabolic power. *J. Exp. Biol.* **115,** 253–262.

Taylor, C. R., and Rountree, V. J. (1973). Running on two or on four legs: Which consumes more energy? *Science* **179,** 186–187.

Taylor, C. R., Schmidt-Nielsen, K., and Raab, J. L. (1970). Scaling of energetic cost of running to body size in mammals. *Am. J. Physiol.* **219,** 1104–1107.

Taylor, C. R., Shkolnik, A., Dmi'el, R., Baharav, D., and Borut, A. (1974). Running in cheetahs, gazelles and goats: Energy cost and limb configuration. *Am. J. Physiol.* **227,** 848–850.

Taylor, C. R., Heglund, N. C., McMahon, T. A., and Looney, T. R. (1980). Energetic cost of generating muscular force during running: A comparison of large and small animals. *J. Exp. Biol.* **86,** 9–18.

Taylor, C. R., Heglund, N. C., and Maloiy, G. M. O. (1982). Energetics and mechanics of terrestrial locomotion. I. Metabolic energy consumption as a function of speed and body size in birds and mammals. *J. Exp. Biol.* **97,** 1–21.

Thompson, S. D., MacMillen, R. E., Burke, E. M., and Taylor, C. R. (1980). The energetic cost of bipedal hopping in small mammals. *Nature (London)* **287,** 223–224.

Weyand, P., Wu, D., Gonzalez, O., Roberts, T., Kennedy, C., and Taylor, C. R. (1993). Efficiency of running uphill: Speed, size and numbers of limbs. *Proc. Int. Union Physiol. Sci., Int. Congr., 32nd* (in press).

Woledge, R. C., Curtin, N. A., and Homsher, E. (1985). "Energetic Aspects of Muscle Contraction." Academic Press, London.

Wu, D. S., and Taylor, C. R. (1992). Running uphill: Why are small animals less efficient? *Physiologist* **35,** 186.

Zuntz, N. (1897). Ueber den Stoffverbrauch des Hundes bei Muskelarbeit. *Arch. Gesamte Physiol. Menschen Tiere* **68,** 191–211.

ADVANCES IN VETERINARY SCIENCE AND COMPARATIVE MEDICINE, VOL. 38A

Circulatory Function during Exercise: Integration of Convection and Diffusion

JAMES H. JONES

Department of Surgical and Radiological Sciences, School of Veterinary Medicine, University of California, Davis, Davis, California 95616

I. Introduction

Exercise places physiological demands on animals that are met by integrated responses of multiple-organ systems. Of particular importance during exercise are the skeletal muscular elements that generate force and produce mechanical power (see papers in this volume by Rome and Taylor), the glycolytic and oxidative enzymes and biochemical pathways that provide metabolic power for sustaining the mechanical function of the muscles (see papers by Conley, Hochachka, and

Billeter *et al.*) and the uptake of oxygen from the atmosphere to meet the demand for oxidative phosphorylation, if the activity is aerobic (see paper by Powell).

The circulatory systems of vertebrates play a central role during aerobic exercise by connecting the sources of substrates—the external gas exchange organs (lungs, gills, or skin) and internal carbon-substrate sources (liver, adipose, and lactate-producing skeletal muscle fibers)—with the sites of substrate delivery and utilization—the mitochondria of working skeletal and cardiac muscle (Fig. 1). As the central convective component of the respiratory system, circulatory function directly influences oxygen transport capacity and, hence, aerobic capacity.

Comparing cardiovascular performance during exercise between different vertebrate taxa is difficult because of inherent differences in the designs of their cardiovascular systems. For example, fish hearts are simple two-chambered structures, most of which lack any coronary circulation, in which cardiac output may be regulated quite differently than in the tetrapods (Farrell, 1984, 1993). The ventricles of amphibian and noncrocodilian reptilian hearts are either single or incompletely separated by a septum, allowing varying amounts of shunting, venous admixture, and reduced arterial saturation and systemic oxygen delivery (White, 1978; van Mierop and Kutsche, 1985). Only the circulatory systems of birds and mammals are completely separated into a low-pressure pulmonary circuit and a high pressure systemic circuit. This separation is essential for achieving high cardiac outputs, as the large driving pressure required for high flow through the systemic circuit would cause marked hydrostatic filtration and pulmonary edema if applied to the thin-walled pulmonary circuit (Schmidt-Nielsen, 1990). The divided circulations of homeothermic birds and mammals enable these animals to achieve much higher cardiac outputs and rates of oxygen transport during exercise than are possible for other vertebrates.

In this paper I will consider the role that the circulatory system plays in both limiting and facilitating aerobic performance. The review is limited to mammals because their circulatory systems are of a single design, they are homeothermic (simplifying metabolic performance comparisons), and the majority of pertinent data have been collected on them. Within the mammals, great variation in aerobic capacity exists owing to differences in body size (*allometric variation*), selection (natural or human) for aerobic performance (*adaptive variation*), and changes within an animal due to prolonged aerobic training

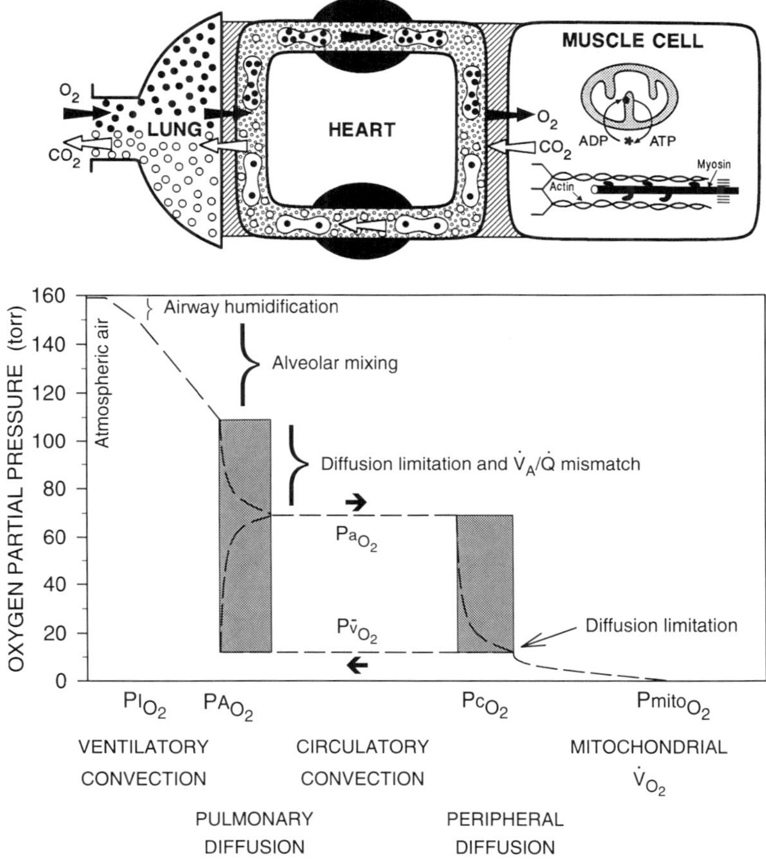

FIG. 1. Schematic of the mammalian respiratory system (*top*) and the partial pressure of O_2 (dashed line, *bottom*) as it passes through the respiratory system from atmospheric air to mitochondrion via convective and diffusive transport steps. Shaded boxes are diffusive exchange between alveolar gas and pulmonary capillary blood (left box) and muscle capillary blood and myocyte (right box). $P\text{I}_{O_2}$ is inspired P_{O_2}, $P\text{A}_{O_2}$ is alveolar P_{O_2}, Pa_{O_2} is arterial P_{O_2}, $P\bar{v}_{O_2}$ is mixed venous P_{O_2}, Pc_{O_2} is muscle capillary P_{O_2}, and $P\text{mito}_{O_2}$ is mitochondrial P_{O_2}, assumed to be near 0 Torr. Values depicted are those measured for a Thoroughbred horse running at $\dot{V}_{O_2\ max}$ and breathing air at sea level.

and development (*induced variation*), all of which will be considered. I will:

1. Review the long-standing argument that circulatory convection limits aerobic performance in mammals;
2. Consider the mechanisms by which circulatory convection interacts

with other components of the respiratory system to take up and deliver oxygen and other substrates during exercise, and by which integrated multiple-step limitations to oxygen flux are established;
3. Discuss how the circulatory system adjusts or fails to adapt to the demands of exercise placed on it;
4. Consider some general principles regarding the design of the respiratory system and the circulatory components within it.

II. An Old Hypothesis: Circulatory Capacity Limits Aerobic Performance

The hypothesis that the circulatory delivery of oxygen to working muscle limits aerobic performance is rooted in some of the earliest studies of human exercise physiology. Following the pioneering work of Zuntz (1897; Zuntz and Hagemann, 1898) on the uptake of oxygen during treadmill exercise in dogs, horses, and humans, Verzar (1912) proposed that oxygen delivery to working muscle is the factor that limits the rate of maximal oxygen consumption during exercise. This concept was further elaborated by Hill and Lupton (1923) and Hill *et al.* (1924), who specifically considered cardiopulmonary transport of oxygen to be *the* limiting factor for aerobic exercise. Since that time a great number of researchers have adopted this paradigm a priori in designing experiments, interpreting data, and formulating conclusions as to factors limiting aerobic performance in a variety of species, including humans (Saltin and Gollnick, 1983; Brooks and Fahey, 1985; Nadel, 1985; Åstrand and Rodahl, 1986; Ekblom, 1986; Cerretelli and di Prampero, 1987), dogs (Musch *et al.*, 1985), rats (Davies *et al.*, 1982), horses (Persson, 1967; Evans and Rose, 1988; Knight *et al.*, 1991), and frogs (Hillman, 1980; Withers and Hillman, 1983; Hillman *et al.*, 1985).

A. Experimental Evidence That Circulation Limits Performance

Perhaps the most commonly cited evidence supporting the hypothesis that circulatory delivery of oxygen limits aerobic capacity is the myriad number of studies in humans that have correlated alterations in oxygen delivery with changes in maximal rates of oxygen consumption ($\dot{V}_{O_2\,max}$). These studies have been performed by increasing circulatory oxygen delivery with infusion of erythrocytes (Buick *et al.*, 1980; Spriet *et al.*, 1986; Celsing *et al.*, 1987) or by breathing inspired gas containing a high partial pressure of oxygen (P_{O_2}) (Powers *et al.*,

1989), or alternatively by decreasing circulatory oxygen delivery with anemia or phlebotomy (Ekblom *et al.*, 1972, 1975, 1976; Woodson *et al.*, 1978; Kanstrup and Ekblom, 1984; Celsing *et al.*, 1986), by reducing blood O_2-binding capacity by breathing CO (Ekblom and Hout, 1972; Raven *et al.*, 1974), or by reducing arterial O_2 saturation by breathing hypoxic gas (Squires and Buskirk, 1982; Young *et al.*, 1985). The general finding of these types of studies cited as evidence that circulatory oxygen delivery limits $\dot{V}_{O_2\ max}$ is that $\dot{V}_{O_2\ max}$ is positively correlated with either blood hemoglobin concentration ([Hb]) or O_2 delivery ($\dot{Q}_T \times Ca_{O_2}$, where $\dot{Q}_T$ is cardiac output and Ca_{O_2} is arterial O_2 concentration).

B. Why the Hypothesis Fails

"Students of statistics are told at an early stage not to confuse significant correlation with causation" (Sokal and Rohlf, 1981). This maxim is well applied when considering the relationship between $\dot{V}_{O_2\ max}$ and the circulatory delivery of O_2 during exercise. The Fick (1870) Principle describes the relationship between O_2 consumption ($\dot{V}_{O_2}$), O_2 delivery, and mixed venous O_2 concentration ($C\bar{v}_{O_2}$):

$$\dot{V}_{O_2} = \dot{Q}_T \times (Ca_{O_2} - C\bar{v}_{O_2}). \tag{1}$$

Inspection of Eq. (1) demonstrates the tautological nature of the argument that correlation of changes in O_2 delivery with changes in $\dot{V}_{O_2 max}$ indicates that circulatory delivery of oxygen is *the* limiting factor for aerobic capacity. Mass balance dictates that it is virtually impossible for the term on the left of the equation, $\dot{V}_{O_2}$, to increase or decrease without the term on the right, circulatory O_2 delivery, increasing or decreasing by an equal amount. Because O_2 flux at every step of the oxygen transport and utilization system (i.e., the respiratory system) must change by exactly the same amount when an animal is in steady state (Section III), it is inevitable that $\dot{V}_{O_2}$ must be highly correlated with circulatory O_2 delivery.

Several papers have considered the weaknesses of the arguments supporting the concept that the cardiovascular system represents a single "weak link" limiting step for O_2 transport and aerobic capacity (Weibel and Taylor, 1981; Taylor *et al.*, 1987; Jones and Karas, 1988; Lindstedt *et al.*, 1988; Wagner, 1988; Jones and Lindstedt, 1993). Lindstedt *et al.* (1988) evaluated changes in $\dot{V}_{O_2\ max}$ that resulted from experimental alterations of Hb concentration or O_2 delivery in a large number of studies performed on humans. Their conclusions (Fig. 2) were that considerable variation existed in the [Hb] that an individual maintained in order to support a given $\dot{V}_{O_2\ max}$, and that changes in

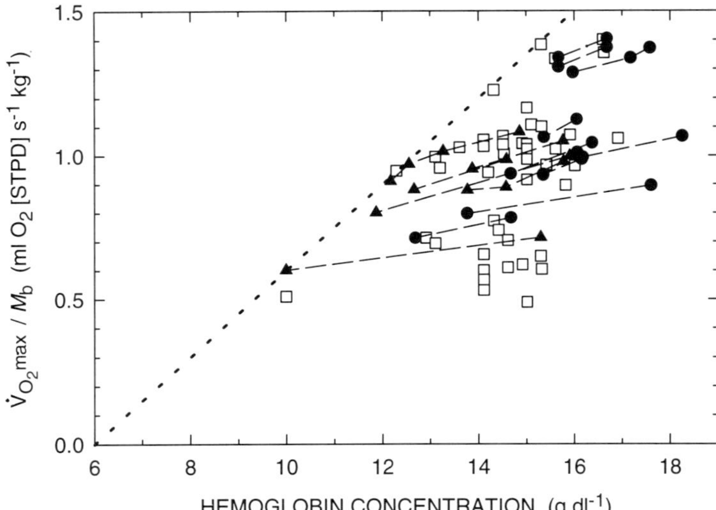

FIG. 2. Mass-specific $\dot{V}_{O_2\ max}$ in human subjects plotted as a function of hemoglobin concentration for a variety of individual experiments (open squares) and for experiments in which hemoglobin concentration was experimentally increased (solid circles) or decreased (solid triangles). Note the great variability in hemoglobin concentration associated with any given $\dot{V}_{O_2\ max}$, although a minimal hemoglobin concentration (dashed line) appears to be required to attain any $\dot{V}_{O_2\ max}$. Also note that in experiments in which hemoglobin concentration was altered, the resulting changes in $\dot{V}_{O_2\ max}$ were not in proportion (parallel) to the line of absolute hemoglobin concentration limitation. Both of these observations argue against the existence of an absolute circulatory limit to O_2 transport at $\dot{V}_{O_2\ max}$. Data from Lindstedt *et al.* (1988).

[Hb] and O_2 delivery were typically disproportionate to the changes in $\dot{V}_{O_2\ max}$ with which they were associated. These findings argue against a simplistic single-step limitation to O_2 transport by the cardiovascular system.

An independent line of reasoning also argues against the idea of a simple circulatory limitation to O_2 transport. Jones *et al.* (1985) and Karas *et al.* (1987b) found that goats maintained identical $\dot{V}_{O_2\ max}$ when running at sea level and breathing air (inspired O_2 fraction, $F\text{I}_{O_2}$ = 0.21) as when breathing gas with $F\text{I}_{O_2}$ = 0.15 (Fig. 3). Because of the significant decrease in arterial O_2 tension (Pa_{O_2}) and saturation (Sa_{O_2}) in hypoxia, the goats had to increase $\dot{Q}_T > 20\%$ to maintain normoxic $\dot{V}_{O_2\ max}$, clearly indicating that O_2 delivery was not limited by circulatory transport in normoxia. Mitochondrial oxidative capacity appeared to be the factor that limited the maximum rate at which O_2 could be used by the caprine respiratory system. A key observation in these

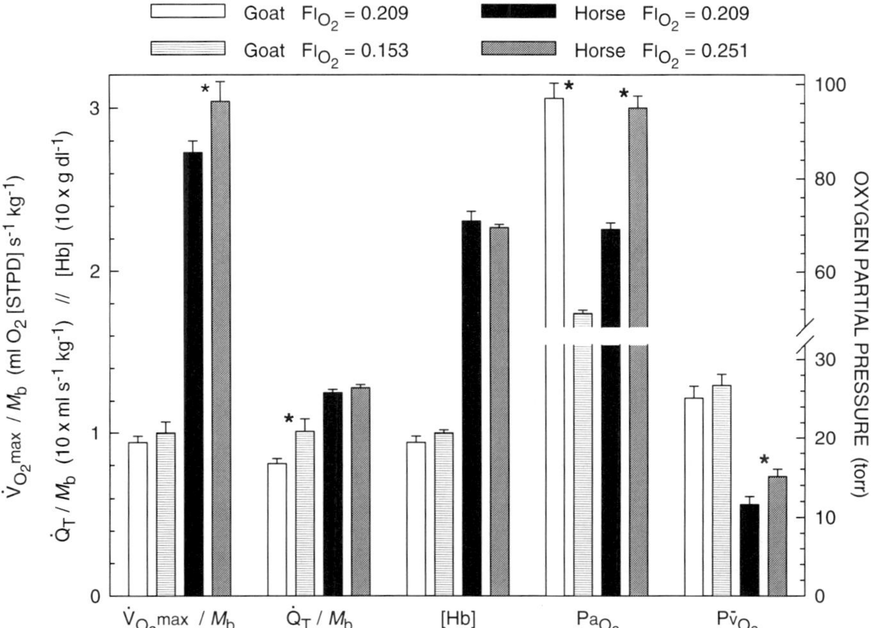

FIG. 3. Values of mass-specific rates of maximal O_2 consumption ($\dot{V}_{O_2\,max}/M_b$), mass-specific cardiac output ($\dot{Q}_T/M_b$), hemoglobin concentration ([Hb]), and arterial and mixed venous O_2 tensions (Pa_{O_2} and $P\bar{v}_{O_2}$, respectively) measured while running at $\dot{V}_{O_2\,max}$ at sea level in goats breathing air (inspired O_2 fraction, FI_{O_2} = 0.209) or hypoxic gas (FI_{O_2} = 0.153), or Thoroughbred horses breathing air or hyperoxic gas (FI_{O_2} = 0.251). Bars are ±1 SE; asterisks denote data for which a significant difference exists between the two gases within a species. Note that $\dot{V}_{O_2\,max}$ did not change for goats in hypoxia because $\dot{Q}_T$ increased to compensate for the significant decrease in Pa_{O_2} (and hemoglobin saturation), indicating that normoxic goats are limited at $\dot{V}_{O_2\,max}$ by mitochondrial oxidative capacity, not their O_2 transport systems. Conversely, $\dot{V}_{O_2\,max}$ in horses increased in hyperoxia with no change in $\dot{Q}_T$, indicating that the O_2 transport system limits normoxic $\dot{V}_{O_2\,max}$ in this species. $P\bar{v}_{O_2}$ (and saturation) increased significantly in horses when $\dot{V}_{O_2\,max}$ increased, in accord with the hypothesis that peripheral tissue diffusion limitation exists at $\dot{V}_{O_2\,max}$. The hypothesis is further supported by the fact that in goats $P\bar{v}_{O_2}$ was the same when $\dot{V}_{O_2\,max}$ did not change, despite a >40 Torr difference in Pa_{O_2}, suggesting that the same end-capillary pressure head is required to overcome peripheral tissue diffusion resistance in order to cause a given O_2 flux. Data from Jones *et al.* (1985, 1989a, 1993) and Karas *et al.* (1987b).

experiments was that mixed venous P_{O_2} ($P\bar{v}_{O_2}$) at $\dot{V}_{O_2\,max}$ was identical in normoxia and hypoxia, suggesting that a similar end-capillary P_{O_2} was needed in both environments to elicit equivalent diffusive flux of O_2 across the peripheral gas exchanger (Section III,A).

A similar line of reasoning has been forwarded by Wagner (1988)

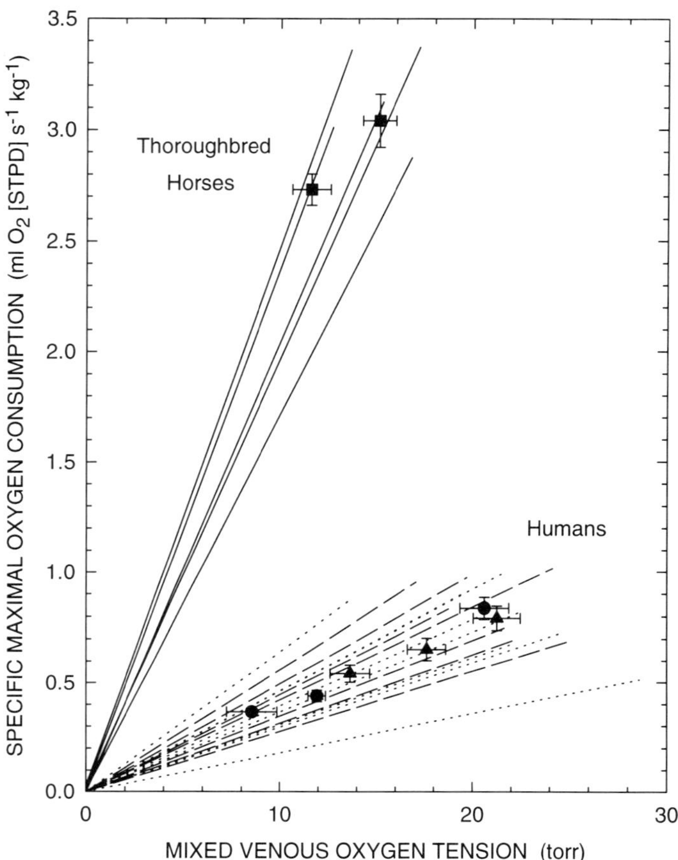

FIG. 4. Relationship between mass-specific rates of maximal O_2 consumption and mixed venous P_{O_2} estimated for human subjects exercising at sea level or during exposure to simulated altitude in a hypobaric chamber, and for Thoroughbred horses running on a treadmill breathing air or hyperoxic gas. Barometric pressures at which the human subjects exercised were 760, 347, and 282 Torr (circles, dashed lines) and 760, 523, and 428 Torr (triangles, dotted lines). Horses breathed air or gas containing 25.1% O_2 at sea level, equivalent to breathing air at pressures of 760 and 911 Torr (squares, solid lines). Lines are drawn between the two or three data points for each subject and the origin; symbols represent the mean ± 1 SE for all subjects within that study. The relationship between $\dot{V}_{O_2\ max}$ and $P\bar{v}_{O_2}$ suggests that peripheral diffusion limitation occurs at $\dot{V}_{O_2\ max}$. The steeper slopes of the equine curves indicate the much greater peripheral tissue diffusion conductance ($D\text{T}_{O_2}$) in horses than in humans. Data from Wagner (1988) and Jones *et al.* (1989a, 1993).

Wagner *et al.* (1991) based on correlated changes in $P\bar{v}_{O_2}$ and $\dot{V}_{O_2\ max}$ in human subjects in which $\dot{V}_{O_2\ max}$ decreased as the subjects were exercised in progressive hypoxia (Fig. 4). Wagner argued that the association between the two variables suggested the existence of peripheral tissue diffusion limitation for oxygen, and that higher fluxes required a larger driving pressure, reflected in the $P\bar{v}_{O_2}$. Studies by Wagner's group with isolated perfused canine limb preparations (Hogan *et al.*, 1988, 1989) support the hypothesis that peripheral tissue diffusion conductance plays a significant role in limiting $\dot{V}_{O_2\ max}$. Some caution must be used in interpreting these results, however, because isolated perfused skeletal muscle preparations inevitably achieve much lower mass-specific $\dot{V}_{O_2\ max}$ (typically less than one-third) than do intact awake animals, and the factors limiting $\dot{V}_{O_2}$ in the preparation may not be the same as those functioning in the whole animal.

Additional evidence arguing against the concept of a single-step cardiovascular limitation to $\dot{V}_{O_2\ max}$ comes from studies of Thoroughbred racehorses exercising in hyperoxia (Jones *et al.*, 1989a, 1993; Birks *et al.*, 1992). When horses running on a treadmill breathed gas with $F\text{I}_{O_2} = 0.25$, $\dot{V}_{O_2\ max}$ increased $> 10\%$ with no increase in $\dot{Q}_T$ or [Hb] (Fig. 3). However, $P\bar{v}_{O_2}$ and $S\bar{v}_{O_2}$ both increased significantly in hyperoxia, again in conformance with the predictions of a peripheral diffusion limitation hypothesis (Fig. 4). In the case of the horses, there is strong evidence not only of peripheral tissue diffusion limitation at $\dot{V}_{O_2\ max}$, but also of pulmonary diffusion limitation (alveolar–arterial P_{O_2} difference > 40 Torr) (Wagner *et al.*, 1989; Jones *et al.*, 1993) and hypoventilation (arterial CO_2 tension, $Pa_{CO_2} > 57$ Torr), suggesting that limitations to O_2 transport occur simultaneously throughout the O_2 transport system in these animals.

III. A New Hypothesis: The Circulation as an Integrated Component of the Respiratory System

Because the circulatory system includes the capillary beds that constitute the pulmonary and peripheral muscle gas exchangers, it is inevitable that changes in circulatory function should affect the transport of O_2 by diffusion in both of these gas exchangers. Furthermore, these two gas exchangers function interdependently with circulatory convection to determine the rate at which O_2 is transported through the respiratory system. Understanding the influence of changes in circulatory convection on the function of the diffusive gas exchangers is essential to understanding why the circulatory system cannot function as a single rate-limiting step for oxygen transport.

A. How Does the Circulation Affect Diffusion and Convection of Oxygen?

Equations (2)–(6) describe the flux of O_2 through the transport steps of the respiratory system at steady state, beginning with the site at which the demand for O_2 is set, the muscle mitochondria:

$$\dot{V}_{O_2} = \frac{\dot{V}_{O_2}}{S\text{mito}} \times \frac{S\text{mito}}{V\text{mito}} \times \frac{V\text{mito}}{V\text{mus}} \times V\text{mus} \tag{2}$$

$$\dot{V}_{O_2} = \int_{t=0}^{t_c} (D\text{T}_{O_2} \times (P\text{c}_{O_2} - P\text{mito}_{O_2}))\, dt \tag{3}$$

$$\dot{V}_{O_2} = \dot{Q}\text{T} \times [\text{Hb}] \times k \times (f(P\text{a}_{O_2}) - f(P\bar{\text{v}}_{O_2})) \tag{4}$$

$$\dot{V}_{O_2} = \int_{t=0}^{t_c} (D\text{L}_{O_2} \times (P\text{A}_{O_2} - P\text{c}_{O_2}))\, dt \tag{5}$$

$$\dot{V}_{O_2} = \dot{V}_{\text{E}} \times (F\text{I}_{O_2} - F\bar{\text{E}}_{O_2}) \tag{6}$$

where Smito is the surface area of inner mitochondrial membranes consuming O_2, Vmito is the volume of mitochondria containing inner mitochondrial membrane, Vmus is the volume of muscle containing mitochondria, t_c is erythrocyte contact (or transit) time through the capillaries of the gas exchangers, $D\text{T}_{O_2}$ is diffusive conductance of the peripheral (muscle) gas exchanger, $P\text{c}_{O_2}$ is O_2 tension in the capillaries in the gas exchangers, $P\text{mito}_{O_2}$ is P_{O_2} in the mitochondria, $D\text{L}_{O_2}$ is pulmonary diffusive conductance, $P\text{A}_{O_2}$ is alveolar P_{O_2}, $\dot{V}_{\text{E}}$ is minute ventilation, and $F\bar{\text{E}}_{O_2}$ is mixed expired gas fraction. At $\dot{V}_{O_2\,\text{max}}$, flux through each of these transport steps is identical, and for $\dot{V}_{O_2}$ to increase, it is necessary for flux at each step to be increased by an identical amount.

In order for O_2 flux at each of the transport steps listed in Eqs. (2)–(6) to be increased, the demands of adjacent gas exchange steps must be met. For each of these transport steps we can identify the factors that must change for O_2 flux through the system to increase.

1. *Mitochondial Oxygen Consumption*

Quantification of mitochondrial structure using morphometric techniques has shown that the surface area of inner membrane per unit volume is nearly constant in mammalian skeletal muscle mitochondria (approximately 30 m^2 cm^{-3}; Schwerzmann *et al.*, 1989) and is independent of variation in whole-animal aerobic capacity owing to allometric, adaptive, or induced variation (Davies *et al.*, 1981; Hoppeler *et al.*, 1981a, 1984a; Mathieu *et al.*, 1981). Furthermore, in experiments in which whole-body $\dot{V}_{O_2\ max}$, *V*mito, and/or *S*mito have been estimated, it has been show that at $\dot{V}_{O_2\ max}$, mitochondria from all mammals consume O_2 on average at a nearly constant rate per unit area of inner mitochondrial membrane (75,000 molecules O_2 s^{-1} μm^{-2}) or per unit volume of mitochondria (3–5 ml [STPD] O_2 min^{-1} cm-3) (Hoppeler and Lindstedt, 1985; Hoppeler, 1986; Lindstedt and Wells, 1988).

This observation suggests that for some mammals (e.g., goats that can maintain normoxic $\dot{V}_{O_2\ max}$ even when convective O_2 transport is acutely stressed by hypoxia, and can increase $\dot{Q}_T$ to compensate; see Section II,B), the surface area of inner mitochondrial membranes (and thus whole-body mitochondrial volume) would have to increase in order for normoxic $\dot{V}_{O_2\ max}$ to increase. Limitations to oxidative capacity of the mitochondria appear to limit $\dot{V}_{O_2\ max}$ in these animals. In contrast, $\dot{V}_{O_2\ max}$ in Thoroughbred horses increases in hyperoxia, indicating that their mitochondria are capable of higher oxidative rates than they sustain at normoxic $\dot{V}_{O_2\ max}$. Oxygen delivery appears to limit normoxic $\dot{V}_{O_2\ max}$ in these animals. Thus, it appears that for a particular species, any of the terms on the right side of Eq. (2) may be required to increase in order to elevate mitochondrial demand for O_2, at least within narrow limits (10–20%).

Evaluation of changes in mitochondrial volume or oxidative enzyme activities in conjunction with changes in whole-body $\dot{V}_{O_2\ max}$ can only be meaningfully made if the mitochondrial/oxidative enzyme changes are evaluated for the whole body as well. Many comparisons of this type have been based on mitochondrial/oxidative enzyme data from biopsies of a single muscle (Saltin and Gollnick, 1983) and have found that whole-body $\dot{V}_{O_2\ max}$ and single-muscle *V*mito may change disproportionately to each other. However, because whole-body $\dot{V}_{O_2\ max}$ reflects global oxygen demand, it should hypothetically be related to the global (i.e., whole-body) mitochondrial pool setting that demand, rather than larger or smaller changes in mitochondrial oxidative capacity that might occur with training in a single muscle or group of muscles. Changes in oxidative capacity of a limited mass of muscle could be offset by changes in muscle mass or mitochondrial volume density of other muscles throughout the rest of the body (Hoppeler *et al.*, 1985). Indeed, if the $\dot{V}_{O_2}$ of a limited group of muscles increases sufficiently to significantly alter the calculated whole-body $V_{O_2\,max}$, *S*mito and *V*mito in that muscle group would have to increase by a proportionately greater amount than in the rest of the body, as the mass of any one muscle group is only a fraction of whole-body muscle mass [Eq. (2)].

2. *Peripheral Tissue Diffusion*

Diffusive flux across the peripheral gas exchanger [Eq. (3)] will increase only if diffusive conductance ($D\text{T}_{O_2}$) or the mean pressure head for O_2 ($P\text{c}_{O_2} - P\text{mito}_{O_2}$) increases (Fig. 1). Peripheral diffusive conductance is a complex function of physical factors such as capillary and mitochondrial surface areas, three-dimensional orientations of capillaries and mitochondria and their resultant effective diffusive distances, capillary volume (V_C), the kinetics of O_2–hemoglobin unbinding, and intramuscular myoglobin concentration (Conley *et al.*, 1987; Hoppeler *et al.*, 1991; Jones and Longworth, 1992). The effective capillary P_{O_2} can only be calculated by integrating the P_{O_2} as an erythrocyte passes down the capillary, for example, with a Bohr integration model (Bohr, 1909; Hill *et al.*, 1973; Karas *et al.*, 1987b; Hempleman and Gray, 1988).

Under steady state conditions at $\dot{V}_{O_2\,max}$, peripheral conductance can be altered with long-term aerobic conditioning to induce structural changes in the muscle, for example, capillary and mitochondrial proliferation, which increase surface areas and decrease distances for diffusion of O_2 (see paper by Billeter *et al.*). These changes are correlated, resulting in a relatively constant ratio to each cubic centimeter of mitochondria of 10 km of capillary length, 1400 cm^2 of capillary surface area, and 0.16 cm^3 of capillary volume (Hoppeler and Billeter, 1991).

Little change is possible in the variables that determine $D\text{T}_{O_2}$ in the short-term unless $\dot{Q}_T$ can be increased to perfuse additional capillary beds, thereby increasing the surface area for diffusion. (Note that this does not indicate that circulatory capacity limits $D\text{T}_{O_2}$ per se, but rather that $\dot{Q}_T$ is one of several structural and kinetic factors that interact to determine its value.) For a species such as the goat, in which mitochondrial oxidative capacity appears to limit $\dot{V}_{O_2\,max}$ in normoxia, $\dot{Q}_T$ could be increased in normoxia, raise $D\text{T}_{O_2}$, and increase the *rate* of O_2 delivery to mitochondria by individual erythrocytes; however, such changes would have no effect on the *quantity* of O_2 delivered over time ($\dot{V}_{O_2}$), as the mitochondrial sink for O_2 is already saturated. Therefore, $P\text{c}_{O_2}$ would reach its equilibrium value sooner, but would be the same P_{O_2}. For species such as the horse and dog that appear to be limited at $\dot{V}_{O_2\,max}$ by oxygen delivery, $\dot{Q}_T$ increases with $\dot{V}_{O_2}$ up to $\dot{V}_{O_2\,max}$, then plateaus with no indication of additional capacity to elevate $\dot{Q}_T$ above its value at $\dot{V}_{O_2\,max}$, at least in normoxia (Karas *et al.*, 1987b; Jones *et al.*, 1989a; Birks *et al.*, 1991; Armstrong *et al.*, 1992). Therefore, $D\text{T}_{O_2}$ is relatively fixed at $\dot{V}_{O_2\,max}$ by structural elements and does not appear to be a mechanism for acutely increasing $\dot{V}_{O_2\,max}$.

Changes in the pressure head for the diffusion of O_2 at $\dot{V}_{O_2\,max}$ are more likely to be responsible for effecting acute changes in O_2 flux. Mitochondria function maximally in working muscle down to $P_{O_2} < 1$ Torr (Gayeski *et al.*, 1987). If $\dot{V}_{O_2\,max}$ were limited by mitochondrial oxidative capacity, as appears to be the case with goats, then mitochondria presumably would be saturated with O_2 at $\dot{V}_{O_2\,max}$ and delivery of "excess" O_2 would cause $P\text{mito}_{O_2}$ to rise above 0 Torr, abolishing the partial pressure difference for O_2, limiting flux, and establishing an end-capillary P_{O_2} that would contribute to a fixed value for $P\bar{v}_{O_2}$. On the other hand, if peripheral tissue diffusion limitation plays a rôle in establishing an O_2 transport limitation to $\dot{V}_{O_2\,max}$, then alterations to systemic O_2 transport that change $\dot{V}_{O_2\,max}$ would result in (or be caused by) correlated changes in Pc_{O_2}, with concomitant changes in the diffusive pressure head. In effect, higher end-capillary P_{O_2} would be required to increase diffusive delivery of O_2 across a relatively fixed conductance, $D\text{T}_{O_2}$. Thus, changes in $P\bar{v}_{O_2}$ should occur in association with changes in $\dot{V}_{O_2\,max}$, as found in the studies by Wagner (1988), Hogan *et al.* (1988, 1989) and Jones *et al.* (1989b, 1993).

3. *Circulatory Convection*

The factors that determine convective delivery of O_2 by the circulation are expressed by the Fick Principle [Eq. (1)] as $\dot{Q}_T$, Ca_{O_2} and $C\bar{v}_{O_2}$. The latter two terms are expanded in Eq. (4) as products of [Hb], k (the O_2-binding capacity of hemoglobin, or Hüfner's constant; Hüfner, 1877), and the function (f) describing the saturation of hemoglobin with O_2 as a function of P_{O_2} (the O_2–Hb equilibrium curve, or OEC; Severinghaus, 1979). Expressed in these terms, it is apparent that for circulatory convective delivery of O_2 to increase at $\dot{V}_{O_2\,max}$, either $\dot{Q}_T$, [Hb], or Pa_{O_2} must increase, or $P\bar{v}_{O_2}$ must decrease.

Cardiac output is a function of two pairs of variables:

$$\dot{Q}_T = f_H \times V_S = \bar{P}_a \times \frac{1}{R_S}, \tag{7}$$

where f_H is heart rate, V_S is cardiac stroke volume, $\bar{P}_a$ is mean systemic arterial blood pressure, and R_S is systemic vascular resistance. Maximum heart rate is a function of body size and cannot be increased with training or increased demand (Baudinette, 1978). Maximum V_S varies with body size and the genetic adaptation for aerobic performance within a species, and increases when an animal's heart hypertrophies with aerobic conditioning. In most mammals, in normoxia, V_S remains constant as the animal progresses from rest to $\dot{V}_{O_2\ max}$, thus V_S appears to have little malleability for acute changes during intense exercise (Karas *et al.*, 1987a; Jones *et al.*, 1989a). Increased $\dot{Q}_T$ associated with training that increases $\dot{V}_{O_2\ max}$ over time is due to the heart becoming a larger pump, not a faster one (Section IV).

The heart's capacity to generate pressure also depends on the extent to which the heart has developed in response to aerobic demand, with mean arterial pressure at $\dot{V}_{O_2\ max}$ relatively constant among mammals and independent of body size, although equids tend to have higher $\bar{P}_a$ at $\dot{V}_{O_2\ max}$ than other families of mammals (Karas *et al.*, 1987a; Jones *et al.*, 1989a, 1992a; Longworth *et al.*, 1989). Systemic vascular resistance at $\dot{V}_{O_2\ max}$ is largely a function of the size of the capillary bed that is perfused, and V_C increases with increased aerobic demand (Hoppeler *et al.*, 1981a,b; Conley *et al.*, 1987; Kayar *et al.*, 1989). The circulatory response to increased aerobic demand in mammals of a given body size is due to a larger, stronger heart that beats at the same maximal f_H, and an increased V_C and lower R_S in conjunction with increased mitochondrial volume (Hoppeler *et al.*, 1981a, 1985, 1987a,b, 1991; Karas *et al.*, 1987a; Jones *et al.*, 1989a; Kayar *et al.*, 1989; Longworth *et al.*, 1989). This is true whether the increased demand is caused by adaptive variation or induced variation.

Although in humans [Hb] changes little with exercise, in most species of mammals this is not the case. As a mammal increases from rest to $\dot{V}_{O_2\ max}$, [Hb] typically increases by 20–50% (Karas *et al.*, 1987a; Jones *et al.*, 1989a; Longworth *et al.*, 1989), presumably as the result of splenic contraction, recruitment of pooled erythrocytes in capacitance vessels, and/or plasma filtration. Maximum [Hb] during exercise increases somewhat with aerobic training, but does not exceed approximately 24 g dl^{-1} (65% hematocrit) in even the most elite of animal athletes (Lindstedt *et al.*, 1991; Jones and Lindstedt, 1993).

4. Pulmonary Diffusion

Equation (5) describes the diffusive transport of O_2 across the pulmonary gas exchanger as the product of a conductance and a pressure head. If one of these variables is fixed, O_2 transport can increase only if the other variable increases. Diffusion limitation of O_2 in the lung, if it exists to a significant extent, will increase the alveolar–arterial P_{O_2} difference (A-aD_{O_2}). Increasing $D\textsc{l}_{O_2}$ will raise Pa_{O_2} and reduce the integrated difference in P_{O_2}, which is the driving pressure for diffusion. However, the product of conductance and pressure head increases, and so does $\dot{V}_{O_2}$.

If little diffusion limitation exists in the lung, increasing $D\text{L}_{O_2}$ will increase the *rate* of O_2 transport to individual erythrocytes, but not the *quantity* of O_2 transported, as Pa_{O_2} will be unaffected (Fig. 5). In effect, the P_{O_2} in the capillary blood will equilibrate with alveolar P_{O_2} ($P\text{A}_{O_2}$) more quickly, however, arterial O_2 saturation of hemoglobin (Sa_{O_2}) will remain the same. The integrated alveolar–capillary P_{O_2} difference will actually decrease for the same change in P_{O_2} from mixed venous to arterial blood if $D\text{L}_{O_2}$ is increased. This occurs because the P_{O_2} difference is nearly zero for a greater portion of the erythrocyte's t_c (Fig. 5). Thus, the increase in $D\text{L}_{O_2}$ is offset by a decrease in the integrated P_{O_2} difference, resulting in almost no change in $\dot{V}_{O_2}$ [Eq. (5)].

Pulmonary diffusing capacity changes nearly in proportion to body size in mammals (Gehr *et al.*, 1981; Taylor and Weibel, 1991; Jones and Longworth, 1992; Longworth, 1993), and not in proportion to $\dot{V}_{O_2\ max}$ (Weibel *et al.*, 1983), although nearly twofold adaptive variation is possible within a body size (Weibel *et al.*, 1987; Constantinopol *et al.*, 1989; Taylor and Weibel, 1991; Longworth, 1993). However, pulmonary structure appears to be largely determined genetically, and little adaptation in $D\text{L}_{O_2}$ occurs in adult animals in response to induced increases in O_2 demand, although transient adaptation may be seen during development (Bartlett, 1970; Bartlett and Remmers, 1971; Burri and Weibel, 1971; Burri *et al.*, 1976; Geelhaar and Weibel, 1971; Gehr *et al.*, 1978; Hugonnaud *et al.*, 1977; Lechner *et al.*, 1982; Weibel *et al.*, 1983).

Pulmonary O_2 transport *rate* will also increase if $P\text{A}_{O_2}$ increases, for example, with hyperventilation. However, this will have relatively little effect on the *quantity* of O_2 transported unless the animal's Pa_{O_2} at $\dot{V}_{O_2\ max}$ normally yields an Sa_{O_2} on the steep portion of the OEC. In this case, small increases in Pa_{O_2} will significantly increase Sa_{O_2} and Ca_{O_2}. This situation exists in athletes that are highly aerobic owing to adaptive variation (Thoroughbred horses; Jones *et al.*, 1989b, 1993) or induced variation (elite human athletes; Dempsey, 1986; Powers *et al.*, 1989), although it may be less important in those that have high mass-specific $\dot{V}_{O_2\ max}$ owing to allometric variation (Longworth *et al.*, 1989).

Equation (5) illustrates that for a given $D\text{L}_{O_2}$ the flux of O_2 increases only if the integrated alveolar–capillary P_{O_2} difference increases. If $P\text{A}_{O_2}$ and $P\bar{v}_{O_2}$ are fixed, this can only occur if Pa_{O_2} decreases, that is, if $\dot{Q}_T$ is sufficiently high relative to $D\text{L}_{O_2}$ to cause significant diffusion limitation. Decreased $P\bar{v}_{O_2}$ will also increase the integrated alveolar–

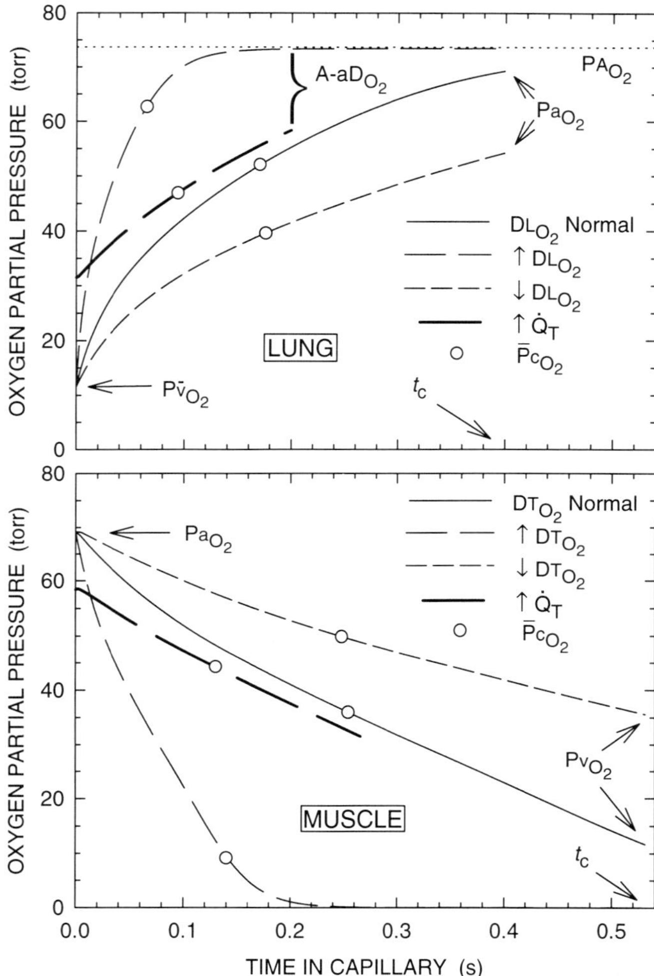

FIG. 5. Bohr integration models of the change in P_{O_2} in an erythrocyte (solid curve) as it passes through the pulmonary capillary (*top*) or muscle capillary (*bottom*) of a Thoroughbred horse exercising at $\dot{V}_{O_2\,max}$. Blood enters the pulmonary capillary with mixed venous P_{O_2} ($P\bar{v}_{O_2}$) and exits with arterial P_{O_2} (Pa_{O_2}); the difference between alveolar P_{O_2} ($P_{A_{O_2}}$) and Pa_{O_2}, the alveolar–arterial P_{O_2} difference (A–aD_{O_2}), is due to diffusion limitation in this simple one-compartment model. Blood enters the muscle capillary with Pa_{O_2} and exits with end-capillary venous P_{O_2} (Pv_{O_2}), which is greater than mitochondrial P_{O_2} (= 0 Torr) as a result of diffusion limitation. The difference between $P_{A_{O_2}}$ (in lung) or mitochondrial P_{O_2} (in muscle) and the integrated mean P_{O_2} during capillary transit ($\bar{P}c_{O_2}$) is the average pressure head for diffusion in that gas exchanger. The capillary transit time (t_c) is calculated as the pulmonary or whole-body muscle capillary volume (estimated morphometrically) divided by cardiac output ($\dot{Q}_T$). The effects of increasing conductance for diffusion ($D_{L_{O_2}}$ in lung, $D_{T_{O_2}}$ in muscle) are shown by the medium-

capillary P_{O_2} difference, however, peripheral gas exchange determines this value, and it is an input variable in terms of the pulmonary gas exchanger.

5. *Ventilatory Convection*

Ventilatory transport of O_2 [Eq. (6)] depends on the product of the quantity of gas ventilated per unit time ($\dot{V}_E$) and the fraction of O_2 extracted from the air that is ventilated ($FI_{O_2} - F\bar{E}_{O_2}$). Analogous to cardiac output, minute ventilation is the product of two pairs of variables:

$$\dot{V}_E = f_R \times V_T = (\bar{P}_R + \bar{P}_I + \bar{P}_T) \times \frac{1}{R_{PUL}}, \tag{8}$$

where f_R is ventilatory frequency, V_T is tidal volume, $\bar{P}_R$, $\bar{P}_I$, and $\bar{P}_T$ are mean pressures to overcome airway resistance, inertia, and tissue viscosity, respectively, and R_{PUL} is total pulmonary resistance. Ventilatory frequency at rest is allometrically related to body size (Stahl, 1967; Schmidt-Nielsen, 1984), such that small mammals breathe more frequently than large ones. During exercise in many mammals ventilation is phase-locked in a 1:1 ratio with stride (Bramble and Carrier, 1983), and stride frequency is also allometrically related to body size (Heglund *et al.*, 1974; Heglund and Taylor, 1988), although with a different mass exponent than resting f_R. Tidal volume at rest is directly related to body size (Stahl, 1967; Schmidt-Nielsen, 1984), however, relatively few data are available from different-sized mammals at $\dot{V}_{O_2\,max}$ to assess the relationship during severe exercise.

Longworth *et al.* (1988) and Jones and Lindstedt (1993) have noted that larger mammals tend to hypoventilate at $\dot{V}_{O_2\,max}$ more than small ones, a fact that may be related to their ventilatory (= stride) frequencies being relatively lower than their metabolic demands for O_2. Highly aerobic mammals may approach maximum expiratory flow limita-

---

dashed lines, and of decreasing conductances by the short-dashed lines. The diffusion limitations experienced by the normal horse (5 Torr in lung, 12 Torr in muscle) are essentially abolished by increasing the conductances fourfold and are exacerbated by halving the conductances. The effects of doubling $\dot{Q}_T$ (with normal DL_{O_2} and DT_{O_2}) are shown by the heavy long-dashed line; note that t_c decreases and diffusion limitation increases in both lung and muscle. As a result, $\dot{V}_{O_2\,max}$ increases by only 22%, although $\dot{Q}_T$ was increased by 100%. The 50% reductions in DL_{O_2} and DT_{O_2} resulted in 22 and 31% reductions in $\dot{V}_{O_2\,max}$, suggesting the horse's respiratory system has equal or greater sensitivity to changes in diffusive conductance as to convective capacity. Models patterned after the approach and data in Karas *et al.* (1987b), Jones and Karas (1988), Constantinopol *et al.* (1989), Kayar *et al.* (1989), and Jones *et al.* (1993).

tion at $\dot{V}_{O_2\ max}$, creating a potential ventilatory limitation in large athletes (e.g., horses) that could explain why they hypoventilate severely at $\dot{V}_{O_2\ max}$. Although allometric effects on airway and pulmonary mechanics during resting ventilation have been considered (Agostini *et al.*, 1959; Stahl, 1967; Tenney and Bartlett, 1967; Lahiri, 1975; Leith, 1976, 1983a,b), few data are available to allometrically evaluate ventilation during maximal exercise.

If minute ventilation is fixed, $\dot{V}_{O_2}$ can only be increased if extraction is increased. Capacitance of ventilated gas is determined by $F\text{I}_{O_2}$, however, hyperoxic gas is only encountered in experimental situations. Barometric pressure (P_B) can also affect inspired P_{O_2} ($P\text{I}_{O_2}$); it is unusual to encounter hyperbaric ventilation except experimentally, although altitude can reduce O_2 availability as a result of hypobaric conditions. Therefore, the only mechanism available to increase ventilatory delivery of O_2 if $\dot{V}_E$ is fixed is to increase extraction by lowering $F\bar{\text{E}}_{O_2}$, and $P\text{A}_{O_2}$ as a consequence.

B. Conflicting Demands: Circulation Does Not Function as an Independent System

1. *Interaction of Circulation and Diffusive Gas Exchangers*

Karas *et al.* (1987b), Jones and Karas (1988), Constantinopol *et al.* (1989), Longworth *et al.* (1989), and Jones *et al.* (1993) have evaluated interactions between circulatory convection and pressure heads for diffusion of O_2 across the pulmonary and/or skeletal muscle gas exchangers using simple one-compartment Bohr integration models. These models assume homogeneity of alveolar ventilation ($\dot{V}_A$) and perfusion in the lung, and homogeneity of perfusion and O_2 demand in the working muscle. They evaluate the effects on O_2 flux of changes in the following variables: ventilation and its effect on $P\text{A}_{O_2}$, diffusive conductance for oxygen ($D\text{L}_{O_2}$ for the lung, $D\text{T}_{O_2}$ for muscle), $\dot{Q}_T$, [Hb], capillary volume (V_C, inversely related to t_c for a given $\dot{Q}_T$), and the OEC.

The findings of these modeling studies are summarized in Fig. 5. For a given diffusive conductance, as $\dot{Q}_T$ increases so does the difference between end-capillary P_{O_2} ($P\text{a}_{O_2}$ in lung, $P\text{v}_{O_2}$ in muscle) and the P_{O_2} with which the blood is equilibrating ($P\text{A}_{O_2}$ in the lung; $P\text{mito}_{O_2}$, assumed to be nearly 0 Torr, in the muscle). Therefore, as $\dot{Q}_T$ increases, t_c becomes insufficient for the available conductance and diffusion limitation becomes more significant. Nevertheless, although diffusion limitation becomes greater and $P\text{a}_{O_2}$ decreases or $P\bar{\text{v}}_{O_2}$ increases, *total* O_2 flux through the system increases. Hence, the tautological nature of the "single-step cardiovascular limitation" hypothesis is realized: increases in circulatory O_2 convection will *always* increase O_2 flux (assuming that mitochondrial oxidative capacity is sufficient to utilize the O_2) even in the face of significant diffusion limitation that contributes significantly to the total resistance to O_2 transport. In this case, increasing or decreasing $\dot{Q}_T$ or [Hb] would raise or lower $\dot{V}_{O_2\ max}$; however, so would raising or lowering $D\text{L}_{O_2}$, $D\text{T}_{O_2}$, or $\dot{V}_A$ (for pulmonary diffusion limitation), *independently* of changes in $\dot{Q}_T$ or [Hb].

The changes in O_2 flux shown in Fig. 5 that result from modification of different components of the respiratory system underscore the interrelationship between circulatory function and O_2 transport. It is evident that increasing either the capacitance of the circulation for O_2 ([Hb]) or the rate of circulatory convection ($\dot{Q}_T$) *always* increases $\dot{V}_{O_2}$ for any given values of $\dot{V}_A$, $D\text{L}_{O_2}$, and $D\text{T}_{O_2}$. However, if significant diffusion limitation exists in either diffusive gas exchanger, O_2 flux can also be increased by raising the appropriate conductance; if pulmonary diffusion limitation exists, $\dot{V}_{O_2}$ can also be increased by raising $\dot{V}_A$ and $P\text{A}_{O_2}$.

It should be noted that for any given $D\text{T}_{O_2}$, O_2 extraction from capillary blood would be greater if t_c were longer, thus lowering $P\bar{v}_{O_2}$ and indicating less diffusion limitation for O_2. However, for any given $D\text{T}_{O_2}$, the systemic transport and delivery of O_2 is greater whenever $\dot{Q}_T$ is increased, t_c is decreased, and $P\bar{v}_{O_2}$ increases, despite the decreased O_2 extraction and greater diffusion limitation for O_2 (Jones and Karas, 1988). In other words, O_2 transport will always increase when circulatory bulk transport ($\dot{Q}_T$) or capacitance ([Hb]) are increased, even if they outstrip conductance and result in reduced extraction of O_2 in the capillaries of the gas exchanger.

Consideration of these factors leads to an interesting conclusion. Table I shows changes in individual variables required to acutely increase $\dot{V}_{O_2\ max}$ for each of the steps of the O_2 transport system. These relationships hold for mammals in which $\dot{V}_{O_2\ max}$ is limited by O_2 transport, not mitochondrial oxidative capacity. (In the goat, e.g., the latter holds, and the requirement for increasing $\dot{V}_{O_2\ max}$ is to increase peripheral demand for O_2, providing a larger sink for it. In such animals $P\text{mito}_{O_2}$ must be lowered to increase the rate of peripheral diffusion.) The right-hand column in the table shows the changes in P_{O_2} at each step that would be required to increase $\dot{V}_{O_2}$ at that step and through the system as a whole. This column is of particular interest because it highlights the fact that the demands of adjacent steps for changes in P_{O_2} conflict:

1. Peripheral tissue diffusive $\dot{V}_{O_2}$ increases if $P\bar{v}_{O_2}$ increases [Eq. (3)],

TABLE I

CHANGES THAT WOULD BE REQUIRED IN VARIABLES AT DIFFERENT STEPS OF THE OXYGEN TRANSPORT SYSTEM TO ACUTELY INCREASE OXYGEN FLUX AT $\dot{V}_{O_2max}$.[a]

Transport step	O_2 flux equation $\dot{V}_{O_2}$ =	Frequency	Volume
Peripheral diffusion	$\int(D\text{T}_{O_2} \times (P\text{c}_{O_2} - P\text{mito}_{O_2}))dt$	—	—
Circulatory convection	$\dot{Q}_\text{T} \times [\text{Hb}] \times (f(P\text{a}_{O_2}) - f(P\bar{\text{v}}_{O_2}))$	f_H ↑	V_S ↑
Pulmonary diffusion	$\int(D\text{L}_{O_2} \times (P\text{A}_{O_2} - P\text{c}_{O_2}))dt$	—	—
Ventilatory convection	$\dot{V}_\text{E} \times (F\text{I}_{O_2} - F\bar{\text{E}}_{O_2})$	f_R ↑	V_T ↑

[a]Symbols as in the text. Boxed pairs in the column on the right highlight conflicting demands of adjacent convective and diffusive transport steps for changes in O_2 partial pressure. * indicates P_{O_2} values that would algebraically increase $\dot{V}_{O_2}$, but cannot be altered in an animal limited at $\dot{V}_{O_2}$max by its O_2 transport system: $P\text{mito}_{O_2}$ is presumably near 0 Torr and is therefore already minimized; $P\text{I}_{O_2}$ is determined by barometric pressure, $F\text{I}_{O_2}$, and the animal's body temperature.

but circulatory convective $\dot{V}_{O_2}$ decreases when $P\bar{\text{v}}_{O_2}$ increases [Eq. (4)];

2. Circulatory convective $\dot{V}_{O_2}$ increases when $P\text{a}_{O_2}$ increases [Eq. (4)], however, pulmonary diffusive $\dot{V}_{O_2}$ decreases when $P\text{a}_{O_2}$ increases [Eq. (5)];
3. Pulmonary diffusive $\dot{V}_{O_2}$ increases when $P\text{A}_{O_2}$ increases [Eq. (5)], however, ventilatory convective $\dot{V}_{O_2}$ decreases when $P\text{A}_{O_2}$ increases [Eq. (6)].

It is impossible for changes in any one step of the O_2 transport system to be truly independent of the adjacent steps. Acute alterations in circulatory O_2 transport will affect diffusive transport in both the pulmonary and peripheral gas exchangers. If these exchangers have sufficient capacity to adjust to the altered pressure head generated by changes in circulatory convection, then O_2 flux can be maintained; otherwise, $\dot{V}_{O_2}$ will change. Note that if significant diffusion limitation exists at either of the diffusive gas exchangers, then O_2 flux through the system can be altered by changing the diffusive conductance of the exchanger at which diffusion limitation occurs with no change in circulatory convection or capacitance.

2. *Symmorphosis*

In 1981 Taylor and Weibel proposed a principle of design for the mammalian respiratory system that they termed "symmorphosis." This principle states that animal design should be economical, with the

TABLE I (*Continued*)

Capacitance	Conductance	O_2 partial pressure		
—	$D_{T_{O_2}}$ ↑	$Pmito_{O_2}$ ↓*		$P\bar{v}_{O_2}$ ↑
[Hb] ↑	—		Pa_{O_2} ↑	$P\bar{v}_{O_2}$ ↓
—	$D_{L_{O_2}}$ ↑	$P_{A_{O_2}}$ ↑	Pa_{O_2} ↓	
$F_{I_{O_2}}$ ↑	—	$P_{A_{O_2}}$ ↓		$P_{I_{O_2}}$ ↑*

formation of structural elements "regulated to satisfy but not exceed the requirements of the functional system." Although the concept of symmorphosis has generated considerable debate (Garland and Huey, 1987; Weibel *et al.*, 1991), it has been suggested that it may be most useful as a working hypothesis for evaluating structural and functional relationships in animals (Lindstedt and Jones, 1987).

The role of cardiovascular components in limiting O_2 flux through the respiratory system can be considered in terms of the concept of symmorphosis. The effects that changes in circulatory convection inherently have on pressure heads for diffusion in adjacent gas exchangers are somewhat analogous to the effects of altering activities of individual enzymes in biochemical systems with multiple catalyzed steps (Kacser and Burns, 1973, 1979; Kacser, 1983). Analyses of these interactions demonstrate the difficulty in defining a single "rate-limiting" step in a system in which components function interdependently. When these concepts are applied to circulatory function during exercise in mammals, we conclude that a single-step, "weak link," absolute limiting step for O_2 flux through the respiratory system can only exist if:

1. Mitochondrial oxidative capacity is exceeded by the quantity of O_2 being transported to the working muscle (mitochondrial oxidative capacity limits $\dot{V}_{O_2\,max}$); or
2. Arterial blood equilibrates with alveolar gas ($Pa_{O_2} = P_{A_{O_2}}$) and O_2 is completely extracted from blood in the muscle capillaries ($P\bar{v}_{O_2}$ = 0 Torr; circulatory convection limits $\dot{V}_{O_2\,max}$); or
3. Oxygen is completely extracted from air in the alveolus (ventilatory convection limits $\dot{V}_{O_2\,max}$).

A mitochondrial oxidative limitation to $\dot{V}_{O_2\,max}$ appears to exist in some species (e.g., goats), demonstrating the first type of limitation.

The latter two limitation patterns are not known to exist in any mammal. In many species of mammals, more than one component of the system may contribute to the limitation (e.g., circulatory O_2 delivery and peripheral diffusion limitation), whereas other components do not. In mammals that demonstrate extreme adaptive variation (Thoroughbred racehorses) or induced variation (elite human athletes) for aerobic performance, significant limitations to O_2 flux may be contributed by *all* components of the O_2 transport system (Section II,B).

The animals that are among the most highly selected for aerobic capacity (horses) demonstrate limitations to O_2 flux throughout their respiratory systems at $\dot{V}_{O_2\ max}$: hypoventilation, pulmonary diffusion limitation, allometrically determined maximal f_H with systemic and pulmonary hypertension (Section IV), and peripheral diffusion limitation. The respiratory systems of these animals have been strongly selected for a phenotypic trait that requires high $\dot{V}_{O_2\ max}$ and have evolved most closely to the pattern predicted by symmorphosis. In horses, all the O_2 transport steps contribute significantly to limiting O_2 flux at $\dot{V}_{O_2\ max}$ simultaneously, indicating no "excess" structure in any of the O_2 transport components. The principle of symmorphosis may be more likely to apply to the equine respiratory system than those of other species simply because in most mammals maximal respiratory function is not a major factor in natural selection. Therefore, in other species, one or more elements of the O_2 transport system may be underdeveloped, relative to the rest of the system, and contribute a proportionately larger share to limiting O_2 flux at $\dot{V}_{O_2\ max}$. However, with the exception of mitochondrial oxidative capacity, they cannot constitute single-step limiting factors for $\dot{V}_{O_2\ max}$. In horses, the more genetically "plastic" respiratory structures (heart, skeletal muscle, capillaries, mitochondria) may have developed to the point where they achieve the O_2 transport capacity of the least malleable element of the system, the lung, resulting in simultaneous matched limitations to O_2 flux that appear to be symmorphotic (Jones and Lindstedt, 1993; Jones *et al.*, 1993).

IV. Circulation at the Upper Limits of Aerobic Performance

A. Circulatory Responses to Allometric, Adaptive, and Induced Demands for Oxygen

The respiratory system contains both malleable and fixed components. The way that malleable components change both structurally and functionally in response to altered demand for O_2 has been the focus of intense research (Weibel and Taylor, 1981; Taylor *et al.*, 1987). The three different mechanisms by which aerobic demand is altered (allometric, adaptive, and induced variation) have different effects on various components of the respiratory system, and particularly the cardiovascular elements. Many of these effects have been mentioned in Section III,A. I will briefly consider the major mechanisms by which cardiovascular function adapts to these demands.

1. *Allometric Effects*

Body size causes larger allometric differences in mass-specific $\dot{V}_{O_2\ max}$ among mammals than are seen within any given size owing to adaptive or induced variation (Fig. 6; Jones and Lindstedt, 1993). These changes in O_2 flux are directly correlated with changes in whole-body mitochondrial and muscle capillary volumes, and are associated with changes in lung and heart volumes, and airway and blood vessel sizes that are either directly or allometrically related to body size (Stahl, 1967; Holt *et al.*, 1968; Leith, 1976; Weibel and Taylor, 1981; Taylor *et al.*, 1987; Jones and Lindstedt, 1993). A number of other variables at $\dot{V}_{O_2\ max}$ (*e.g.*, [Hb], $\bar{P}_a$, and the kinetics of O_2 binding and unbinding with Hb) are independent of body size.

Because mass-specific $\dot{V}_{O_2\ max}$ increases as body size decreases ($\propto M_b^{-0.14}$, where M_b is body mass) (Fig. 6; Jones and Lindstedt, 1993), and [Hb], Sa_{O_2}, and $S\bar{v}_{O_2}$ are independent of size (Karas *et al.*, 1987a; Jones *et al.*, 1989a; Longworth *et al.*, 1989), Eqs. (1) and (4) dictate that mass-specific $\dot{Q}_T$ must increase similarly to the allometric change in $\dot{V}_{O_2\ max}$. Confirmation of this relationship is shown in Fig. 6, with $\dot{V}_{O_2\ max}/M_b$ and $\dot{Q}_T/M_b$ changing with similar allometric exponents. The mechanism by which $\dot{Q}_T$ changes with body size is also evident in Fig. 6, as data for maximal f_H of mammals of different size show a similar relationship. Indeed, the highest $\dot{V}_{O_2}/M_b$ reported in a mammal (uncertain to be $\dot{V}_{O_2\ max}$) of 6.7 ml (STPD) s^{-1} kg^{-1} in the 2-g Etruscan shrew is associated with the highest f_H reported, 1300 min^{-1} (Fig. 6; Fons and Sicart, 1976; Weibel, 1984). Thus, it is apparent that body size determines the maximal f_H that a mammal can achieve, and maximal f_H determines the maximal $\dot{Q}_T$, depending on the size of the heart and the V_S it can generate [Eq. (7)].

The constancy of $\bar{P}_a$ at $\dot{V}_{O_2\ max}$ as body size changes indicates that as body size decreases, capillary density in muscle must increase, a relationship originally recognized by Krogh (1919). Body size is, therefore, a prime determinant of the upper limit of cardiovascular capacity, exerting its effect directly on absolute heart size (V_S) and differentially

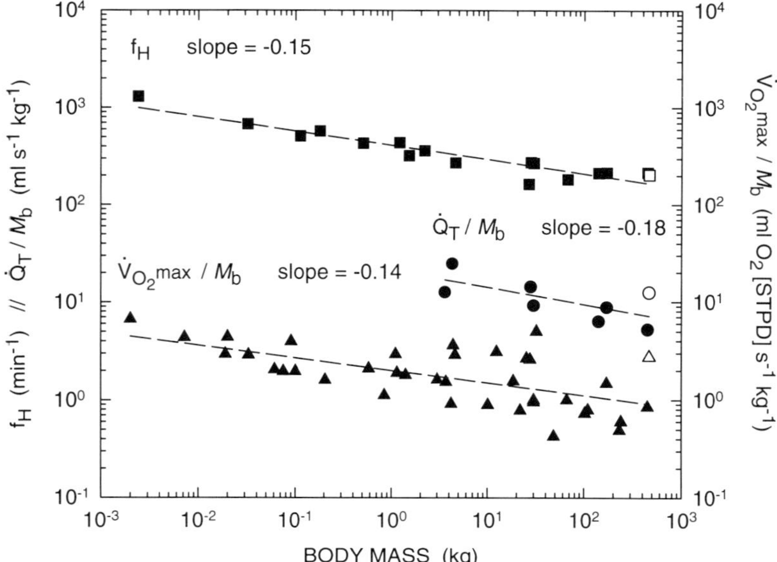

FIG. 6. Allometric relationships between body mass and mass-specific maximal rate of O_2 consumption ($\dot{V}_{O_2\ max}/M_b$, triangles), mass-specific cardiac output ($\dot{Q}_T/M_b$, circles), and maximal heart rate (f_H, squares) in mammals running at $\dot{V}_{O_2\ max}$. Least-squares linear regressions calculated on the log-transformed data have similar slopes and are shown. Open symbols are values for 465-kg Thoroughbred racehorses. Data from Baudinette (1978), Karas *et al.* (1987a), Jones *et al.* (1989b, 1993), Longworth *et al.* (1989), and Jones and Lindstedt (1993).

on vascular resistance (V_C) and cardiac frequency (f_H). Size-dependent differences in biological frequencies are typical of allometric relationships (Schmidt-Nielsen, 1984).

2. *Adaptive Variation*

Within a body size, differences in maximal cardiovascular performance are due to two factors shown in Eq. (4): $\dot{Q}_T$ and [Hb]. The higher maximal $\dot{Q}_T$ in species that are more highly adapted for aerobic performance is entirely due to larger V_S, as maximal f_H is identical (Karas *et al.*, 1987a; Jones *et al.*, 1989a; Longworth *et al.*, 1989). The hearts of highly aerobic species are typically twice as large as those of less aerobic species, although in cases of extreme adaptation (e.g., the pronghorn antelope) the difference may be larger (Lindstedt *et al.*, 1991).

Although both athletic (highly aerobic) and average species increase [Hb] with exercise intensity, the value of [Hb] is typically as much as 50% higher at $\dot{V}_{O_2\ max}$ in the highly athletic species (Karas *et al.*, 1987a; Jones *et al.*, 1989a; Lindstedt *et al.*, 1991). Because the effects of $\dot{Q}_T$ and [Hb] on circulatory O_2 transport are multiplicative [Eq. (4)], the combination of these adaptations enables athletic species to achieve much greater increases in $\dot{V}_{O_2}$ than the amount that either of the individual variables changes.

Mean arterial pressure at $\dot{V}_{O_2\ max}$ appears to be relatively independent of aerobic adaptation, although equids tend to have pressures that are 25–100 Torr higher than other mammals that have been studied (Karas *et al.*, 1987a; Jones *et al.*, 1989a; Longworth *et al.*, 1989). Because $\bar{P}_a$ at $\dot{V}_{O_2\ max}$ is similar in athletic and average species, yet $\dot{Q}_T$ is higher, Eq. (7) dictates that V_C must be larger in athletic species (Hoppeler *et al.*, 1981a; Conley *et al.*, 1987; Kayar *et al.*, 1989; Hoppeler and Billeter, 1991). Capillary volume is larger in athletic species because of both higher capillary density in their muscles and a higher proportion of skeletal muscle in their bodies (Snow, 1985; Karas *et al.*, 1987a; Jones *et al.*, 1989a; Kayar *et al.*, 1989; Lindstedt *et al.*, 1991).

Differences in circulatory function due to adaptive variation result from quantitative and qualitative structural changes in cardiac and skeletal muscle. The relative amounts of both types of muscle are larger in athletic species; capillary (and mitochondrial) densities are higher in the skeletal muscles of athletic species, although cardiac muscle is qualitatively similar in both groups (Hoppeler *et al.*, 1984b).

3. *Induced Variation*

Changes in circulatory structures and function resulting from training within an individual animal are constrained by allometric and adaptive considerations. As described earlier, maximum f_H is determined by body size, and training does not affect it. Heart size (V_S) increases with training, as do skeletal muscle capillary and mitochondrial densities (Saltin and Rowell, 1980; Blomqvist and Saltin, 1983; Hoppeler *et al.*, 1985; Birks *et al.*, 1992). As a result, maximal $\dot{Q}_T$ increases with training, as does [Hb] at $\dot{V}_{O_2\ max}$; the latter change does not occur in human athletes, who are unusual among mammals in not increasing [Hb] as exercise intensity increases from rest to $\dot{V}_{O_2\ max}$. Therefore, changes in cardiovascular structures resulting from training mirror those due to adaptive variation. Animals of athletic species amplify their adaptive aerobic advantage in the size of these structures when the structures hypertrophy further during training.

B. Extreme Demand for Circulatory Function May Exceed Structural Adaptation

Exercise-Induced Pulmonary Hemorrhage in Horses

If the demands of allometric, adaptive, and induced variation for enhanced circulatory function converge, it is possible for the forces exerted by malleable elements of the cardiovascular system to exceed the tolerances of some of the fixed structures, leading to catastrophic failure of these components. An example of this type of failure may be the phenomenon of exercise-induced pulmonary hemorrhage in racehorses. Although pulmonary hemorrhage is extremely rare in mammals, it occurs nearly ubiquitously in Thoroughbred horses that exercise strenuously (Whitwell and Greet, 1984; Sweeney, 1991). These animals have been highly selected by humans for hundreds of generations for phenotypic traits requiring high aerobic performance (Snow, 1985) and are trained to achieve peak performance for competition. Although they are relatively large mammals (*ca.* 500 kg), horses have mass-specific $\dot{V}_{O_2\,max}$ that is nearly half that of the 2-g Etruscan shrew (Fig. 6). Extreme demands are placed on the circulatory components of the equine respiratory system to enable high $\dot{Q}_T/M_b$ and O_2 flux; maximum f_H for horses is only one-sixth that of the shrew (220 min^{-1} versus 1300 min^{-1}; Fig. 6) (Weibel, 1984; Jones *et al.*, 1989a, 1993; Birks *et al.*, 1991).

As mentioned previously, $\bar{P}_a$ is higher for equids than for other mammals (240 versus 150 Torr at $\dot{V}_{O_2\,max}$), and it has been well documented that mean pulmonary artery pressure ($\bar{P}_{PA}$) is also unusually high, approaching values of 120 Torr when measured at the height of the heart (Birks *et al.*, 1991; Jones *et al.*, 1992a, 1993). These high values appear to result from high mean left atrial pressure ($>$ 70 Torr at $\dot{V}_{O_2\,max}$) caused by low left ventricular compliance, rather than from high pulmonary vascular resistance (Jones *et al.*, 1992a,b). Because the harmonic mean thickness (0.6 μm) of the equine alveolar–capillary membrane is relatively thick compared to that of other mammals (Gehr *et al.*, 1981; Constantinopol *et al.*, 1989), there is no evidence that the pulmonary vasculature in horses is particularly weak and therefore, subject to stress failure (West *et al.*, 1993). Nevertheless, microsphere data indicate that blood originating from the pulmonary circulation contributes to exercise-induced pulmonary hemorrhage (Jones *et al.*, 1992c), suggesting that high transmural pressures in the pulmonary vessels are responsible for causing the hemorrhage to occur (Jones *et al.*, 1992a; West *et al.*, 1993).

Taken together, these data suggest that the heavy selective pressure that humans have placed on Thoroughbred horses for high aerobic performance may have led to an extreme in adaptive variation that is pathologically amplified when training induces additional hypertrophy of cardiovascular structures. The demands for high $\dot{Q}_T/M_b$, constrained allometrically by a relatively low maximal f_H, may have led to the development of a cardiac pump that is stiff because of the relatively thicker ventricular walls required by the Law of Laplace to generate high pressures in a large chamber (large V_S) (Martin and Haines, 1970). The conflicting demands of allometric constraints versus those of adaptive and induced variation may lead to structural failure when a large animal, such as the horse, pushes the design of its cardiovascular system to attain high $\dot{Q}_T/M_b$, whereas smaller species can utilize higher f_H to achieve high $\dot{Q}_T/M_b$ at lower ventricular volumes without raising ventricular filling pressures.

V. Summary

The cardiovascular system has frequently been hypothesized to be *the* limiting step for O_2 transport that determines $\dot{V}_{O_2\ max}$ in many species of mammals. Careful analysis of the factors that determine how O_2 is transported by the circulation demonstrate that such a single-step limitation cannot exist. Evaluation of the results of experiments in which circulatory O_2 transport capacity was experimentally altered demonstrates no direct or absolute relationship between changes in O_2 transport capacity and changes in $\dot{V}_{O_2\ max}$. Furthermore, experimental evidence collected during maximal exercise in hypoxia and hyperoxia supports the concept that multiple components of the O_2 transport system contribute to limiting O_2 flux at $\dot{V}_{O_2\ max}$.

Consideration of the basic equations that describe O_2 transport through the respiratory system shows that changes in P_{O_2} at each step of the system required to increase O_2 flux through that step conflict with the changes in P_{O_2} required to increase flux through adjacent steps. Changes in convection, capacitance, or conductance at one step affect gas transport through the adjacent steps. Hence, no single-step limitation to O_2 transport is possible, because the convective and diffusive gas exchangers are interdependent. Increasing $\dot{Q}_T$ at $\dot{V}_{O_2\ max}$ always increases O_2 flux (although not necessarily in proportion to the increase in $\dot{Q}_T$), unless $\dot{V}_{O_2\ max}$ is limited by mitochondrial oxidative capacity, as in goats.

Cardiovascular structure and function in mammals reflects allometric, adaptive and induced variation. Maximal heart rate is determined strictly by body size, thus maximal $\dot{Q}_T/M_b$ is inevitably lower in larger mammals. Adaptive and induced variation elicit hypertrophy of muscle, capillaries, and mitochondria, increasing circulatory capacity and $\dot{V}_{O_2\ max}$. When selection for maximal respiratory function is weak, as in most species of mammals, any component(s) of the respiratory system may be underdeveloped, relative to other structures in the system, and contribute disproportionately to limiting O_2 flux. When selection for aerobic capacity is strong, as in racehorses, malleable elements of the respiratory system, including the cardiovascular structures, may hypertrophy until their capacities for O_2 transport match that of the least malleable structure, the lung. Amplifying circulatory function so greatly in a large animal may lead to functional demand exceeding structural capacity, resulting in the nearly ubiquitous occurrence of exercise-induced pulmonary hemorrhage in racehorses.

Acknowledgements

Support for work presented in this paper was provided by the Equine Research Laboratory of the University of California, Davis, with contributions from the California Satellite Wagering Fund, the Oak Tree Racing Foundation, and private contributors, and the Grayson-Jockey Club Research Foundation.

References

Agostoni, E., Thimm, F. F., and Fenn, W. O. (1959). Comparative features of the mechanics of breathing. *J. Appl. Physiol.* **14,** 679–683.

Armstrong, R. B., Essen-Gustavsson, B., Hoppeler, H., Jones, J. H., Kayar, S. R., Laughlin, M. H., Lindholm, A., Longworth, K. E., Taylor, C. R., and Weibel, E. R. (1992). O_2 delivery at $\dot{V}_{O_2\ max}$ and oxidative capacity in muscles of Standardbred horses. *J. Appl. Physiol.* **73,** 2274–2282.

Åstrand, P., and Rodahl, K. (1986). "Textbook of Work Physiology," 3rd ed. McGraw Hill, New York.

Bartlett, D. (1970). Postnatal growth of the mammalian lung: Influence of low and high oxygen tensions. *Respir. Physiol.* **9,** 58–64.

Bartlett, D., and Remmers, J. E. (1971). Effects of high altitude exposure on the lungs of young rats. *Respir. Physiol.* **13,** 116–125.

Baudinette, R. V. (1978). Scaling of heart rate during locomotion in mammals. *J. Comp. Physiol.* **127,** 337–342.

Birks, E. K., Jones, J. H., and Berry, J. D. (1991). Plasma lactate kinetics during exercise. *In* "Equine Exercise Physiology III" (S. G. B. Persson, A. Lindholm, and L B. Jeffcott, eds.), pp. 179–187. Almqvist & Wiksell, Stockholm.

Birks, E. K., Jones, J. H., Hughes, T. R., Smith, B. L., and Edinger, A. L. (1992). Training effects on glucose and lactate kinetics in exercising horses. *Physiologist* **35,** 209.

Blomqvist, C. G., and Saltin, B. (1983). Cardiovascular adaptation to physical training. *Annu. Rev. Physiol.* **45,** 169–189.

Bohr, C. (1909). Ueber die spezifische Taetigkeit der Lungen bei der respiratorischen Gasaufnahme und ihr Verhalten zu der durch die Alveolarwand stattfindenden Gasdiffusion. *Skand. Arch. Physiol.* **22,** 221–280.

Bramble, D. M., and Carrier, D. R. (1983). Running and breathing in mammals. *Science* **219,** 251–256.

Brooks, G. A., and Fahey, T. D. (1985). "Exercise Physiology." Macmillan, New York.

Buick, F. J., Gledhill, N., Froese, A. B., Spriet, L., and Meyers, E. C. (1980). Effect of induced erythrocythemia on aerobic work capacity. *J. Appl. Physiol.* **48,** 636–642.

Burri, P. H., and Weibel, E. R. (1971). Morphometric estimation of pulmonary diffusion capacity. II. Effect of P_{O_2} on the growing lung. Adaptation of the growing rat lung to hypoxia and hyperoxia. *Respir. Physiol.* **11,** 247–264.

Burri, P. H., Gehr, P., Mueller, K., and Weibel, E. R. (1976). Adaptation of the growing lung to increased $\dot{V}_{O_2}$. I. IDPN as an inducer of hyperactivity. *Respir. Physiol.* **28,** 129–140.

Celsing, F., Nystrom, J., Pihlstedt, P., Werner, B., and Ekblom, B. (1986). Effect of long-term anemia and retransfusion on central circulation during exercise. *J. Appl. Physiol.* **61,** 1358–1362.

Celsing, F., Svedenhag, J., Pihlstedt, P., and Ekblom, B. (1987). Effects of anemia and stepwise-induced polycythemia on maximal aerobic power in individuals with high and low hemoglobin concentrations. *Acta Physiol. Scand.* **129,** 47–54.

Cerretelli, P., and di Prampero, P. E. (1987). Gas exchange in exercise. *In* "Handbook of Physiology" (L. E. Farhi and S. M. Tenney, eds.), Sect. 3, 2 pp. 297–340. Am. Physiol. Soc., Bethesda, MD.

Conley, K. E., Kayar, S. R., Rosler, K., Hoppeler, H., and Weibel, E. R. (1987). Adaptive variation in the mammalian respiratory system in relation to energetic demand. IV. Capillaries and their relationship to oxidative capacity. *Respir. Physiol.* **69,** 47–64.

Constantinopol, M., Jones, J. H., Weibel, E. R., Taylor, C. R., Lindholm, A., and Karas, R. H. (1989). Oxygen transport during exercise in large mammals. II. Oxygen uptake by the pulmonary gas exchanger. *J. Appl. Physiol.* **67,** 871–878.

Davies, K. J. A., Packer, L., and Brooks, G. A. (1981). Biochemical adaptation of mitochondria, muscle, and whole-animal respiration to endurance training. *Arch. Biochem. Biophys.* **209,** 539–554.

Davies, K. J. A., Maguire, J. J., Brooks, G. A., Dallman, P. A., and Packer, L. (1982). Muscle mitochondrial bioenergetics, oxygen supply, and work capacity during dietary iron deficiency and repletion. *Am. J. Physiol.* **242,** E418–E427.

Dempsey, J. A. (1986). Is the lung built for exercise? *Med. Sci. Sports Exercise* **18,** 143–155.

Ekblom, B. (1986). Factors determining maximal aerobic power. *Acta Physiol. Scand.* **128,** 15–19.

Ekblom, B., and Hout, R. (1972). Response to submaximal and maximal exercise at different levels of carboxyhemoglobin. *Acta Physiol. Scand.* **86,** 474–482.

Ekblom, B., Goldbarg, A. N., and Gullbring, B. (1972). Response to exercise after blood loss and reinfusion. *J. Appl. Physiol.* **33,** 175–180.

Ekblom, B., Hout, R., Stein, E. M., and Thorstensson, A. T. (1975). Effect of changes in arterial oxygen content on circulation and physical performance. *J. Appl. Physiol.* **39,** 71–75.

Ekblom, B., Wilson, G., and Åstrand, P.-O. (1976). Central circulation during exercise after venesection and reinfusion of red blood cells. *J. Appl. Physiol.* **40,** 379–383.

Evans, D. L., and Rose, R. J. (1988). Cardiovascular and respiratory responses to submaximal exercise training in the thoroughbred horse. *Pflüegers Arch.* **411,** 316–321.

Farrell, A. P. (1984). A review of cardiac performance in the teleost heart: Intrinsic and humoral regulation. *Can. J. Zool.* **62,** 523–536.

Farrell, A. P. (1993). Cardiac output in fish: Regulation and limitations. *In* "The Vertebrate

Gas Transport Cascade: Adaptations to Environment and Mode of Life" (J. E. P. W. Bicudo, ed.), pp. 208–214. CRC Press, Boca Raton, FL.

Fick, A. (1870). Ueber die Messung des Blutquantums in den Herzventrikeln. *Sitzungsber. Phys.-Med. Ges. Wurzburg* **16.**

Fons, R., and Sicart, R. (1976). Contribution à la connaissance du métabolisme energetique chez deuz crocidurinae: *Suncus etruscus* (Savi, 1822) et *Crocidura russala* (Hermann, 1780) (Insectivora, Soricidae). *Mammalia* **40,** 299–311.

Garland, T. J., and Huey, R. B. (1987). Testing symmorphosis: Does structure match functional requirements? *Evolution (Lawrence, Kans.)* **41,** 1404–1409.

Gayeski, T. E., Connett, R. J., and Honig, C. R. (1987). Minimum intracellular P_{O_2} for maximum cytochrome turnover in red muscle *in situ. Am. J. Physiol.* **252,** H906-H915.

Geelhaar, A., and Weibel, E. R. (1971). Morphometric estimation of pulmonary diffusing capacity. III. The effect of increased oxygen consumption in Japanese waltzing mice. *Respir. Physiol.* **11,** 354–366.

Gehr, P., Hugonnaud, C., Burri, P. H., Bachofen, H., and Weibel, E. R. (1978). Adaptation of the growing lung to increased $\dot{V}_{O_2}$. III. The effect of exposure to cold environment in rats. *Respir. Physiol.* **32,** 345–353.

Gehr, P., Mwangi, D. K., Ammann, A., Maloiy, G. M. O., Taylor, C. R., and Weibel, E. R. (1981). Design of the mammalian respiratory system. V. Scaling morphometric diffusing capacity to body mass: Wild and domestic mammals. *Respir. Physiol.* **44,** 61–86.

Heglund, N. C., and Taylor, C. R. (1988). Speed, stride frequency and energy cost per stride: How do they change with body size and gait? *J. Exp. Biol.* **138,** 301–318.

Heglund, N. C., Taylor, C. R., and McMahon, T. A. (1974). Scaling stride frequency and gait to animal size: Mice to horses. *Science* **186,** 1112–1113.

Hempleman, S. C., and Gray, A. T. (1988). Estimating steady-state DL_{O_2} with nonlinear dissociation curves and $\dot{V}_A/\dot{Q}$ inequality. *Respir. Physiol.* **73,** 279–288.

Hill, A. V., and Lupton, H. (1923). Muscular exercise, lactic acid, and the supply and utilization of oxygen. *Q. J. Med.* **16,** 135–171.

Hill, A. V., Long, C. N. H., and Lupton, M. (1924). Muscular exercise, lactic acid and supply and utilisation of oxygen. IV. Method of studying the respiratory exchanges in man during rapid alterations produced by muscular exercise and while breathing various gas mixtures. *Proc. R. Soc. London, Ser. B* **97,** 84–95.

Hill, E. P., Power, G. G., and Longo, L. D. (1973). Mathematical simulation of pulmonary O_2 and CO_2 exchange. *Am. J. Physiol.* **224,** 904–917.

Hillman, S. S. (1980). The effect of anemia on metabolic performance in the frog, *Rana pipiens. J. Exp. Zool.* **211,** 107–111.

Hillman, S. S., Withers, P. C., Hedrick, M. S., and Kimmel, P. B. (1985). The effects of erythrocytemia on blood viscosity, maximal systemic oxygen transport capacity and maximal rate of oxygen consumption in an amphibian. *J. Comp. Physiol. B* **155B,** 577–582.

Hogan, M. C., Roca, J., Wagner, P. D., and West, J. B. (1988). Limitation of maximal O_2 uptake and performance by acute hypoxia in dog muscle *in situ. J. Appl. Physiol.* **65,** 815–821.

Hogan, M. C., Roca, J., West, J. B., and Wagner, P. D. (1989). Dissociation of maximal O_2 uptake from O_2 delivery in canine gastrocnemius *in situ. J. Appl. Physiol.* **66,** 1919–1926.

Holt, J. P., Rhode, E. A., and Kines, H. (1968). Ventricular volumes and body weight in mammals. *Am. J. Physiol.* **215,** 704–715.

Hoppeler, H. (1986). Exercise-induced ultrastructural changes in skeletal muscle. *Int. J. Sports Med.* **7,** 187–204.

Hoppeler, H., and Billeter, R. (1991). Conditions for oxygen and substrate transport in muscles in exercising mammals. *J. Exp. Biol.* **160,** 263–283.

Hoppeler, H., and Lindstedt, S. L. (1985). Malleability of skeletal muscle tissue in overcoming limitations: Structural elements. *J. Exp. Biol.* **115,** 355–364.

Hoppeler, H., Mathieu, O., Krauer, R., Claassen, H., and Armstrong, R. B. (1981a). Design of the mammalian respiratory system. VI. Distribution of mitochondria and capillaries in various muscles. *Respir. Physiol.* **44,** 87–111.

Hoppeler, H., Mathieu, O., Weibel, E. R., Krauer, R., and Lindstedt, S. L. (1981b). Design of the mammalian respiratory system. VIII. Capillaries in skeletal muscles. *Respir. Physiol.* **44,** 129–150.

Hoppeler, H., Lindstedt, S. L., Uhlmann, E., Niesel, A., and Cruz-Orive, L. M. (1984a). Oxygen consumption and the composition of skeletal muscle tissue after training and inactivation in the European woodmouse (*Apodemus sylvaticus*). *J. Comp. Physiol. B* **155B,** 51–61.

Hoppeler, H., Lindstedt, S. L., Claasen, H., Taylor, C. R., and Mathieu, O. (1984b). Scaling mitochondrial volume in heart to body mass. *Respir. Physiol.* **55,** 131–137.

Hoppeler, H., Howald, H., Conley, K. E., Lindstedt, S. L., and Claasen, H. (1985). Endurance training in humans: Aerobic capacity and structure of skeletal muscle. *J. Appl. Physiol.* **59,** 320–327.

Hoppeler, H., Hudlicka, O., and Uhlmann, E. (1987a). Relationship between mitochondria and oxygen consumption in isolated cat muscles. *J. Physiol. (London)* **385,** 661–675.

Hoppeler, H., Jones, J. H., Lindstedt, S. L., Claasen, H., Longworth, K. E., Taylor, C. R., Straub, R., and Lindholm, A. (1987b). Relating $\dot{V}_{O_2\ max}$ to skeletal muscle mitochondria in horses. *In* "Equine Exercise Physiology II" (J. R. Gillespie and N E. Robinson, eds.), pp. 278–289. ICEEP Publ., Davis CA.

Hoppeler, H., Mathieu-Costello, O., and Kayar, S. R. (1991). Mitochondria and microvascular design. *In* "The Lung: Scientific Foundations" (R. G. Crystal and J. B. West, eds.), Vol. 2, pp. 1467–1477. Raven Press, New York.

Hüfner, C. G. (1877). Uber die Quantitat Sauerstoff welche 1 gramm Haemoglobin zu binden vermag. *Z. Physiol. Chem.* **1–2,** 317.

Hugonnaud, C., Gehr, P., Weibel, E. R., and Burri, P. H. (1977). Adaptation of the growing lung to increased oxygen consumption. II. Morphometric analysis. *Respir. Physiol.* **29,** 1–10.

Jones, J. H., and Karas, R. H. (1988). Structural *vs.* functional limitations to oxygen transport: Is there a difference? *Adv. Exp. Med. Biol.* **227,** 293–300.

Jones, J. H., and Lindstedt, S. L. (1993). Limits to maximal performance. *Annu. Rev. Physiol.* **55,** 547–569.

Jones, J. H., and Longworth, K. E. (1992). Gas exchange at rest and during exercise in mammals. *In* "Comparative Biology of the Normal Lung" (R. A. Parent, ed.), pp. 271–307. CRC Press, Boca Raton, FL.

Jones, J. H., Lindstedt, S. L., Longworth, K. E., Karas, R. H., and Taylor, C. R. (1985). Muscle respiration limits aerobic capacity in goats. *Physiologist* **28,** 342.

Jones, J. H., Birks, E. K., and Berry, J. D. (1989a). Does oxygen transport limit $\dot{V}_{O_2\ max}$ in horses? *FASEB J.* **3,** A234.

Jones, J. H., Longworth, K. E., Lindholm, A., Conley, K. E., Karas, R. H., Kayar, S. R., and Taylor, C. R. (1989b). Oxygen transport during exercise in large mammals. I. Adaptive variation in oxygen demand. *J. Appl. Physiol.* **67,** 862–870.

Jones, J. H., Smith, B. L., Birks, E. K., Pascoe, J. R., and Hughes, T. R. (1992a). Left atrial and pulmonary arterial pressures in exercising horses. *FASEB J.* **6,** A2020.

Jones, J. H., Smith, B. L., Pascoe, J. R., Tyler, W. S., and Thomas, W. P. (1992b). Why are left atrial pressures high in exercising horses? *Physiologist* **35,** 225.

Jones, J. H., Tyler, W. S., Pascoe, J. R., Smith, B. L., and Birks, E. K. (1992c). Origin of airway blood in exercise-induced pulmonary hemorrhage in race horses. *Physiologist* **35,** 225.

Jones, J. H., Birks, E. K., and Pascoe, J. R. (1993). Factors limiting aerobic performance. *In* "The Vertebrate Gas Transport Cascade: Adaptations to Environment and Mode of Life" (J. E. P. W. Bicudo, ed.), pp. 169–178. CRC Press, Boca Raton, FL.

Kacser, H. (1983). The control of enzymes *in vivo*: Elasticity analysis of the steady state. *Biochem. Soc. Trans.* **11,** 35–40.

Kacser, H., and Burns, J. (1973). The control of flux. *Symp. Soc. Exp. Biol.* **27,** 65–104.

Kacser, H., and Burns, J. A. (1979). Molecular democracy: Who shares the controls? *Biochem. Soc. Trans.* **7,** 1149–1160.

Kanstrup, I. L., and Ekblom, B. (1984). Blood volume and hemoglobin concentration as determinants of maximal aerobic power. *Med. Sci. Sports Exercise* **16,** 256–263.

Karas, R. H., Taylor, C. R., Rösler, K., and Hoppeler, H. (1987a). Adaptive variation in the mammalian respiratory system in relation to energetic demand. V. Limits to oxygen transport by the circulation. *Respir. Physiol.* **69,** 65–80.

Karas, R. H., Taylor, C. R., Jones, J. H., Lindstedt, S. L., and Reeves, R. B. (1987b). Adaptive variation in the mammalian respiratory system in relation to energetic demand. VII. Flow of oxygen across the pulmonary gas exchanger. *Respir. Physiol.* **69,** 101–115.

Kayar, S. R., Hoppeler, H., Lindstedt, S. L., Claasen, H., Jones, J. H., and Essen-Gustavsson. B. (1989). Total muscle mitochondrial volume and capillary length in relation to aerobic capacity of horses and steers. *Pflüegers. Arch.* **413,** 343–347.

Knight, P. K., Sinha, A. K., and Rose, R. J. (1991). Effect of training intensity on maximum oxygen uptake. *In* "Equine Exercise Physiology III" (S. G. B. Persson, A. Lindholm, and L. B. Jeffcott, eds.), pp. 77–82. Almqvist & Wiksell, Stockholm.

Krogh, A. (1919). The number and distribution of capillaries in muscles with calculations in the oxygen pressure head necessary for supplying the tissue. *J. Physiol. (London)* **52,** 405–415.

Lahiri, S. (1975). Blood oxygen affinity and alveolar ventilation in relation to body weight in mammals. *Am. J. Physiol.* **229,** 529–536.

Lechner, A. J., Grimes, M. J., Aquin, L., and Banchero, N. (1982). Adaptive lung growth during cold plus hypoxia is age-dependent. *J. Exp. Zool.* **219,** 285–291.

Leith, D. E. (1976). Comparative mammalian respiratory mechanics. *Physiologist* **19,** 485–510.

Leith, D. E. (1983a). Mammalian tracheal dimensions: Scaling and physiology. *J. Appl. Physiol.* **55,** 196–200.

Leith, D. E. (1983b). Comparative mammalian respiratory mechanics. *Am. Rev. Respir. Dis.* **128,** S77-S82.

Lindstedt, S. L., and Jones, J. H. (1987). Symmorphosis: The concept of optimal design. *In* "New Directions in Ecological Physiology" (M. E. Feder, A. F. Bennett, W. W. Burggren, and R. B. Huey, eds.), pp. 289–309. Cambridge Univ. Press, Cambridge, UK.

Lindstedt, S. L., and Wells, D. J. (1988). Skeletal muscle mitochondria: The aerobic gate? *Adv. Exp. Med. Biol.* **227,** 207- 214.

Lindstedt, S. L., Wells, D. J., Jones, J. H., Hoppeler, H., and Thronson, H. A. (1988).

Limitations to aerobic performance in mammals: Interaction of structure and demand. *Int. J. Sports Med.* **9,** 210–217.

Lindstedt, S. L., Hokanson, J. F., Wells, D. J., Swain, S. D., Hoppeler, H., and Navarro, V. (1991). Running energetics in the pronghorn antelope. *Nature (London)* **353,** 748–750.

Longworth, K. E. (1993). Scaling of respiratory variables. *In* "The Vertebrate Gas Transport Cascade: Adaptations to Environment and Mode of Life" (J. E. P. W. Bicudo, ed.), pp. 162–168. CRC Press, Boca Raton, FL.

Longworth, K. E., Jones, J. H., Karas, R. H., and Bicudo, J. E. P. W. (1988). Scaling of blood gases in athletic mammals at $\dot{V}_{O_2\ max}$. *FASEB J.* **2,** A499.

Longworth, K. E., Jones, J. H., Bicudo, J. E. P. W., Taylor, C. R., and Weibel, E. R. (1989). High rate of O_2 consumption in exercising foxes: Large P_{O_2} difference drives diffusion across the lung. *Respir. Physiol.* **77,** 263–276.

Martin, R. R., and Haines, H. (1970). Application of Laplace's law to mammalian hearts. *Comp. Biochem. Physiol.* **34,** 959–962.

Mathieu, O., Krauer, R., Hoppeler, H., Gehr, P., and Lindstedt, S. L. (1981). Design of the mammalian respiratory system. VII. Scaling mitochondrial volume in skeletal muscle to body mass. *Respir. Physiol.* **44,** 113–128.

Musch, T. I., Haidet, G. C., Ordway, G. A., Longhurst, J. C., and Mitchell, J. H. (1985). Dynamic exercise training in foxhounds. I. Oxygen consumption and hemodynamic responses. *J. Appl. Physiol.* **59,** 183–189.

Nadel, E. R. (1985). Physiological adaptations to aerobic exercise. *Am. Sci.* **73,** 334–343.

Persson, S. (1967). On blood volume and working capacity in horses. *Acta Vet. Scand.*, Suppl. **19.**

Powers, S. K., Lawler, J., Dempsey, J. A., Dodd, S., and Landry, G. (1989). Effects of incomplete pulmonary gas exchange on $\dot{V}_{O_2\ max}$. *J. Appl. Physiol.* **66,** 2491–2495.

Raven, P. B., Drinkwater, B. L., Ruhling, R. O., Bolduan, N., and Taguchi, S. (1974). The effect of carbon monoxide and peroxyacetylnitrate on man's maximal aerobic capacity. *J. Appl. Physiol.* **26,** 288–293.

Saltin, B., and Gollnick, P. D. (1983). Skeletal muscle adaptability: Significance for metabolism and performance. *In* "Handbook of Physiology" (L. D. Peachy, R. H. Adrian, and S. R. Geiger, eds.), Sect. 10, pp. 555–631. Williams & Wilkins, Baltimore, MD.

Saltin, B., and Rowell, L. B. (1980). Functional adaptation to physical activity and inactivity. *Fed. Proc., Fed. Am. Soc. Exp. Biol.* **39,** 1506–1513.

Schmidt-Nielsen, K. (1984). "Scaling: Why Is Animal Size So Important?" Cambridge Univ. Press, New York.

Schmidt-Nielsen, K. (1990). "Animal Physiology: Adaptation and Environment." Cambridge Univ. Press, New York.

Schwerzmann, K., Hoppeler, H., Kayar, S. R., and Weibel, E. R. (1989). Oxidative capacity of muscle and mitochondria: Correlation of physiological, biochemical and morphometric characteristics. *Proc. Natl. Acad. Sci. U.S.A.* **86,** 1583–1587.

Severinghaus, J. W. (1979). Simple, accurate equations for human blood O_2 dissociation computations. *J. Appl. Physiol.* **46,** 599–602.

Snow, D. H. (1985). The horse and dog, elite athletes—Why and how. *Proc. Nutr. Soc.* **44,** 267–272.

Sokal, R. R. and Rohlf, F. J. (1981). "Biometry," 2nd ed. Freeman, New York.

Spriet, L. O., Gledhill, N., Froese, A. B., and Wilkes, D. L. (1986). Effect of graded erythrocythemia on cardiovascular and metabolic responses to exercise. *J. Appl. Physiol.* **61,** 1942–1948.

Squires, R. W., and Buskirk, E. R. (1982). Aerobic capacity during acute exposure to simulated altitude, 914 to 2286 meters. *Med. Sci. Sports Exercise* **14,** 36–40.

Stahl, W. R. (1967). Scaling of respiratory variables in mammals. *J. Appl. Physiol.* **22,** 453–460.

Sweeney, C. R. (1991). Exercise-induced pulmonary hemorrhage. *Vet. Clin. North Am.* **7,** 93–104.

Taylor, C. R., and Weibel, E. R. (1981). Design of the mammalian respiratory system. I. Problem and strategy. *Respir. Physiol.* **44,** 1–10.

Taylor, C. R., and Weibel, E. R. (1991). Learning from comparative physiology. *In* "The Lung: A Scientific Foundation" (R. G. Crystal and J. B. West, eds.), pp. 1595–1607. Raven Press, New York.

Taylor, C. R., Karas, R. H., Weibel, E. R., and Hoppeler, H. (1987). Adaptive variation in the mammalian respiratory system in relation to energetic demand. *Respir. Physiol.* **69,** 1–127.

Tenney, S. M., and Bartlett, D. (1967). Comparative quantitative morphology of the mammalian lung: Trachea. *Respir. Physiol.* **3,** 130–135.

van Mierop, L. H. S. and Kutsche, L. M. (1985). Some aspects of comparative anatomy of the heart. *In* "Cardiovascular Shunts: Phylogenetic, Ontogenetic and Clinical Aspects" (K. Johansen and W. W. Burggren, eds.), pp. 38–56. Munksgaard, Copenhagen.

Verzar, F. (1912). The gaseous metabolism of striated muscle in warm-blooded animals. *J. Physiol. (London)* **44,** 243–258.

Wagner, P. D. (1988). An integrated view of the determinants of maximum oxygen uptake. *Adv. Exp. Med. Biol.* **227,** 245–256.

Wagner, P. D., Gillespie, J. R., Landgren, G. L., Fedde, M. R., Jones, B. W., DeBowes, R. M., Pieschl, R. L., and Erickson, H. H. (1989). Mechanism of exercise-induced hypoxemia in horses. *J. Appl. Physiol.* **66,** 1227–1233.

Wagner, P. D., Hoppeler, H., and Saltin, B. (1991). Determinants of maximal oxygen uptake. *In* "The Lung: Scientific Foundations" (R. G. Crystal and J. B. West, eds.), Vol. 2, pp. 1585–1593. Raven Press, New York.

Weibel, E. R. (1984). "The Pathway for Oxygen." Harvard Univ. Press, Cambridge, MA.

Weibel, E. R., and Taylor, C. R. (1981). Design of the mammalian respiratory system. *Respir. Physiol.* **44,** 1–164.

Weibel, E. R., Taylor, C. R., O'Neil, J. J., Leith, D. E., and Gehr, P. (1983). Maximal oxygen consumption and pulmonary diffusing capacity: A direct comparison of physiologic and morphometric measurements in canids. *Respir. Physiol.* **54,** 173–188.

Weibel, E. R., Marques, L. B., Constantinopol, M., Doffey, F., and Gehr, P. (1987). Adaptive variation in the mammalian respiratory system in relation to energetic demand. VI. The pulmonary gas exchanger. *Respir. Physiol.* **69,** 81–100.

Weibel, E. R., Taylor, C. R., and Hoppeler, H. (1991). The concept of symmorphosis: A testable hypothesis of structure–function relationship. *Proc. Natl. Acad. Sci. U.S.A.* **88,**10357–10361.

West, J. B., Mathieu-Costello, O., Jones, J. H., Birks, E. K., and Logemann, R. B. (1993). Stress failure of pulmonary capillaries in racehorses with exercise-induced pulmonary hemorrhage. *J. Appl. Physiol.* **75,** 1097–1109.

White, F. N. (1978). Circulation: A comparison of reptiles, mammals and birds. *In* "Respiratory Function in Birds, Adult and Embryonic" (J. Piiper, ed.), pp. 51–60. Springer-Verlag, Berlin.

Whitwell, K. E., and Greet, T. R. C. (1984). Collection and evaluation of tracheobronchial washes in the horse. *Equine Vet. J.* **16,** 499–508.

Withers, P. C., and Hillman, S. S. (1983). The effects of hypoxia on pulmonary function and maximal rates of oxygen consumption in two anuran amphibians. *J. Comp. Physiol.* **152,** 125–129.

Woodson, R. D., Wills, R. E., and Lenfant, C. (1978). Effect of acute and established anemia on O_2 transport at rest, submaximal and maximal workload. *J. Appl. Physiol.* **44,** 36–43.

Young, A. J., Cymerman, A., and Burse, R. L. (1985). The influence of cardiorespiratory fitness on the decrement in maximal aerobic power at high altitude. *Eur. J. Appl. Physiol.* **54,** 12–15.

Zuntz, N. (1897). Ueber den Stoffverbrauch des Hundes bei Muskelarbeit. *Arch. gesamte Physiol. Menschen Tiere* **68,** 191–211.

Zuntz, N., and Hagemann, O. (1898). Untersuchungen über den Stoffwechsel des Pferdes. *Landwirtsch. Jahrb.* **27,** Suppl. III, 1–438.

Respiratory Gas Exchange during Exercise

FRANK L. POWELL, JR.

Physiology Division, Department of Medicine, University of California, San Diego, La Jolla, California 92093

I. Introduction

A. Objective

At the most fundamental level, exercise physiology is the study of elevated energy exchange between organisms and their environment. Of the several physiological systems that could be used to study this basic process, the respiratory system offers several advantages. An experimental advantage is that oxygen uptake simply and completely describes the physiological exchange process and it usually occurs at a single anatomical site. This contrasts, for example, with measures of physical performance that must account for both mechanical and thermal work done on the environment by multiple muscles in an organism. Another advantage is the great diversity of respiratory organs corresponding to the diversity of vertebrate behaviors and ecologies. However, all vertebrate respiratory organs use the same basic processes of passive diffusion and convection by muscular pumps for gas exchange. The goal of this chapter is to examine the diverse applications of these processes for common themes to help us understand:

1. The physiological basis for differences in exercise capabilities of various animals;
2. What selective forces might have been most important in the evolution of vertebrate lungs and gills.

The relevance of these themes to organismal–environmental energy exchange in general will also be considered.

B. Scope

Oxygen gas exchange will be considered for the lungs, gills, and skin of vertebrates. The focus is on exercise, but resting data will also be considered to provide a baseline and make extrapolations where good exercise data are not available. Representative examples will be selected for each type of gas exchange situation for which there is adequate experimental data for a model analysis. Peripheral gas exchange in the tissues is only considered as necessary to understand the model analyses. The effects of body size are not considered.

This focus on exchange in the respiratory organ considers ventilation and blood flow as inputs to the system being studied. Accordingly, the comparative physiology of respiratory mechanics, cardiovascular physiology, and cardiopulmonary control are not covered here, except as necessary to explain a gas exchanger's performance. Readers are referred to the papers in this volume on exercise performance in the different groups of animals. Diffusing capacity and spatial matching of ventilation, blood flow, and diffusing capacity will be considered in more detail as processes intrinsic to the gas exchangers.

Gas tensions in the respiratory medium and blood entering the exchangers are also considered as inputs, so factors such as cardiac shunts that cause differences between mixed venous and pulmonary arterial tensions will not be discussed. Reasons for differences in blood–gas dissociation curves will not be considered either, although their influence on gas exchanger performance is discussed.

C. Model Approach

The diversity of vertebrate respiratory organs requires that different models be used to examine their gas exchange performance in exercise. Different sizes and metabolic capacities of different animals (e.g., endotherms versus ectotherms) will obviously have different absolute capacities for gas exchange. However, any given model can also be evaluated in terms of how it performs in nature relative to ideal conditions. *Ideal conditions* generally refer to:

1. *diffusion equilibrium* between blood and respiratory medium across the exchange barrier;
2. *homogeneity* of spatial distributions of ventilation, blood flow, and diffusing capacity so the exchanger behaves like a one-compartment system.

The *performance* of the models is generally quantified by their outputs, namely, *gas flux* and *gas tensions of blood and respiratory medium leaving the exchanger*. This paper focuses on the ways in which diffusion limitation and heterogeneity affect oxygen uptake and expired and arterial P_{O_2} in the different models of vertebrate gas exchange relative to their ideal performance.

Often it is difficult to distinguish the factors that reduce performance from ideal value, so simplifying assumptions (e.g., homogeneity) must be made to compare data from multiple experiments and draw general conclusions. This approach is used first to establish a framework. Then more detailed experiments determining the specific mechanisms of gas exchange limitations are considered for each type of gas exchange model or vertebrate class where possible. For many animals, it is difficult to attain maximal exercise intensities in experiments that collect the necessary data for a model analysis. In these cases, data from submaximal exercise intensities will be used, and the significance for maximal exercise will be discussed later.

II. Homogeneous Models

A. Analytical Methods

Piiper and Scheid (1972, 1975) have developed models to analyze the efficiency of gas exchange in four different types of *homogeneous* respiratory organs:

1. Countercurrent model of fish gills;
2. Crosscurrent model of avian parabronchial lungs;

3. Ventilated pool model of mammalian alveolar lungs, unicameral and multicameral reptilian and amphibian lungs;
4. Infinite pool model of cutaneous exchange.

Inputs to the models include ratios of conductances for ventilation with the respiratory medium, that is, air or water (*G*vent), respiratory organ blood flow (*G*perf), and diffusive transport between the respiratory medium and blood (*G*diff), equal to diffusing capacity. Conductances are defined as the factor in the Fick equation necessary to produce a given gas flux ($\dot{M}$) for a given partial pressure difference (cf. Piiper *et al.*, 1971):

$$\dot{M} = G \cdot \Delta P. \tag{1}$$

*G*diff is the diffusing capacity. *G*vent and *G*perf are the products of convective flow rates ($\dot{V}$ and $\dot{Q}$, respectively) and their respective capacitance coefficients (β; cf. Piiper *et al.*, 1971):

$$G\text{vent} = \dot{V}\beta \tag{2a}$$

$$G\text{perf} = \dot{Q}\beta. \tag{2b}$$

β is the slope of the curve relating concentration to partial pressure in air, water, or blood:

$$\beta = \Delta C/\Delta P. \tag{3}$$

The models also include the following assumptions:

1. steady state, or constancy of all partial pressures and conductances in time;
2. linear blood capacitance coefficients, that is, linear dissociation curves and constant β;
3. spatial homogeneity, that is, no mismatch of conductances.

Deviations from these assumptions have been discussed previously (Hammond and Hempleman, 1987; Hempleman and Gray, 1988; Piiper and Scheid, 1975, 1977) and will be considered in the following, as necessary, to explain differences between diffusion and heterogeneity limitations.

Outputs from the models include *relative*, or normalized, partial pressure differences of inspired and expired respiratory medium (*P*i and *P*e) and blood entering and exiting the exchanger (*P*v and *P*a):

$$\Delta P\text{vent} = (P\text{i-}P\text{e})/(P\text{i-}P\text{v}), \tag{4}$$

$$\Delta P\text{perf} = (P\text{a-}P\text{v})/(P\text{i-}P\text{v}), \tag{5}$$

$$\Delta P\text{tr} = (P\text{e-}P\text{a})/(P\text{i-}P\text{v}). \tag{6}$$

Note that these ΔP are dimensionless, as were the input conductance ratios.

This chapter will emphasize ΔPtr as an extremely useful index of gas exchange efficiency, it being a normalized form of the alveolar–arterial difference, a standard index for alveolar gas exchange. In an alveolar model, ΔPtr can be interpreted as the fraction of resistance to gas transfer between inspired air and venous blood attributed to alveolar gas–capillary blood exchange (Piiper and Scheid, 1977). However, in the more efficient crosscurrent and countercurrent models, ΔPtr cannot be easily interpreted as a fractional resistance, because Pa_{O_2} can exceed Pe_{O_2}, and ΔPtr can assume negative values. A negative ΔPtr emphasizes that the geometrical arrangement of these more efficient models results in performance only possible in alveolar exchange by active secretion.

Piiper and Scheid (1977) derived the contribution of a model's spatial arrangement to ΔPtr as $\Delta P\text{tr} = \Delta P\text{o}$ when Gdiff is infinite (i.e., ideal conditions). More generally, they define

$$\Delta P\text{tr} = \Delta P\text{o} + \Delta P\text{diff}. \tag{7}$$

ΔPdiff can be greater or less than ΔPtr, but it is always greater than zero.

Because these models assume homogeneity, all increases in ΔPtr from ideal values will be attributed to diffusion limitation. To explore the physiology of diffusion limitation in more depth, Powell and Hempleman (1993) calculated another ΔP, the average blood–gas (or water) partial pressure difference across the diffusion barrier:

$$\Delta P_{\text{D}} = (\overline{P\text{m} - P\text{b}})/(P\text{i} - P\text{v}). \tag{8}$$

Note that ΔP_{D} is different from ΔPdiff; ΔPdiff represents the *increase from the ideal expired–arterial gradient* caused by a finite Gdiff, whereas ΔP_{D} is the *average gradient across the diffusion barrier* with a finite Gdiff. With knowledge of the normalization value $P\text{i} - P\text{v}$, ΔP_{D} is useful for comparing diffusing capacity requirements between different models.

Another model output used by Piiper and Scheid (Piiper *et al.*, 1976) to compare performance is the total conductance:

$$G\text{tot} = \dot{M}/(P\text{i} - P\text{v}). \tag{9}$$

Gtot is the product of a conductance and its associated ΔP, that is, $G\text{tot} = G\text{diff} \cdot P_{\text{D}} = G\text{vent} \cdot \Delta P\text{vent} = G\text{perf} \cdot \Delta P\text{perf}$.

Finally, to quantify the effects of different conductance values, Piiper and Scheid (1975) defined limitations for the four models, again as functions of conductance ratios:

$$L = (\dot{M}\text{max} - \dot{M}\text{act})/\dot{M}\text{max} \tag{10}$$

where $\dot{M}$max is maximal $\dot{M}$ with infinite conductance and $\dot{M}$act is an actual value with a finite conductance. *L*diff, calculated with infinite and measured or estimated diffusing capacity, is emphasized here.

B. Application to Experimental Data

The models are now used to explore how the various values for diffusing capacity estimated for different types of animals affect O_2 exchange efficiency during exercise. Physiological estimates of *G*diff, corrected for heterogeneity where possible, are used (Powell and Hempleman, 1993). Consequently, the model results here can indicate better performance than is actually measured in some cases (*e.g.*, a smaller or more negative ΔPtr), and *L*diff will be smaller than the value calculated with uncorrected diffusing capacities. This differs from the approach of Piiper and Scheid that used uncorrected *G*diff, and helps distinguish between heterogeneity and diffusion limitations.

Data sets were selected for completeness of the necessary variables collected under *hypoxic conditions.* In hypoxia, the oxyhemoglobin equilibrium curve is approximately linear (cf. assumption 2 of constant β). Also, *G*diff is measured most accurately in hypoxia because inhomogeneity effects are minimized (cf. Piiper and Scheid, 1972). Thus, effects of hypoxia per se should be due to its influence on *G*vent and *G*perf. Ventilations are corrected for dead space, so as to focus on processes within the exchange organ, as explained earlier.

Using the foregoing criteria, Powell and Hempleman (1993) applied the analysis to experimental data for yellowfin tuna (*Thunnus albacares*; Bushnell and Brill, 1992), white Pekin ducks (*Anas platyrhynchos*; Kiley *et al.*, 1985), and humans (Hammond and and Hempleman, 1987). Men were sitting or pedaling on a bicycle ergometer. Ducks were standing or running on a treadmill. Tuna were ram ventilated in flow-through tanks at 25°C with their spinal nerves blocked to prevent locomotion. Exercising tuna data were extrapolated from measurements on alert resting and swimming dogfish (Piiper and Scheid, 1977). I am not aware of any experimental data meeting these requirements for the infinite pool model (i.e., cutaneous gas exchange), so it is not compared (but see Section II,D).

The measured and calculated variables are summarized in Table I.

TABLE I

COMPARISON OF HOMOGENEOUS GAS EXCHANGE MODELS FOR VENTILATED POOL (HUMAN), CROSSCURRENT (DUCK), AND COUNTERCURRENT (TUNA) EXCHANGERS AT REST AND DURING EXERCISE IN HYPOXIA[a]

		Human		Duck		Tuna	
Variable	Units	Rest	Exer	Rest	Exer	Rest	Exer
$\dot{M}_{O_2}$	μmol min^{-1} kg^{-1}	0.32	1.2	0.48	1.7	0.47	0.69[b]
$\dot{V}$[c]	liters min^{-1} kg^{-1}	0.10	0.62	0.47	2.1	6.9	—
$\dot{Q}$	liters min^{-1} kg^{-1}	0.14	0.25	0.28	0.60	0.12	—
Gdiff[d]	μmol min^{-1} Torr^{-1} kg^{-1}	27	53	38	68	18	—
Gvent/Gperf	—	0.21	0.68	0.68	1.4	0.99	1.5[b]
Gdiff/Gperf	—	1.1	1.1	1.1	1.0	1.6	31.9
ΔPtr	—	0.08	0.17	−0.05	0.13	−0.23	−0.21
Ldiff	—	0.08	0.17	0.19	0.29	0.38	0.27
Lvent	—	0.76	0.50	0.36	0.20	0.23	0.15
Lperf	—	0.09	0.20	0.22	0.29	0.23	0.33

[a] $\dot{M}_{O_2}$ is oxygen flux, $\dot{V}$ is ventilation rate, $\dot{Q}$ is perfusion rate, G is conductance and L conductance ratios for ventilation (vent), perfusion (perf), and diffusion (diff), and ΔPtr is the normalized partial pressure difference of medium and blood entering and exiting the gas exchanger.

[b] Not measured in tuna but increased from resting values by factors measured in dogfish on transition from alert rest to swimming (Piiper *et al.*, 1977).

[c] $\dot{V}_A$ calculated from $\dot{M}_{CO_2}$ and Pa_{CO_2} in human; parabronchial $\dot{V}$ calculated from $\dot{V}_I$ and measured dead space in ducks; gill $\dot{V}$ measured with dye dilution technique.

[d] DL_{O_2} (pulmonary diffusing capacity for oxygen) corrected for heterogeneity ($\dot{Q}$ weighted) in human; D_{O_2} corrected for heterogeneity and cardiac output in 2.4kg ducks: D_{O_2} (μmol min^{-1} Torr^{-1} kg^{-1}) = $\dot{Q}$ (liters min^{-1}) × 97 + 25 (Hempleman *et al.*, 1986, and unpublished data); D_{O_2} calculated countercurrent model, that is, not T_{O_2}.

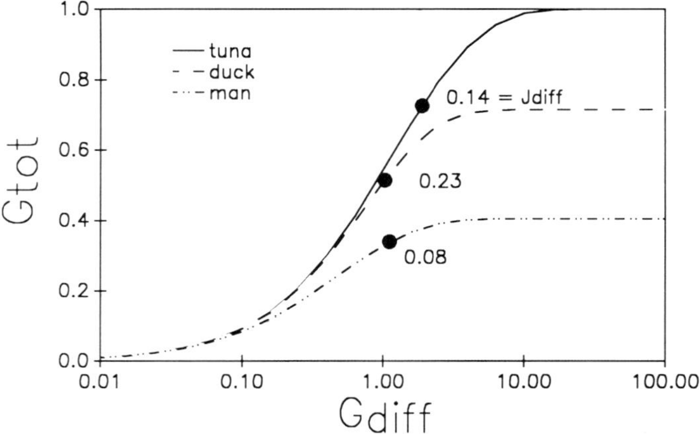

FIG. 1. *G*tot as a function of *G*diff for tuna (countercurrent model), duck (crosscurrent model) and human (ventilated pool model) for the measured hypoxic exercise conditions in Table I. *G*perf is normalized to a common value of 1 for all cases. All of the animals show significant diffusion limitations at these intensities of hypoxic exercise and would benefit from increased diffusing capacity, as shown by *J*diff [cf. Eq. (11)], but note that the tuna would have to increase *G*diff to much higher values than human or duck to reach the *G*tot plateau.

Weight-specific $\dot{M}_{O_2}$ is relatively similar at rest, and the values on exercise simply reflect the degree of activity that the experimenters could induce in the subjects. In resting conditions, the models follow the order of their ideal efficiency: countercurrent > crosscurrent > ventilated pool. Countercurrent ΔPtr is mostly negative, and ventilated pool ΔPtr is mostly positive. Piiper and Scheid have pointed out that actual efficiencies are usually less than ideal values because of diffusion limitations and deviations of *G*vent/*G*perf from the ideal value near unity. However, this comparative set of data shows that the ideal theoretical order is still followed in nature, because *G*vent/*G*perf is near unity in all cases and *G*diff is large enough, when corrected for heterogeneity, to allow expression of the models' inherent efficiencies. Note the negative ΔPtr in crosscurrent and countercurrent models.

Resting *G*tot also follows the same order as the intrinsic efficiencies of the models if *G*perf is normalized to 1 for all cases so they can be compared (Fig. 1); *G*tot is 0.16 in human, 0.43 in duck, and 0.62 in tuna (Powell and Hempleman, 1993). However, the degree to which the different models actually approach their ideal performance follows the inverse order. *G*tot in the tuna is only 62% of the countercurrent ideal, duck *G*tot is 81% of the crosscurrent ideal, and human *G*tot is 92% of the ventilated pool ideal (i.e., *L*diff = 0.38, 0.19, and 0.08, respectively). In this sense, it is clear that actual efficiencies are more similar than optimal values.

Exercise decreases the species differences observed at rest. At exercise versus rest, both ducks and humans show larger diffusion limitation (*L*diff = 0.29 and 0.17, respectively), whereas the tuna shows a smaller diffusion limitation (*L*diff = 0.27). *G*diff increases with exercise in all cases, but so does *G*perf. Thus, *G*diff/*G*perf (i.e., the form in which *G*diff is input to the models) does not increase significantly. The result is that all three species are on steep parts of the curve relating *G*tot to *G*diff in exercise (Fig. 1). The physiological significance of this figure is that it shows all three species and models could have better O_2 exchange with higher diffusing capacities.

However, Fig. 1 shows a fundamental difference between the countercurrent and other models: maximum *G*tot is not achieved until very high *G*diff values. The infinite pool also behaves like crosscurrent and ventilated pool models with a *G*tot plateau at *G*diff ≈ 3 (Powell and Hempleman, 1993). One way to quantify this situation is with *J*diff, the sensitivity of *G*tot to *G*diff (Powell and Hempleman, 1993):

$$J\text{diff} = \partial G\text{tot}/\partial G\text{diff}. \tag{11}$$

For human and duck, exercise *J*diff (0.08 and 0.23, respectively) is greater than resting *J*diff (0.02 and 0.12), as exercise moves them farther from the *G*tot versus *G*diff plateau. Reasonable increases in *G*diff could increase *G*tot in resting or exercising human and duck to the plateau where *J*diff is ≈ 0. However, *J*diff is essentially unchanged in tuna (0.15 at rest versus 0.14 in exercise) because the *G*tot curve does not plateau until very high *G*diff values. *G*diff would have to increase about fivefold more in tuna than in duck or human to maximize gas flux at the measured *G*vent, *G*perf, Pi and Pv (Powell and Hempleman, 1993).

C. Cutaneous Gas Exchange

As mentioned earlier, I am not aware of any infinite pool experimental data that are suitable for this homogeneous model analysis using the same criteria as applied to the examples here. However, the analysis has been applied to inert gas data in normoxia and hyperoxia from exclusively skin-breathing salamanders, *Desmognathus* (Piiper *et al.*, 1976). Inert gas data were used to calculate *G*diff and *G*perf values, which were then used to estimate the conductances for O_2. Resting data indicate that O_2 exchange is predominantly diffusion limited (*L*diff ≈ 0.80). The sensitivity of *G*tot to *G*diff, that is, *J*diff, is also large (0.80 calculated using an average value for *G*diff/*G*perf = 0.22). Although this large diffusion limitation and sensitivity is similar to the situation seen in the countercurrent gill data, a plot of *G*tot versus *G*diff (similar to Fig. 1) appears more like the ventilated pool and crosscurrent models in that it shows a plateau at *G*diff/*G*perf values < 10.

Feder *et al.* (1988) used a similar approach to study *Desmognathus* at rest and during walking to exhaustion on a treadmill. O_2 exchange showed large diffusion limitations at rest and exercise, but inert gas *G*diff increased with exercise. The authors conclude that this may be important for enhancing O_2 exchange during the twofold increase in $\dot{M}_{O_2}$ observed by others in the same species. Interestingly, capillary recruitment may not be the main mechanism of increasing cutaneous *G*diff in frogs; instead, changes in local hematocrit could be important (Malvin and Wood, 1992). However, cutaneous exchange in these salamanders is qualitatively similar to aquatic countercurrent exchange in gills in being largely diffusion limited under all conditions.

D. Significance of Countercurrent Gills

The unique behavior of countercurrent exchange in gills invites further investigation. One question is: "How did the tuna evolve with the intrinsically most efficient gas exchange model operating farthest from ideal conditions? More specifically, how does the tuna's evolutionary solution compare with other possibilities, for example, (a) more optimal performance in less efficient models or, (b) smaller but more ideally matched conductances in the countercurrent model? The models have been used to explore the physiological consequences of these alternatives for resting fish (Powell and Hempleman, 1993), and are expanded to include exercise data here.

1. Larger Conductances in Less Efficient Models

To consider the first alternative, *G*tot is calculated for the three finite ventilation models, given the conductances measured in the exercising tuna (cf. Table I). *G*tot is 0.65 and 0.55 in crosscurrent and ventilated pool models, respectively, compared to 0.73 in the countercurrent model (i.e., the value shown in Fig. 1). Next, the conductances required to increase the crosscurrent or ventilated pool model's *G*tot to 0.73 are calculated. Increasing gill ventilation was not considered, because the increased work of breathing water with high density and viscosity and low O_2 capacitance would presumably be selected against in nature. Increasing *G*vent would also require increasing *G*perf to maintain *G*vent/*G*perf near the ideal value of one. Assuming the tuna *G*vent/*G*perf value, increasing *G*diff is not a viable solution with the ventilated pool model; maximum *G*tot with infinite *G*diff is only 0.60. However, a crosscurrent model could reach a *G*tot of 0.73 if *G*diff/*G*perf increased two- to threefold. Reasons why this may not have occurred are considered later.

2. *Better Conductance Matching in the Most Efficient Model*

Figure 1 is useful for considering the alternative of more ideal conductance ratios in the countercurrent model. *G*vent/*G*perf is already very near the ideal value of 1, so again we focus on *G*diff/*G*perf. We calculated that *G*diff/*G*perf would have to increase almost 10-fold in the tuna to come as close to optimal performance as humans do, and *G*diff/*G*perf is already greatest in the tuna. For tuna to increase *G*diff 10-fold, respiratory organ capillary volume, surface/volume ratio, and/or exchange barrier thickness would have to change significantly from the general vertebrate plan. The strength of capillaries and the developmental program for vascular branching may preclude such evolutionary changes. Furthermore, the gills are involved in numerous physiological functions in addition to oxygen exchange (McDonald *et al.*, 1991), so increasing their surface area may compromise other vital processes, like CO_2, water, or ion homeostasis.

On the other hand, increasing *G*diff/*G*perf by 10-fold through decreasing *G*perf would not only move *G*vent/*G*perf farther from the ideal value of 1, but would require tissue oxygen delivery and exchange in fish to be much more efficient than in other vertebrates. This would presumably necessitate a different and more efficient design of systemic oxygen transport than is found in the other vertebrates. Perfusion limitations are not larger in fish than in terrestrial vertebrates (Piiper and Scheid, 1975), so the required increase in perfusive oxygen delivery could not likely be achieved through separation of gas exchanger and systemic circulations.

E. Significance of Crosscurrent Lungs

Differences in efficiency of gas exchange models used in air breathing are not as dramatic as the differences between air and water breathers. However, the analysis in Table I and Fig. 1 clearly shows that crosscurrent gas exchange is more efficient than ventilated pool gas exchange. This is not always the case (e.g., in normoxic rest), but it is certainly true in hypoxia when effects of heterogeneity are reduced (see Section III,B). However, Shams and Scheid (1989) used the model analysis to examine pulmonary gas exchange efficiency in resting, unrestrained, awake ducks exposed to different inspired O_2 concentrations at sea level, and they reached different conclusions. They concluded that very high ventilation in deeply hypoxic ducks effectively eliminated any O_2 gradients along the parabronchi, that is, $P\text{I}_{O_2}$ approached $P\text{E}_{O_2}$ and Pa_{O_2}. This eliminates the advantage of a crosscurrent arrangement, and pulmonary O_2 transport is reduced to the efficiency of an alveolar lung.

Shams and Scheid (1989) quantified the efficiency of avian versus alveolar O_2 exchange as the difference between the Pa_{O_2} measured in ducks and that predicted for an alveolar lung with the same conductances and metabolic rate. This Pa_{O_2} difference was 5, 3, and 0 Torr at $P\text{I}_{O_2}$ = 144, 49, and 30 Torr, respectively. For $P\text{I}_{O_2}$ = 49 Torr, this translates into an altitude advantage allowing the duck to have the same Pa_{O_2} at 8000 m that it would have at only 7000 m with an alveolar lung. In contrast, the difference of 0 Torr at the simulated altitude of 10,900 m means the bird is no better off with a crosscurrent, instead of an alveolar, lung. However, these predictions are critically dependent on the ventilatory, perfusive, and diffusive conductances, and there is some uncertainty about them in birds.

Figure 2 shows how this conclusion depends on the conductance ratios used in the analysis. The upper right panel shows that ΔPtr for ventilated pool lungs is 0 with infinite Gdiff (i.e., ideal conditions) at all Gvent/Gperf. However, note on the upper left panel of Fig. 2 that maximum efficiency of an ideal crosscurrent system, in terms of Pa_{O_2} exceeding $P\text{E}_{O_2}$ and causing negative ΔPtr, occurs only when Gvent/Gperf is near unity. As Piiper and Scheid (1975) pointed out, deviations from ideal conditions, like decreases in Gdiff/Gperf or Gvent/Gperf straying from unity, reduce the efficiency of a given model and differences between models with the same conductances.

The lower panel of Fig. 2 compares crosscurrent and ventilated pool models at different Gvent/Gperf for Gdiff/Gperf = 2, a value similar to that in Shams and Scheid's (1989) ducks with $P\text{I}_{O_2}$ < 40 Torr. The shaded region of Fig. 2 shows the range of Gvent/Gperf reported in the literature (Shams and Scheid, 1989; Black and Tenney, 1980) for awake, unrestrained, deeply hypoxic ducks. Note that over this range there is no significant difference in the ventilated pool ΔPtr, but there is a notable difference for the crosscurrent ΔPtr, ranging from 0 to −0.1 for the data of Shams and Scheid (1989) and Black and Tenney (1980), respectively. In both sets of data, (Pi − Pv) was about 20 Torr, so the difference in (Pe − Pa) is only a couple of Torr.

Two Torr may not seem like a large difference, but Pa_{O_2} is <30 Torr

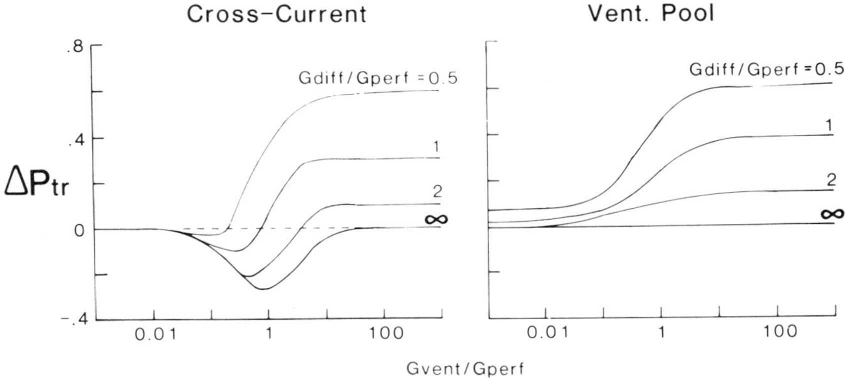

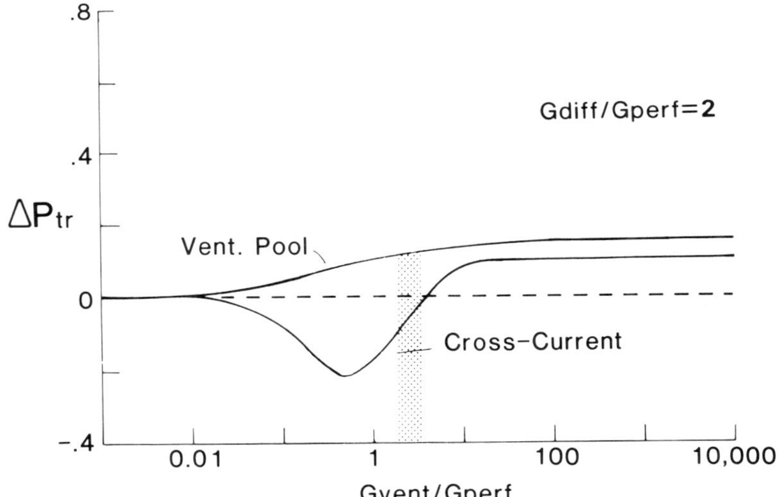

FIG. 2. Gas exchange efficiency in crosscurrent and ventilated pool models in terms of the normalized expired–arterial difference, ΔPtr. ΔPtr is a function of both Gvent/Gperf and Gdiff/Gperf. Lower panel compares the two models over a range of Gvent/Gperf measured in ducks at measured Gdiff/Gperf. With Gvent/Gperf = 3.4, crosscurrent ΔPtr = 0 (i.e., indistinguishable from ideal ventilated pool), but with Gvent/Gperf = 2.7, ΔPtr is negative (i.e., clearly crosscurrent).

in these conditions, so even a 2-Torr increase could result in a significant advantage for crosscurrent exchange, compared to alveolar lungs. West and Wagner (1980) considered that barometric pressure was the most important variable (that could reasonably be expected to change) for humans climbing Mt. Everest without supplemental O_2. The expected daily variation of 4 Torr was predicted to change $\dot{M}_{O_2}$ by 10%, mainly by increasing Pa_{O_2}. Thus, conclusions about the advantages of crosscurrent gas exchange for birds at altitude depend on changes that may be small and difficult to measure but physiologically very significant.

F. Summary

All the animals examined would benefit from increased diffusing capacity, especially in hypoxic exercise. (The significance of hypoxia to these conclusions is considered in Section IV.) However, there are differences between the models in terms of their approach to ideal conditions. Generally, the more efficient models operate further from ideal conditions in nature; this is most clearly seen in hypoxic rest (Table I). The more efficient models also tend to be more sensitive to changes in diffusing capacity. These observations suggest that the sensitivity of physiological function to structural capacity may influence the evolution of exchange organs.

The analysis also suggests that evolutionary compromise is a common cause of nonideal gas exchange. Diffusion limitation was largest in cutaneous and gill gas exchange. In both cases, the exchange organ has to transfer gas *and* operate as an ionic barrier to the environment. In the case of the fish requiring high O_2 exchange, it was shown that the countercurrent model was the only reasonable solution in light of the evolutionary history of the vertebrate cardiovascular and respiratory systems.

III. Heterogeneous Models

A. Methods of Study

The previous analysis with homogeneous models indicated that all the examples studied would benefit from increased diffusing capacities. However, there is a potential physiological cost to this benefit that should be considered. Respiratory organ exchange surface area increased by partitioning of the exchangers with the evolution of the capacity to attain high O_2 consumption. This partitioning not only increases the diffusing capacity, but it also increases the probability of nonuniform spatial distributions of ventilatory, perfusive, and diffusive conductances. Such *spatial mismatching of conductances* within a respiratory organ is termed *heterogeneity*. The most common example of heterogeneity discussed in the literature is ventilation/perfusion ($\dot{V}/\dot{Q}$) heterogeneity. Note that technically it is ventilatory/perfusive conductance heterogeneity that is important for gas exchange (cf. *G*vent/*G*perf in homogeneous model analysis in the foregoing), but it is often physiologically meaningful to study simply $\dot{V}/\dot{Q}$ heterogeneity, that is, nonuniform *spatial* distributions of the convective flow rates. Such heterogeneity can be distinguished from temporal inhomogeneities in convective flows or mismatching of the *overall* $\dot{V}/\dot{Q}$ ratio by increasing the expired–arterial P_{O_2} difference sampled at any point in time above ideal values. Because limitation to diffusion can also increase the expired–arterial P_{O_2} difference, distinguishing the effects of heterogeneity from diffusion limitation has been a significant challenge to respiratory physiologists.

1. *Inert Gas Studies*

There have been several methods developed to determine the mechanisms responsible for positive exp–art P_{O_2}. The multiple inert gas elimination technique (abbreviated MIGET) for measuring $\dot{V}/\dot{Q}$ distributions (Hlastala, 1984) has proven useful for separating the effects of $\dot{V}/\dot{Q}$ *heterogeneity* and *blood–gas diffusion limitation* on exp–art P_{O_2}. Inert gases are not diffusion limited (Piiper and Scheid, 1989), so differences between the measured exp–art P_{O_2} and that predicted for a degree of measured $\dot{V}/\dot{Q}$ heterogeneity are presumably caused by diffusion limitation. An O_2 diffusing capacity for the lung ($D\text{L}_{O_2}$ = *G*diff) can be calculated with assumptions about *G*diff/*G*perf distribution; these are considered in Section III,B. An advantage of this model is that it allows for nonlinear blood–gas dissociation curves and CO_2–O_2 interactions in blood, so hypoxia is not a requirement. However, it is a difficult experimental technique, so it has not been applied to many potentially interesting comparative cases.

2. *Comparison with Homogeneous Models*

There is a fundamental difference between this approach of partitioning the increase in a limitation index from ideal values (*e.g.*, exp–art P_{O_2}) between two limitations and the homogeneous model approach used in the foregoing. By analogy with the *L*diff approach assuming homogeneity (see Section II,A), *L*inhomo can be calculated as the effect of $\dot{V}/\dot{Q}$ inhomogeneity on gas flux assuming diffusion equilibrium (Powell and Hempleman, 1988). Note that this is the case for inert gases and is the physiological basis of the MIGET. *L*diff and *L*inhomo can be used to compare the effects of *independent* changes in $D\text{L}_{O_2}$ or $\dot{V}/\dot{Q}$ heterogeneity on O_2 exchange. This differs from the approach described earlier that implicitly considers the interactive effects of these two limitations.

An example using the *L*inhomo approach is shown for crosscurrent

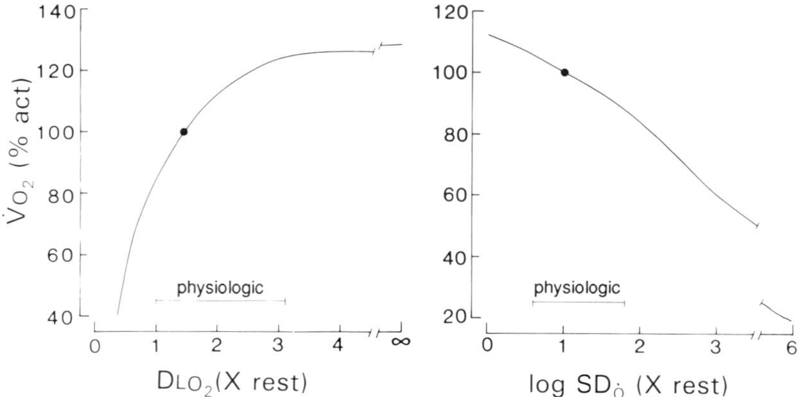

FIG. 3. Modeled independent effects of changes in diffusing capacity (assuming homogeneity, left) or $\dot{V}/\dot{Q}$ heterogeneity (assuming diffusion equilibrium, right) on crosscurrent O_2 exchange in hypoxic ducks. Points indicate measured or estimated values for moderate running exercise (see text). Over the expected physiological range (see text), changes in $D\text{L}_{O_2}$ could have a larger effect on gas exchange than would changes in heterogeneity (log SD_Q).

gas exchange in Fig. 3. The data from hypoxic ducks used in the preceding homogeneous model analysis (Table I) are also used here (Kiley *et al.*, 1985). The range of $D\text{L}_{O_2}$ extends from physiological estimates that were not corrected for heterogeneity (Burger *et al.*, 1979) to maximum morphometric estimates (Maina and King, 1982). The range of $\dot{V}/\dot{Q}$ heterogeneity extends from a negligible amount that could not be distinguished from homogeneity by the MIGET to the maximum value measured in hypoxic anesthetized ducks (Powell and Hastings, 1983). Figure 3 indicates that the effects on gas exchange of changes in $D\text{L}_{O_2}$ are potentially much larger than effects of changes in $\dot{V}/\dot{Q}$ heterogeneity. Physiological increases or decreases in heterogeneity could only change $\dot{V}_{O_2}$ ± 10%, whereas changes in diffusion could change $\dot{V}_{O_2}$ by 20% or more over a similar, that is, physiological, range. However, this finding can be misleading, because the actual effects of a given change in heterogeneity on O_2 exchange would be larger if $D\text{L}_{O_2}$ were smaller and there were more diffusion limitation. The general conclusion that diffusion limitation is more important in this case is reasonable, however, because the effects of heterogeneity are relatively small in hypoxia (see Section III,D).

A more meaningful approach would be to take *measured* $D\text{L}_{O_2}$ *and* $\dot{V}/\dot{Q}$ heterogeneity as a starting point, instead of assuming one or the other is ideal. Then one could model the effects of changes in heterogeneity given a physiological diffusion limitation, or vice versa. Assumptions are required about the distribution of *G*diff, *G*vent, and *G*perf, but there is a good physiological basis for making these assumptions in many cases. The most important argument for this approach is that the effects of heterogeneity and diffusion are not always additive (Powell and Hempleman, 1988). Finally, conclusions based on models of the interactive effects are probably more relevant to evolution, as natural selection does not usually start with one factor idealized.

3. *Respiratory Gas Studies*

The first experimental methods developed to estimate $\dot{V}/\dot{Q}$ heterogeneity were based on measurements of O_2 and CO_2 exchange. Physiological dead space and venous admixture are examples of such measurements that have been applied to a few cases of nonalveolar gas exchange (e.g., Hastings and Powell, 1986). In general, they have not been used to quantitatively distinguish between diffusion and heterogeneity as O_2 exchange limitations. Estimates of parallel $\dot{V}/\dot{Q}$ heterogeneity in artificially unidirectionally ventilated avian lungs have been made from CO_2 data (Burger *et al.*, 1979), but this requires the use of an inert gas or special bronchial catheters (Hempleman and Powell, 1986) to distinguish true dead space from high $\dot{V}/\dot{Q}$ regions. Oxygen exchange data are frequently useful for estimating the relative importance of heterogeneity by providing evidence for a diffusion limitation, however.

B. Mammals

Mammals are the best model to consider first, for several reasons. First, they have been studied most, so we are more certain of the mechanisms that determine the efficiency of alveolar gas exchange. Second, the ventilated pool model is easiest to discuss because the alveolar–arterial P_{O_2} difference is always zero in the ideal case; this is to be compared with the potentially negative values of ΔPtr for cross-current and countercurrent models discussed earlier. To facilitate comparison of mammalian data with other groups, however, I will hereafter refer to the alveolar–arterial P_{O_2} difference as the exp–art P_{O_2} difference.

1. *Human MIGET Studies*

Studies with the MIGET have quantified the role of diffusion and heterogeneity in the well-known increase in exp–art P_{O_2} in humans with exercise at altitude. Wagner and co-workers (1986) measured $\dot{V}/\dot{Q}$

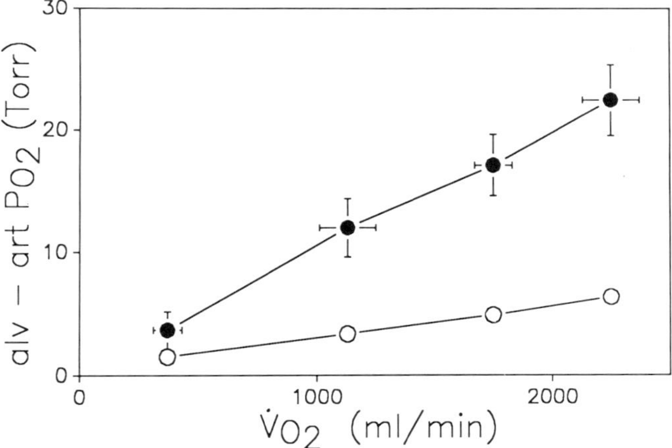

FIG. 4. Expired–arterial P_{O_2} difference (alv–art P_{O_2}) at different intensities of hypoxic ($PI_{O_2} \approx 80$ Torr) bicycle ergometer exercise in humans. The difference between measured values (solid symbols) and values predicted for the amount of $\dot{V}/\dot{Q}$ heterogeneity measured (open symbols) indicates diffusion limitation. Diffusion limitation was significant at all intensities above rest and is increasingly more important that heterogeneity at widening alv–art P_{O_2} with increased $\dot{V}_{O_2}$ (Wagner *et al.*, 1986).

distributions in human subjects exercising up to near maximal $\dot{V}_{O_2}$ at sea level and simulated altitudes to 15,000 ft (429 Torr in a hypobaric chamber). Figure 4 shows exp–art P_{O_2} measured in their subjects and those predicted for the degree of $\dot{V}/\dot{Q}$ heterogeneity measured. The predicted exp–art P_{O_2} is relatively small at rest because the effects of $\dot{V}/\dot{Q}$ heterogeneity on O_2 are reduced in hypoxia; it increases with exercise because of increased heterogeneity. Regional differences in blood flow over large vertical distances are improved in exercise with elevated cardiac output and pulmonary artery pressure, so increased heterogeneity represents changes in $\dot{V}/\dot{Q}$ on a smaller scale, perhaps between acini. The amount of $\dot{V}/\dot{Q}$ heterogeneity can be quantified as the standard deviation of the distribution of blood flow versus log $\dot{V}/\dot{Q}$, which has nearly a log-normal distribution (Wagner *et al.*, 1986). The log SD_Q increased only a small amount from 0.3 at rest to 0.5 on exercise. Possible causes of increased heterogeneity include pulmonary edema, nonuniform pulmonary vasoconstriction, and/or ventilatory time constant inequality.

The difference between the predicted and measured exp–art P_{O_2} was significant in every case except rest, indicating significant diffusion limitation in hypoxic exercise. In fact, diffusion limitation explains about 70% of the 22 Torr exp–art P_{O_2} at the highest intensity of hypoxic exercise. The difference between measured and predicted exp–art P_{O_2} can be used to estimate an O_2 diffusing capacity ($D_{L_{O_2}}$) if assumptions are made about the regional distribution of $D_{L_{O_2}}$ to blood flow (Hammond and Hempleman, 1987). This calculated $D_{L_{O_2}}$ increases with exercise, doubling from rest to heavy exercise. In exercise, it "comes close to morphometric predictions of maximal oxygen diffusing capacity in man" (Hempleman *et al.*, 1986).

It is reasonable to assume uniform $D/\dot{Q}$ in alveolar lungs because capillary surface area is recruited with blood flow. However, note that this assumption does not mean conductance, that is, *G*diff/*G*perf, homogeneity. Recall that *G*perf is the product of $\dot{Q}$ and the effective O_2–blood solubility, β_{O_2}, [Eq. (2a)]. $\dot{V}/\dot{Q}$ heterogeneity results in $P_{A_{O_2}}$ heterogeneity, and therefore nonuniform β_{O_2}, because of nonlinearity of the O_2 equilibrium curve; this increases *G*diff/*G*perf heterogeneity. In hypoxia, the O_2 equilibrium curve is more linear and the effects of *G*diff/*G*perf heterogeneity are small, so the estimates of $D_{L_{O_2}}$ should be accurate (Hempleman and Gray, 1988). The agreement between $D_{L_{O_2}}$ estimated by this method and by morphometric methods is interesting, considering the different conclusion of the morphometric (symmorphosis) analysis that the lung has excess diffusing capacity (Taylor *et al.*, 1987). Note that a limited diffusing capacity explains most of the measured exp–art P_{O_2} with exercise in this analysis.

The contribution of $\dot{V}/\dot{Q}$ heterogeneity to exp–art P_{O_2} in normoxic exercising humans has also been studied and reviewed by Wagner (1992). In normal subjects, exp–art P_{O_2} can reach 20 to 30 Torr at maximal $\dot{V}_{O_2}$ of over 3 liters min^{-1}. Only half of this difference can be explained by the relatively modest increase in $\dot{V}/\dot{Q}$ heterogeneity; log SD_Q increased from 0.4 at rest to 0.6 at maximal exercise. However, there is large variation between subjects, and the effects of these small changes in heterogeneity on arterial oxygenation are minor (Wagner, 1992). Hence, diffusion limitation is also important in normoxic exercising humans.

2. *Animal Studies*

Another interesting case of alveolar gas exchange is the exercising horse. Wagner and co-workers (1989) found exp–art P_{O_2} increasing to almost 30 Torr in horses running at 14 m s^{-1}, and 70% of this difference was attributed to diffusion limitation. This is essentially the same result as that obtained in the human study discussed in the foregoing (Wagner *et al.*, 1986). However, in contrast to the human case, $\dot{V}/\dot{Q}$ heterogeneity did not worsen with increasing exercise in the horse, even though pulmonary artery pressures are very high and pulmonary hemorrhage is common in exercising horses. The authors also noted differences from their earlier results from dogs (Schumacker *et al.*, 1985). Dogs, unlike humans and horses, showed no increase in exp–art P_{O_2} with heavy normoxic exercise, indicating no significant $\dot{V}/\dot{Q}$ heterogeneity or diffusion limitation. These differences are difficult to explain, but the authors speculate that they might involve species differences in *G*diff/*G*perf heterogeneity.

3. *Summary*

For alveolar gas exchange, most of the large exp–art P_{O_2} measured in normoxic or hypoxic heavy exercise is caused by diffusion limitation. There are species differences in effects of exercise on $\dot{V}/\dot{Q}$ heterogeneity. However, $\dot{V}/\dot{Q}$ heterogeneity never improves with exercise, in contrast to diffusing capacity, which does increase with exercise.

C. Reptiles

Consideration of the lungs of the living reptiles provides an evolutionary perspective on respiratory system function, although it must be appreciated that they are probably highly evolved for specialized functions (Perry, 1989). Furthermore, only a few species have been studied, and the diversity of this vertebrate class has barely been tapped. Unfortunately, very few of these studies consider gas exchange during exercise, and modern methods have not yet been used to distinguish mechanisms of change in gas exchange efficiency with exercise in reptiles.

1. *Reptilian MIGET Studies*

Hlastala and co-workers (1985) first applied the MIGET to a reptile in their study of the Tegu lizard (*Tupinambis nigropunctatus*). This unicameral lung is similar to the simplest vertebrate lungs of amphibians and is described by the ventilated pool model. It is a paired ovoid hollow organ lined by 1- to 4-mm-thick *faveolar* gas exchange tissue. Amazingly, even in this simple baglike lung, there was more $\dot{V}/\dot{Q}$ heterogeneity measured by the MIGET than in similar studies on dogs or humans. Furthermore, the authors note that gas mixing in the central cavity probably caused heterogeneity in the faveoli to be underestimated.

A full analysis, like the preceding one for the mammals with predicted exp–art P_{O_2}, was not carried out, but crude estimates of heterogeneity effects can be made using an *effective* blood–O_2 solubility (i.e., by assuming a linear dissociation curve, effective β_{O_2} = 0.05 mmol liter^{-1} Torr^{-1}) from air breathing data (Table 2 in Hlastala *et al.*, 1985). $\dot{V}/\dot{Q}$ heterogeneity would cause an expired–arterial partial pressure difference equal to over 20% of the total inspired–mixed venous difference ($\approx$100 Torr for O_2). Using effective β_{O_2} in normoxia underestimates the contribution, so heterogeneity could increase exp–art P_{O_2} over 20 Torr. This is still only about half of the measured expired–arterial P_{O_2} difference of 47 Torr. The balance of the difference includes effects of significant right-to-left shunts in addition to diffusion.

The Crocodilia, with completely separate right and left ventricles, provide a reptilian model that reduces the complicating effects of extrapulmonary shunts. Using the MIGET on alligators (*Alligator mississippiensis)*, Powell and Gray (1989) found less $\dot{V}/\dot{Q}$ heterogeneity than reported for Tegu lizards, even though alligators have more complex multicameral lungs. The log SD_Q was 0.47 in alligators, compared to 0.90 in Tegu lizards and normal values just under 0.4 in healthy humans. In normoxic awake alligators at 25°C, the P_{O_2} difference between end-tidal gas and pulmonary venous blood (which is not sensitive to extrapulmonary shunts) is 18 Torr (J. W. Hicks, personal communication). Using the same data to calculate an effective β_{O_2} allows one to predict a 5-Torr expired–arterial P_{O_2} difference for these alligators from measurements of $\dot{V}/\dot{Q}$ heterogeneity (Powell and Gray, 1989). Again, the effective β_{O_2} approach in normoxia underestimates the effect of heterogeneity on exp–art P_{O_2}, but the exp–art P_{O_2} is still relatively small.

2. *Gas Phase Diffusion Limitation*

The discrepancy between the measured and predicted exp–art P_{O_2} difference must be caused by diffusion limitation, and in this case, it can be shown to be at least partially a gas phase diffusion limitation. Gas phase diffusion limitations can also be detected by MIGET because of different diffusivities of the inert gases with different molecular weights (Hlastala, 1984). In the alligator, there was a systematic molecular weight-dependent difference between the retentions of inert gases, consistent with limited gas phase diffusivity in the relatively large air spaces of the crocodilian lung (Powell and Gray, 1989). Such differences are not generally seen in alveolar lungs with their small diffusion distances in the terminal air spaces (Hlastala, 1984).

This molecular weight-dependent effect in the alligator can be modeled as if it were stratification between two hypothetical intrapulmonary gas compartments separated by a diffusion barrier (cf. Piiper and Scheid, 1989). The conductance across this barrier is a simplified lumped parameter index of intrapulmonary gas mixing efficiency, comparable to the diffusive conductance across the blood–gas tissue barrier. It can be estimated by finding a stratification diffusing capacity (Dstrat) that predicts the molecular-weight-dependent behavior of selected inert gases in a lung model with measured $\dot{V}/\dot{Q}$ heterogeneity. Dstrat averaged 6 μmol min^{-1} Torr^{-1} kg^{-1} for the alligators and is predicted to cause a 10-Torr exp–art P_{O_2} difference in the normoxic alligators discussed here. The 3-Torr balance of the measured exp–art P_{O_2} must be membrane diffusion limitation so, collectively, diffusion limitations are more important than heterogeneity.

This model of stratification is certainly an oversimplification of the complicated interaction of convection and diffusion at asymmetric airway branches that results in heterogeneous alveolar gas tensions (Engel, 1983). The fact that molecular-weight-dependent effects were not detected with the MIGET in Tegu lizards (Hlastala *et al.*, 1985) supports the importance of bronchial branching and heterogeneity. The model is useful, however, for comparing the effects of this class of diffusion limitation with other determinants of exp–art P_{O_2}. Dstrat/$D\text{L}_{O_2}$ is estimated to be between 1.6 and 10 in mammals (Piiper and Scheid, 1989). Alligators are near the upper end of this range, with Dstrat being 7.5 times physiological estimates of $D\text{L}_{O_2}$ (calculated as 1.23 × rebreathing $D\text{L}_{O_2}$; A. T. Gray and F. L. Powell, unpublished observations). Molecular-weight-dependent effects must be easier to detect in alligators because absolute mass-specific Dstrat and $D\text{L}_{O_2}$ are lower.

The model is also useful for comparing the effects of gas phase and membrane diffusion limitations on exp–art P_{O_2}. For a case like the alligator, in which Dstrat is similar to morphometric estimates of $D\text{L}_{O_2}$ (Perry, 1989, 1990), gas phase diffusion limitations may have much larger effects on exp–art P_{O_2}. Stratification causes the entire partial pressure difference driving gas phase diffusion to be reflected in exp–art P_{O_2}. However, the gradient driving diffusion across the blood–gas barrier is an *average* of the gradient along the capillary that exceeds exp–art P_{O_2}. Initially, this gradient is venous–expired P_{O_2}, but as O_2 is taken up by the blood, the gradient decreases to exp–art P_{O_2} by the end of the capillary. Thus, the average gradient is larger than exp–art P_{O_2} (similar to ΔP_D and ΔPtr, respectively, see earlier). Regardless of the accuracy of this model, however, diffusion limitations collectively explain most of the measured exp–art P_{O_2} in resting alligators.

3. *Reptilian O_2 Exchange Studies*

The MIGET has not been used on exercising reptiles, but Mitchell and co-workers (1981) studied effects of exercise on O_2 exchange in

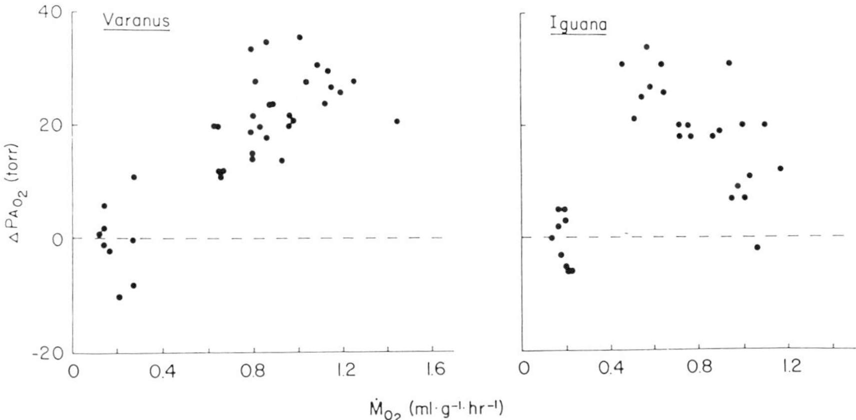

FIG. 5. Change in expired–arterial P_{O_2} difference [Δ(A − a)P_{O_2}] in lizards induced by treadmill running at 35°C, relative to resting differences (i.e., 0 = 14 and 19 Torr in *Varanus* and *Iguana*, respectively). Pa_{O_2} does not change with exercise, and the increased difference is most likely caused by diffusion limitation (Mitchell *et al.*, 1981).

savannah monitor lizards (*Varanus exanthematicus*) and green iguanas (*Iguana iguana*). Exp–art P_{O_2} was greater in the paucicameral-lunged iguana (19 Torr) than in the multicameral-lunged monitors (14 Torr), and started to increase at lower $\dot{V}_{O_2}$ in iguanas. Fivefold increases in $\dot{V}_{O_2}$ increased exp–art P_{O_2} over 30 Torr in both species (Fig. 5).

The authors concluded that diffusion limitation was the most likely cause of increased exp–art P_{O_2}. Expired P_{O_2} increased in the face of constant Pa_{O_2}, whereas, Pa_{O_2} would be expected to decrease if a shunt explained the results. Other evidence against a shunt is the elimination of the right-to-left cardiac shunt in turtles by epinephrine (Hicks and Malvin, 1992); this same mechanism might be expected during exercise. Hyperventilation increased the pressure head for O_2 diffusion and may have offset effects of $\dot{V}/\dot{Q}$ heterogeneity. However, changes in heterogeneity cannot be reliably estimated from the data available.

4. Summary

At rest, $\dot{V}/\dot{Q}$ heterogeneity in multicameral lungs is similar to that in alveolar lungs. However, most of the measured exp–art P_{O_2} differences in resting reptiles are explained by shunts or diffusion limitation, in contrast to heterogeneity in mammals. In reptilian lungs with both large air spaces and bronchial branching (e.g., Crocodilia), much of the diffusion limitation is in the gas phase. In exercising reptiles, diffusion limitation increases, and presumably explains, most of the measured exp–art P_{O_2}, similar to in mammals. However, measurements of heterogeneity in reptilian lungs during exercise remain to be made.

D. Birds

The MIGET has not been used to study exercising birds. However, it has been used on resting birds and can be considered in conjunction with studies of O_2 exchange in exercising birds to estimate the importance of heterogeneity versus diffusion limitation.

1. MIGET Studies on Resting Birds

The major factor increasing ΔPtr above ideal values in normoxic birds at rest is heterogeneity. In anesthetized ducks and geese artificially ventilated with normoxic gas, the ideal crosscurrent predicted Pa_{O_2} averages 25 Torr less than experimentally measured values, and over half of this discrepancy can be explained by $\dot{V}/\dot{Q}$ heterogeneity measured with the MIGET (Powell, 1992). Usually, a difference between the actual Pa_{O_2} and the value predicted for measured $\dot{V}/\dot{Q}$ heterogeneity is interpreted as a diffusion impairment for O_2 in mammals (see Section III,B). However, there is no evidence for diffusion limitation in normoxic resting birds. Physiological estimates of DL_{O_2} corrected for $\dot{V}/\dot{Q}$ heterogeneity agree well with morphometric estimates, and diffusion equilibrium is predicted for such high diffusing capacities (Hempleman and Powell, 1986; Powell and Scheid, 1989).

The most likely factor explaining lower than ideal crosscurrent Pa_{O_2} values in the anesthetized artificially ventilated birds is the extrapulmonary shunt. This can average 5% of cardiac output in birds under these conditions, or very much larger than in mammals, and it is apparently explained by vertebral–pulmonary venous circulatory connections (Bickler *et al.*, 1986). It is not known how this shunt is controlled, or if it occurs under more natural conditions, but measured Pa_{O_2} in these experiments was similar to that reported for undisturbed awake birds.

In hypoxia, birds show obvious crosscurrent behavior, with negative exp–art P_{O_2} (cf. ΔPtr in Table I), and there is much better agreement between measured and predicted Pa_{O_2} (Powell, 1992). $\dot{V}/\dot{Q}$ heterogeneity is not changed by hypoxia, but it increases exp–art P_{O_2} less because gas exchange is operating on the steep part of the O_2–blood dissociation curve (Powell, 1992). The extrapulmonary shunt described here must not occur in hypoxia. Half of the 4-Torr difference between measured and ideal predicted Pa_{O_2} in hypoxic resting birds is explained by $\dot{V}/\dot{Q}$ heterogeneity, and the other half is consistent with a small diffusion limitation, based on estimated DL_{O_2} (Powell, 1992). Stratification, that is, diffusion resistances in the air capillaries, and serial heterogeneity of perfusion along the parabronchi are not predicted to be important at rest (reviewed by Powell and Scheid, 1989).

2. *Oxygen Exchange in Exercising Birds*

O_2 exchange has been studied in ducks resting and running on a treadmill in normoxia and hypoxia (Kiley *et al.*, 1985); the hypoxic data were used earlier for the analysis of homogeneous crosscurrent exchange (see Section II,B). Assuming homogeneity, the authors used their data to calculate DL_{O_2}, and found that it increased with exercise. Recruitment of diffusing capacity with elevated cardiac output or reduced $\dot{V}/\dot{Q}$ heterogeneity were considered the two most likely physiological mechanisms for the increased DL_{O_2}. Diffusing capacity recruitment was the most likely explanation, according to another study of DL_{O_2} changes with $\dot{V}_{O_2}$ increased by 2,4-dinitrophenol in anesthetized hypoxic ducks (Hempleman and Powell, 1986). This latter study corrected DL_{O_2} for $\dot{V}/\dot{Q}$ heterogeneity (estimated from CO_2 exchange) and found no evidence for changes in heterogeneity with $\dot{V}_{O_2}$. Also, there was good agreement between DL_{O_2} estimates in the two studies, indicating that the effects of heterogeneity are minor.

Unfortunately, these data do not address normoxic exercise. Geiser *et al.* (1984) studied changes in DL_{O_2} with 2,4-dinitrophenol hypermetabolism in *normoxic* ducks and found increases similar to those reported for hypoxic ducks by Powell and Hempleman (1993). The normoxic study did not correct for heterogeneity, and the increased DL_{O_2} could have been only *apparent* owing to decreased heterogeneity. However, it seems unlikely that the same magnitude of increase in DL_{O_2} in normoxia and hypoxia would be due to an apparent increase in one case and a true increase in the other.

The increased DL_{O_2} observed by Hempleman and Powell (1986) was shown to be most closely correlated with increased cardiac output. It is reasonable to assume that this also explains the other increases reported, but evidence for recruitment of pulmonary capillary volume and surface area with increased cardiac output in birds is not as convincing as it is for mammals. The avian lung is relatively rigid, and pulmonary vascular resistance measurements indicate limited capacity for capillary distension and recruitment (Powell *et al.*, 1985). However, evidence of potential recruitment (i.e., folds) is observed in avian pulmonary capillaries (Powell and Mazzone, 1983), and recruitment seems to be a general feature of vertebrate gas exchange organs, as it also occurs in fish gills (Farrell *et al.*, 1980).

3. *Summary*

In normoxia, $\dot{V}/\dot{Q}$ heterogeneity has a larger effect on O_2 exchange in crosscurrent avian lungs than it does in ventilated pool mammalian or reptilian lungs. This is because crosscurrent exchange is more sensitive to heterogeneity, not because heterogeneity is much greater in birds. In hypoxia, $\dot{V}/\dot{Q}$ heterogeneity is not decreased, but it affects O_2 exchange less, and diffusion limitation becomes equally important. In hypoxic exercise, diffusion is the dominant limitation. The relative roles of heterogeneity and diffusion limitations in normoxic exercise are unknown, and MIGET studies remain to be done on exercising birds. However, available evidence indicates $D\text{L}_{O_2}$ is recruited by exercise in both normoxia and hypoxia, whereas heterogeneity is relatively insensitive to exercise or O_2 tension, similar to findings in mammals.

E. Fishes

Countercurrent gas exchange in fish gills has the highest intrinsic efficiency of any vertebrate model, as discussed in Section II,D. However, as we also saw earlier, O_2 exchange in fish operates far from optimal countercurrent predictions, with significant positive exp–art P_{O_2} differences (Piiper *et al.*, 1977). Part of this deviation is explained by *overall* $\dot{V}/\dot{Q}$ mismatch and "water and blood mismatch shunts" that change the Gvent/Gperf from the ideal value of 1 (Piiper and Scheid, 1975, 1989). This is different than the spatial heterogeneity discussed earlier. Unfortunately, we have no measures of $\dot{V}\dot{Q}$ heterogeneity in fish. The MIGET is theoretically applicable to countercurrent exchange, but it is not practical, because solubilities of inert gases in blood and water are essentially identical. Hence, gills cannot be exposed to the wide range of $\dot{V}/\dot{Q}$ conductance ratios necessary for the MIGET.

Malte and Weber (1989) have modeled the effects of parallel heterogeneity on gill O_2 exchange as being potentially significant, and notably more important than for CO_2 exchange. Exp–art P_{O_2} is increased in their model by heterogeneity for three reasons. First, expired P_{O_2} is increased by flow-weighting from high $\dot{V}/\dot{Q}$ compartments, and Pa_{O_2} is decreased by flow-weighting from low $\dot{V}/\dot{Q}$ compartments, that is, the usual explanation for alveolar lungs. Second, Gvent/Gperf values not equaling 1 in individual compartments increase exp–art P_{O_2} relative to the ideal value, unlike in the ventilated pool case, in which the ideal value equals 0 for all *G*vent/*G*perf. Third, diffusing capacity is not redistributed with blood flow in the model, so the effects of *G*diff/*G*perf heterogeneity are included. The importance of these factors cannot be judged without knowledge of the degree of heterogeneity, however.

Indirect evidence of the amount of $\dot{V}/\dot{Q}$ heterogeneity has been obtained by digital imaging techniques applied to leopard shark (*Triakis semifasciata*) gills (Bhargava *et al.*, 1992). They show considerable heterogeneity in capillary transit times, which may reflect differences in perfusion that, if not matched by water flow, will cause $\dot{V}/\dot{Q}$ heterogeneity. They may also cause *G*diff/*G*perf heterogeneity. Transit times were relatively long (20 times greater than alveolar transit times in similarly sized mammals), but if there is diffusion disequilibrium, then this may cause heterogeneity in end-capillary P_{O_2} and contribute to exp–art P_{O_2} increases.

Consideration of diffusing capacity estimates for fish leads to a different conclusion about the importance of heterogeneity. Morphometric and physiological estimates of $D\text{L}_{O_2}$ agree well for dogfish (*Scyliorhinus stellaris*), and this can be interpreted to indicate minimal heterogeneity (Piiper and Scheid, 1989). Exp–art P_{O_2} increases with exercise in normoxic dogfish (Fig. 6), similar to the observation in lizards (see earlier). Diffusion resistances can explain a -8-Torr exp–art P_{O_2} difference at rest and +39-Torr difference in swimming exercise (Piiper *et al.*, 1977). Interestingly, Pa_{O_2} changes relatively little with such exercise (64 Torr at rest versus 57 Torr in exercise), but expired P_{O_2} increases significantly. This is more like the lizards than the mammals discussed earlier, but it may relate to lower intensities of exercise in the ectotherms. Calculated $D\text{L}_{O_2}$ values increase a modest 10 to 20% with such exercise.

As mentioned in the foregoing Piiper and Scheid analysis, fish have large diffusion resistances in their respiratory medium. Over 50% of the total diffusive resistance to O_2 uptake is calculated as being in the interlamellar water space (Piiper and Scheid, 1989). The effect of such stratification on exp–art P_{O_2} is more proportional in fish gills than in ventilated pool models, however. Countercurrent exchange with membrane diffusion limitations like those occurring in fish results in an average *water–blood* P_{O_2} gradient that is more uniform than the average *alveolar–blood* P_{O_2} gradient. Consequently, water and membrane diffusion limitations contribute similar amounts to exp–art P_{O_2} in fish, in contrast to gas phase and membrane diffusion limitations in alligators.

In summary, the MIGET is not practical for aquatic O_2 exchange,

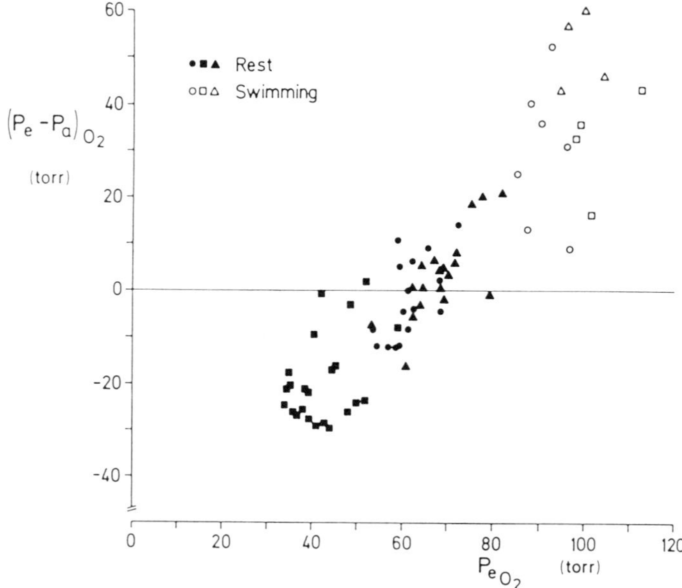

FIG. 6. Changes in expired–arterial P_{O_2} in dogfish going from alert rest to swimming as a function of expired P_{O_2}. Negative exp–art P_{O_2} differences are explained by counter-current exchange. Pe_{O_2} increased with O_2 consumption whereas Pa_{O_2} was relatively constant, indicating increased diffusion limitation (Piiper *et al.*, 1977).

and other methods are not consistent in estimating how much heterogeneity may exist in gills. However, diffusion resistances are substantial and can explain the large increase in exp–art P_{O_2} with exercise in fish. The similarity of O_2 exchange in fish and lizards is striking, and they both may have significant stratification, although they have very different lung structures.

IV. Synthesis and Conclusions

The first goal of this paper was largely descriptive, namely, to review the physiological basis for differences in exercise capabilities of various animals. Unfortunately, data are not available for most vertebrate classes to accurately answer the question in terms of *maximum* capabilities. However, submaximal exercise data indicate that increased diffusing capacity is the most important feature associated with increased metabolism during exercise, both between different vertebrate groups and within a species. For example, if mammalian lungs evolved from anything resembling the modern reptilian lungs studied, then the primary change was increased $D_{L_{O_2}}$ and an ability to recruit more $D_{L_{O_2}}$ upon exercise. In every case examined, however, the increase in diffusing capacity with exercise is insufficient to prevent decreased O_2 exchange efficiency.

In contrast, $\dot{V}/\dot{Q}$ heterogeneity either does not increase with exercise or is less important than diffusion at increased O_2 uptake. There are fewer quantitative data on which to base this conclusion, and it represents an important area for future research. Especially lacking are inert gas studies to determine the effects of heterogeneity versus diffusion on high $\dot{V}_{O_2}$ in birds and reptiles. Until such studies are done, the conclusion that diffusion is a more important O_2 uptake limitation for exercising vertebrates is only preliminary. However, it does provide a theoretical framework for future experiments.

The preliminary conclusion that diffusion is a more important O_2 exchange limitation than heterogeneity can also be used to address the second goal of this paper, that of understanding the evolution of gas exchange organs. This conclusion suggests that there is small evolutionary risk of increased heterogeneity when increasing diffusing capacity by subdividing lungs into smaller compartments. Comparative $\dot{V}/\dot{Q}$ heterogeneity data also support this conclusion by showing very similar degrees of inhomogeneity in reptilian, mammalian, and avian lungs. This is especially interesting for birds, because they followed a unique evolutionary pathway to increasing gas exchange surface area (Duncker, 1978). Perhaps the amount of $\dot{V}/\dot{Q}$ heterogeneity measured in most animals is an inevitable consequence of the morphogenesis of bronchial branching and the scaling of diffusive and convective gas transport.

Another interesting evolutionary problem is determining the factors that influence the natural selection of one gas exchange model over another. Sometimes the data and model analyses provide a clear rationale for a given evolutionary solution, and other times they do not. For example, it is clear that a ventilated pool model could not support the $\dot{V}_{O_2}$ measured in tuna, which use the intrinsically more efficient countercurrent model (Section II,D). In contrast, the significance of the crosscurrent model for explaining avian versus mammalian differences in exercise performance is not clear (see Section II,E). The superior performance of birds, compared to mammals, is probably a preadaptation to hypoxia conferred by crosscurrent gas exchange that evolved for other reasons; there is no compelling evidence that the high-altitude ecological niche confers any avian evolutionary advantage.

In general, the intrinsically more efficient models operate farther from their optimal behavior in nature. This inverse relation should be tested in many more cases, however. The relation held strictly in terms of the homogeneous model for diffusion limitation in hypoxic rest, but it was less clear in hypoxic exercise (Section II,B). The relevance of exercise versus rest when considering evolution of gas exchange organs is obvious in terms of providing a predator or prey with a survival advantage. There is no similar rationale for studying hypoxia; it is merely an experimental convenience to simplify model analyses. Normoxic exercise data that can be analyzed with models that do not require simplifying assumptions are necessary to further test the generality of this inverse relationship.

The ability to collect such experimental data and analyze them in a theoretical framework supports the introductory statement that comparative respiratory gas exchange is a good model to study general problems of exchange processes between organisms and their environment. It is reasonable to hypothesize that the inverse relation between intrinsic efficiency and actual performance (relative to ideal) applies to other biological exchange systems. Observations from the respiratory system can be used to generate further hypotheses about factors governing the evolution of such processes. For example, the sensitivity of countercurrent O_2 exchange to changes in diffusing capacity indicates that significant gains could be achieved by relatively small structural changes. However, the increase needed to actually reach ideal performance is probably not reasonable, given the evolutionary history of the structures. Similarly, avian crosscurrent gas exchange also appears very sensitive to heterogeneity in normoxia, especially when compared to the less efficient ventilated pool model. However, heterogeneity is similar in all vertebrates, suggesting an evolutionary history that may limit the approach to ideal homogeneous performance. These observations suggest that attainment of ideal performance in physiological exchange processes may require an ideal performance plateau within a certain evolutionary distance from the ancestral conditions. Sensitivity of function to structural changes might even be postulated to hinder the evolutionary approach to ideal performance if equivalent gains and losses could result from random mutations in opposite directions. This is a mechanistic interpretation of optimal design in biological systems, which in itself is not a goal of evolution.

Finally, it should be emphasized that such interpretations depend on the biological relevance of the performance indices used. Measures of O_2 flux across the exchanger are obviously relevant. However, limitation indices, like *L*diff and *L*inhomo, have the disadvantage of not accounting for interactions between limitations (see Section III,A,2). The limitation interactions can be modeled, but this has not been done systematically for comparative gas exchange. Instead, interactions have been minimized with experimental conditions that reduce the influence of one limitation (e.g., hypoxia reducing $\dot{V}'$ heterogeneity effects), but these conditions may not be biologically relevant. Indices based on O_2 tensions, like exp–art P_{O_2}, clearly have biological meaning, but they can be very complex and represent several interrelationships. Expired P_{O_2} is a function of ventilatory mechanical efficiency, and Pa_{O_2} is a function of gas exchange efficiency in a respiratory organ. Also, Pa_{O_2} is a critical determinant of circulatory O_2 delivery in tissues and is under the control of cardiopulmonary chemoreflexes. The significance of increased exp–art P_{O_2} with exercise may differ if it is caused by increased expired P_{O_2} (e.g., exercising lizards and fish) instead of decreased Pa_{O_2} (e.g., horses). Hence, exp–art P_{O_2} is not the ideal measure of gas exchange performance, but any index will have to consider interactions to be meaningful.

References

Bhargava, V., Lai, N. C., Graham, J. B., Hempleman, S. C., and Shabetai, R. (1992). Digital image analysis of shark gills: Modelling of oxygen transfer in the domain of time. *Am. J. Physiol.* **32,** R741–R746.

Bickler, P. E., Maginniss, L. A., and Powell, F. L. (1986). Intrapulmonary and extrapulmonary shunt in ducks. *Respir. Physiol.* **63,** 151–160.

Black, C. P., and Tenney, S. M. (1980). Oxygen transport during progressive hypoxia in high-altitude and sea-level waterfowl. *Respir. Physiol.* **39,** 217–239.

Burger, R. E., Meyer, M., Graf, W., and Scheid, P. (1979). Gas exchange in the parabronchial lung of birds: Experiments in unidirectionally ventilated ducks. *Respir. Physiol.* **36,** 19–37.

Bushnell, P. G., and Brill, R. W. (1992). Oxygen transport and cardiovascular responses in skijack tuna (*Katsuwonus pelamis*) and yellowfin tuna (*Thunnus albacares*) exposed to acute hypoxia. *J. Comp. Physiol.* **162,** 131–143.

Duncker, H. R. (1978). Morphological and functional aspects of respiration and circulation. *Verh. Dtsch. Zool. Ges.* **1978,** 99–132.

Engel, L. A. (1983). Gas mixing within the acinus of the lung. *J. Appl. Physiol.* **4,** 609–618.

Farrell, A. P., Sobin, S. S., Randall, D. J., and Crosby, S. (1980). Intralamellar blood flow patterns in fish gills. *Am. J. Physiol.* **239,** R428–R436.

Feder, M. E., Full, R. J., and Piiper, J. (1988). Elimination kinetics of acetylene and Freon 22 in resting and active lungless salamanders. *Respir. Physiol.* **72,** 229–240.

Geiser, J., Gratz, R. K., Hiramoto, T., and Scheid, P. (1984). Effects of increasing metabolism by 2,4-dinitrophenol on respiration and pulmonary gas exchange in the duck. *Respir. Physiol.* **57,** 1–14.

Hammond, M. D., and Hempleman, S. C. (1987). Oxygen diffusing capacity estimates derived from measured $\dot{V}A/\dot{Q}$ distributions in man. *Respir. Physiol.* **69,** 129–147.

Hastings, R. H., and Powell, F. L. (1986). Physiological dead space and effective parabronchial ventilation in ducks. *J. Appl. Physiol.* **60,** 85–91.

Hempleman, S. C., and Gray, A. T. (1988). Estimating steady-state DL_{O_2} with nonlinear dissociation curves and $\dot{V}A/\dot{Q}$ inequality. *Respir. Physiol.* **73,** 279–288.

Hempleman, S. C., and Powell, F. L. (1986). Influence of pulmonary blood flow and O_2 flux on D_{O_2} in avian lungs. *Respir. Physiol.* **63,** 285–292.

Hempleman, S. C., Adamson, T. P., and Burger, R. E. (1986). Sensitivity of avian intrapulmonary chemoreceptors to venous CO_2 load. *Respir. Physiol.* **66,** 53–60.

Hicks, J. W., and Malvin, G. M. (1992). Mechanisms of infracarchiac shunting in the turtle *Pseudemys scripta. Am. J. Physiol.* **262,** R986–R992.

Hlastala, M. P. (1984). Multiple inert gas elimination technique. *J. Appl. Physiol.* **56,** 1–7.

Hlastala, M. P., Standaert, T. A., Pierson, D. J., and Luchtel, D. L. (1985). The matching of ventilation and perfusion in the lung of the tegu lizard, *Tupinambis nigropunctatus. Respir. Physiol.* **60,** 277–294.

Kiley, J. P., Faraci, F. M., and Fedde, M. R. (1985). Gas exchange during exercise in hypoxic ducks. *Respir. Physiol.* **59,** 105–115.

Maina, J. N., and King, A. S. (1982). Morphometrics of the avian lung. 2. The wild mallard (*Anas platyrhynchos*) and greylag goose (*Anser anser*). *Respir. Physiol.* **50,** 299–310.

Malte, H., and Weber, R. E. (1989). Gas exchange in fish gills with parallel in homogeneities. *Respir. Physiol.* **76,** 129–138.

Malvin, G. M., and Wood, S. C. (1992). Effects of capillary red cell density on gas conductance of frog skin. *J. Appl. Physiol.* **73,** 224–233.

McDonald, D. G., Cavdek, V., and Ellis, R. (1991). Gill design in freshwater fishes: Interrelationships among gas exchange, ion regulation, and acid–base regulation. *Physiol. Zool.* **64,** 103–123.

Mitchell, G. S., Gleeson, T. T., and Bennett, A. F. (1981). Pulmonary oxygen transport during activity in lizards. *Respir. Physiol.* **43,** 365–375.

Perry, S. F. (1989). Structure and function of the reptilian respiratory system. *In* "Comparative Pulmonary Physiology: Current Concepts" (S. C. Woods, ed.), pp. 369–416. Dekker, New York.

Perry, S. F. (1990). Gas exchange strategy in the Nile crocodile: A morphometric study. *J. Comp. Physiol. B* **159B,** 761–769.

Piiper, J., and Scheid, P. (1972). Maximum gas transfer efficacy of models for fish gills, avian lungs and mammalian lungs. *Respir. Physiol.* **14,** 115–124.

Piiper, J., and Scheid, P. (1975). Gas transport efficacy of gills, lungs and skin: Theory and experimental data. *Respir. Physiol.* **23,** 209–221.

Piiper, J., and Scheid, P. (1977). Comparative physiology of respiration: Functional analysis of gas exchange organs in vertebrates. *Int. Rev.* **14,** 219–253.

Piiper, J., and Scheid, P. (1989). Gas exchange: Theory, models, and experimental data. *In* "Comparative Pulmonary Physiology: Current Concepts" (S. C. Wood, ed.), pp. 369–416. Dekker, New York.

Piiper, J., Dejours, P., Haab, P., and Rahn, H. (1971). Concepts and basic quantities in gas exchange physiology. *Respir. Physiol.* **13,** 292–304.

Piiper, J., Gatz, R. N., and Crawford, E. C., Jr. (1976). Gas transport characteristics in an exclusively skin-breathing salamander, *Desmognathus fuscus* (Plethodontidae). *In* "Respiration in Amphibious Vertebrates" (G. M. Hughes, ed.), pp. 339–356. Academic Press, San Diego.

Piiper, J., Meyer, M., Worth, H., and Willmer, H. (1977). Respiration and circulation using swimming activity in the dogfish *Scyliorhinus stellaris*. *Respir. Physiol.* **30,** 221–239.

Powell, F. L. (1992). Birds at altitude. *In* "Funktionanalyse biologischer Systeme" (P. Scheid, ed.), Vol. 23, pp. 351–358. Fischer, Stuttgart.

Powell, F. L., and Gray, A. T. (1989). Ventilation–perfusion relationships in alligators. *Respir. Physiol.* **78,** 83–94.

Powell, F. L., and Hastings, R. H. (1983). Effects of hypoxia on ventilation—perfusion matching in birds. *Physiologist* **26,** A50 (abstr.).

Powell, F. L., and Hempleman, S. C. (1988). Comparative physiology of oxygen transfer in lungs. *In* "Oxygen Transfer from Atmosphere to Tissues" (N. C. Gonzalez and M. R. Fedde, eds.), pp. 53–65. Plenum, New York and London.

Powell, F. L., and Hempleman, S. C. (1993). Diffusion limitation in comparative models of gas exchange. *Respir. Physiol.* **91,** 17–29.

Powell, F. L., and Mazzone, R. W. (1983). Morphometrics of rapidly frozen goose lungs. *Respir. Physiol.* **51,** 319–332.

Powell, F. L., and Scheid, P. (1989). Physiology of gas exchange in the avian respiratory system. *In* "Form and Function in Birds" (A. S. King and J. McLelland, eds.), Vol. 4, pp. 393–437. Academic Press, San Diego.

Powell, F. L., Hastings, R. H., and Mazzone, R. W. (1985). Pulmonary vascular resistance during unilateral pulmonary arterial occlusion in ducks. *Am. J. Physiol.* **249,** R39–R43.

Schumacker, P. T., Guth, B., Suggett, A. J., Wagner, P. D., and West, J. B. (1985). Effects of transfusion-induced polycythemia on O_2 transport during exercise in the dog. *J. Appl. Physiol.* **58,** 749–758.

Shams, H., and Scheid, P. (1989). Efficiency of parabronchial gas exchange in deep hypoxia: Measurements in the resting duck. *Respir. Physiol.* **77,** 135–146.

Taylor, C. R., Weibel, E. R., Karas, R. H., and Hoppeler, H. (1987). Adaptive variation in the mammalian respiratory system in relation to energetic demand. VIII. Structural and functional design principles determining the limits to oxidative metabolism. *Respir. Physiol.* **69,** 117–127.

Wagner, P. D. (1992). Ventilation–perfusion matching during exercise. *Chest* **101,** 192S–198S.

Wagner, P. D., Gale, G. E., Moon, R. E., Torre-Bueno, J. R., Stolp, B. W., and Saltzman, H. A. (1986). Pulmonary gas exchange in humans exercising at sea level and simulated altitude. *J. Appl. Physiol.* **61,** 260–270.

Wagner, P. D., Gillespie, J. R., Landgren, G. L., Fedde, M. R., Jones, B. W., DeBowes, R. M., Pieschl, R. L., and Erickson, H. H. (1989). Mechanism of exercise-induced hypoxemia in horses. *J. Appl. Physiol.* **66,** 1227–1233.

West, J. B., and Wagner, P. D. (1980). Predicted gas exchange on the summit of Mt. Everest. *Respir. Physiol.* **42,** 1–16.

Index

E

F

G

H

L

M

N

O

P

R

S

T

V

W

Z

ISBN 0-12-039238-0